Lecture Notes in Computer Scien

T0238256

Commenced Publication in 1973
Founding and Former Series Editors:
Gerhard Goos, Juris Hartmanis, and Jan van Leeuwen

Jano van Hemert   Carlos Cotta (Eds.)

Evolutionary Computation in Combinatorial Optimization

8th European Conference, EvoCOP 2008
Naples, Italy, March 26-28, 2008
Proceedings

 Springer

Volume Editors

Jano van Hemert
University of Edinburgh, National e-Science Centre
15 South College Street, Edinburgh EH8 9AA, UK
E-mail: jano@vanhemert.co.uk

Carlos Cotta
Universidad de Málaga, Dept. Lenguajes y Ciencias de la Computación
ETSI Informática, Campus de Teatinos, 29071 Málaga, Spain
E-mail: ccottap@lcc.uma.es

Cover illustration: "Ammonite II" by Dennis H. Miller (2004-2005)
www.dennismiller.neu.edu

Library of Congress Control Number: 2008922955

CR Subject Classification (1998): F.1, F.2, G.1.6, G.2.1, G.1

LNCS Sublibrary: SL 1 – Theoretical Computer Science and General Issues

ISSN   0302-9743
ISBN-10   3-540-78603-1 Springer Berlin Heidelberg New York
ISBN-13   978-3-540-78603-0 Springer Berlin Heidelberg New York

Springer is a part of Springer Science+Business Media

springer.com

© Springer-Verlag Berlin Heidelberg 2008
Printed in Germany

Typesetting: Camera-ready by author, data conversion by Scientific Publishing Services, Chennai, India
Printed on acid-free paper   SPIN: 12239986   06/3180   5 4 3 2 1 0

Preface

Metaheuristics have been shown to be effective for difficult combinatorial optimization problems appearing in various industrial, economical, and scientific domains. Prominent examples of metaheuristics are evolutionary algorithms, tabu search, simulated annealing, scatter search, memetic algorithms, variable neighborhood search, iterated local search, greedy randomized adaptive search procedures, ant colony optimization and estimation of distribution algorithms. Problems solved successfully include scheduling, timetabling, network design, transportation and distribution, vehicle routing, the travelling salesman problem, packing and cutting, satisfiability and general mixed integer programming.

EvoCOP began in 2001 and has been held annually since then. It was the first event specifically dedicated to the application of evolutionary computation and related methods to combinatorial optimization problems. Originally held as a workshop, EvoCOP became a conference in 2004. The events gave researchers an excellent opportunity to present their latest research and to discuss current developments and applications. Following the general trend of hybrid metaheuristics and diminishing boundaries between the different classes of metaheuristics, EvoCOP has broadened its scope over the last years and invited submissions on any kind of metaheuristic for combinatorial optimization.

This volume contains the proceedings of EvoCOP 2008, the 8th European Conference on Evolutionary Computation in Combinatorial Optimization. It was held in Naples, Italy, on 26–28 March 2008, jointly with EuroGP 2008, the Eleventh European Conference on Genetic Programming, EvoBIO 2008, the Sixth European Conference on Evolutionary Computation and Machine Learning in Bioinformatics, and EvoWorkshops 2008, which consisted of the following nine individual workshops: EvoCOMNET, the Fifth European Workshop on the Application of Nature-Inspired Techniques to Telecommunication Networks and other Connected Systems; EvoFIN, the Second European Workshop on Evolutionary Computation in Finance and Economics; EvoHOT, the Fourth European Workshop on Bio-inspired Heuristics for Design Automation; EvoIASP, the Tenth European Workshop on Evolutionary Computation in Image Analysis and Signal Processing; EvoMUSART, the Sixth European Workshop on Evolutionary Music and Art; EvoNUM, the First European Workshop on Bio-inspired Algorithms for Continuous Parameter Optimization; EvoPhD, the Third European Graduate Student Workshop on Evolutionary Computation; EvoSTOC, the Fifth European Workshop on Evolutionary Algorithms in Stochastic and Dynamic Environments, and EvoTransLog, the Second European Workshop on Evolutionary Computation in Transportation and Logistics. Since 2007, all these events have been grouped under the collective name EvoStar, and together they constitute Europe's premier event on evolutionary computation.

The papers presented at previous EvoCOP events have been published by Springer in the Lecture Notes in Computer Science series. Below we report their statistics.

EvoCOP	submitted	accepted	acceptance ratio	LNCS Volume
2001	31	23	74.2%	2037
2002	32	18	56.3%	2279
2003	39	19	48.7%	2611
2004	86	23	26.7%	3004
2005	66	24	36.4%	3448
2006	77	24	31.2%	3906
2007	81	21	25.9%	4446
2008	69	24	34.8%	4972

The rigorous, double-blind reviewing process of EvoCOP 2008 resulted in a strong selection among the submitted papers; the acceptance rate was 34.8%. Each paper was reviewed by at least three members of the international program committee. All accepted papers were presented orally at the conference and are included in this proceedings volume. We would like to thank the members of our program committee, to whom we are very grateful for their thorough work. EvoCOP 2008 contributions present new algorithms together with insight into how well these algorithms can solve prominent example problems from the literature or selected real-world problems.

We would like to express our sincere gratitude to the two internationally renowned invited speakers, who gave the keynote talks at the conference: Professor Emeritus H.-P. Schwefel from the University of Dortmund, Germany, IEEE Fellow for contributions to evolutionary computation, and Stefano Nolfi, head of the Laboratory of Artificial Life and Robotics of the Institute of Cognitive Science and Technologies, National Research Council (CNR), Rome, Italy.

The success of the conference resulted from the input of many people to whom we would like to express our appreciation. The local organizers, Ivanoe De Falco, ICAR-CNR; Antonio Della Cioppa, University of Salerno; Ernesto Tarantino, ICAR-CNR, and Giuseppe Trautteur, University of Naples Federico II, did an extraordinary job for which we are very grateful. Gratitude goes to Naples City Council for supporting the local organization and for their patronage of the event. We thank Marc Schoenauer from INRIA in France for his support with the MyReview conference management system. Thanks are also due to Jennifer Willies and the Centre for Emergent Computing at Napier University in Edinburgh, Scotland, for administrative support and event coordination. Last, but not least, we would specially like to thank Jens Gottlieb and Günther Raidl for their support and guidance; due to their hard work and dedication, EvoCOP has now become one of the reference events in evolutionary computation.

March 2008

Jano van Hemert
Carlos Cotta

Organization

EvoCOP 2008 was organized jointly with EuroGP 2008, EvoBIO 2008, and EvoWorkshops 2008.

Organizing Committee

Chairs	Jano van Hemert, University of Edinburgh, UK
	Carlos Cotta, Universidad de Málaga, Spain
Local Chair	Ivanoe De Falco, ICAR-CNR, Italy
Publicity Chair	Anna Isabel Esparcia-Alcázar, Universitat Politècnica de València, Spain

EvoCOP Steering Committee

Carlos Cotta, Universidad de Málaga, Spain
Jens Gottlieb, SAP AG, Germany
Jano van Hemert, University of Edinburgh, UK
Günther Raidl, Technische Universität Wien, Austria

Program Committee

Adnan Acan, Middle East Technical University, Ankara, Turkey
Hernán Aguirre, Shinshu University, Nagano, Japan
Enrique Alba, Universidad de Málaga, Spain
Francisco Almeida, Universidad de Murcia, Spain
Mehmet Emin Aydin, University of Bedfordshire, UK
Ruibin Bai, University of Nottingham, UK
Christian Bierwirth, Universität Bremen, Germany
Maria Biesa, Universitat Politècnica de Catalunya, Spain
Christian Blum, Universitat Politècnica de Catalunya, Spain
Peter Brucker, Universität Osnabrück, Germany
Edmund Burke, University of Nottingham, UK
Rafael Caballero, Universidad de Málaga, Spain
Pedro Castillo, Universidad de Granada, Spain
David W. Corne, Heriot-Watt University, Edinburgh, UK
Ernesto Costa, Universidade de Coimbra, Portugal
Carlos Cotta, Universidad de Málaga, Spain
Peter Cowling, University of Bradford, UK
Bart Craenen, Napier University, Edinburgh, UK
Keshav Dahal, University of Bradford, UK

Karl Dörner, Universität Wien, Austria
Jeroen Eggermont, Leiden University Medical Center, The Netherlands
Anton V. Eremeev, Omsk Branch of Sobolev Institute of Mathematics, Russia
Richard F. Hartl, Universität Wien, Austria
Antonio J. Fernández, Universidad de Málaga, Spain
Francisco Fernández de Vega, Universidad de Extremadura, Spain
Bernd Freisleben, Universität Marburg, Germany
José Enrique Gallardo, Universidad de Málaga, Spain
Michel Gendreau, Université de Montréal, Canada
Jens Gottlieb, SAP, Germany
Walter Gutjahr, Universität Wien, Austria
Jin-Kao Hao, Université d'Angers, France
Emma Hart, Napier University, Edinburgh, UK
Geir Hasle, SINTEF Applied Mathematics, Norway
Juhos István, University of Szeged, Hungary
Bryant Arthur Julstrom, St. Cloud State University, MN, USA
Mario Köppen, Fraunhofer IPK, Berlin, Germany
Graham Kendall, University of Nottingham, UK
Joshua Knowles, University of Manchester, UK
Gary Kochenberger, University of Colorado, Denver, USA
Jozef Kratica, University of Belgrade, Serbia
Rhyd Lewis, Cardiff University, UK
Andrea Lodi, Universitá degli Studi di Bologna, Italy
José Antonio Lozano, University of the Basque Country, Spain
Vittorio Maniezzo, Universitá degli Studi di Bologna, Italy
Jose Marcos Moreno, Universidad de Murcia, Spain
Dirk C. Mattfeld, Technische Universität Braunschweig, Germany
Helmut Mayer, Universität Salzburg, Austria
Barry McCollum, Queen's University Belfast, UK
Juan Julián Merelo, Universidad de Granada, Spain
Daniel Merkle, Universität Leipzig, Germany
Peter Merz, Technische Universität Kaiserslautern, Germany
Martin Middendorf, Universität Leipzig, Germany
Julian Molina, Universidad de Málaga, Spain
Pablo Moscato, University of Newcastle, Australia
Christine L. Mumford, Cardiff University, UK
Nysret Musliu, Technische Universität Wien, Austria
Gabriela Ochoa, University of Nottingham, UK
Joaquin Pacheco, Universidad de Burgos, Spain
Francisco J. B. Pereira, Universidade de Coimbra, Portugal
Adam Prügel-Bennett, University of Southampton, UK
Jakob Puchinger, National ICT Australia, Melbourne, Australia
Günther Raidl, Technische Universität Wien, Austria
Marcus Randall, Bond University, Queensland, Australia
Marc Reimann, Institute for Operations Research, Switzerland

Table of Contents

Adaptive Tabu Tenure Computation in Local Search

I. Devarenne, H. Mabed, and A. Caminada

UTBM, SET Lab, 90010 Belfort Cedex, France
{isabelle.devarenne, hakim.mabed, alexandre.caminada}@utbm.fr

Abstract. Optimization methods based on complete neighborhood exploration such as Tabu Search are impractical against large neighborhood problems. Strategies of candidate list propose a solution to reduce the neighborhood exploration complexity. We propose in this paper a generic Tabu Search algorithm using adaptive candidate list strategy based on two alternate candidate lists. Each candidate list strategy corresponds to a given search phase: intensification or diversification. The optimization algorithm uses a Tabu list containing the variables causing loops. The paper proposes a classification of Tabu tenure managing in the literature and presents a new and original Tabu tenure adaptation mechanism. The generic method is tested on the k-coloring problem and compared with some best methods published in the literature. Obtained results show the competitiveness of the method.

1 Introduction

In this paper, we propose a generic Tabu Search based on Adaptive Candidate List strategy, noted ACL_TS. The method alternates the use of two candidate lists [22] corresponding to intensification and diversification phases of search. During the intensification phase, a variable v is chosen from the candidate list CL_I and moved. However, when a loop is detected, a variable belonging to an extended candidate list CL_D is selected during the next iteration. A loop appears when a given variable is chosen more then a given threshold α during the last M iterations. After loop detection, the variable causing the loop is made Tabu. The choice of the loop variable is then forbidden (neighborhood restriction) for a given number of iterations named Tabu tenure. Alternation of intensification and diversification operators allows controlling the concentration and the repartition of visited solutions in certain areas of search. The intensification operator is based on the selection of a variable from a limited set of critical variables. The diversification operator selects a different category of variables covering a broader set of variables of the problem. The working scheme of the generic method is given in figure 1.

ACL_TS is applied to the k-coloring problem [6, 7]. The problem consists, given an undirected graph, in coloring the nodes using only k colors in such manner to assign different colors to adjacent nodes. From optimization point of view, the objective is to minimize the number of conflicts due to reusing the

J. van Hemert and C. Cotta (Eds.): EvoCOP 2008, LNCS 4972, pp. 1–12, 2008.

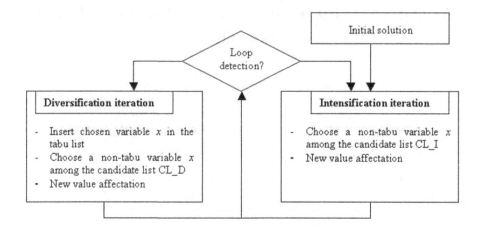

Fig. 1. Adaptive candidate list strategy

same color on linked nodes. During ordinary iteration, intensification operator chooses a node from the most conflicted nodes set. In the opposite, when a loop is detected, a conflicted node is randomly chosen. A node is said in conflict when it is colored with the same color than an adjacent node. In both cases the new assigned color is chosen among the best ones (less used color by the adjacent nodes).

Study performed in [6] shows that the method performances are very impacted by the Tabu tenure value. The Tabu notion was introduced first by Tabu Search method. Glover [13, 14] and Hansen and Jaumard [15], introduced a heuristic using a memory structure to exclude certain choices and restrict the neighborhood of the current solution to a subset of $V(s)$ also called accessible solutions. The Tabu list structure memorizes some information on past moves such as: solutions components, some moves or complete solutions. A Tabu tenure parameter has been introduced to prohibit some actions for a given number of iterations. In the literature there are different methods used to specify its value and its evolution during the search.

The rest of the paper is organized into 4 sections. In the first section, we present a classification of Tabu tenure specification approaches. Then in the second section, we present an adaptive mechanism for Tabu tenure calculation based on the number of visits of each variable. In the third section, we give analysis of the method results on the well-known DIMACS instances. Finally, a comparison with other famous works is made in section 4 and we conclude in the last section.

2   Tabu Tenure in the Literature

The Tabu tenure is a critical parameter that greatly influences the performance of the method. The duration of the prohibition period can be either static or dynamic. When the Tabu tenure is static, the value of Tabu duration is fixed throughout the search. In other words, the number of iterations for which the move is prohibited is fixed as in [16]. In the opposite, dynamic Tabu tenure varies during the execution. Several studies, such as those conducted by Hao et al. [17], have shown that a dynamic Tabu tenure could be more interesting than a static value.

We present different dynamic Tabu tenure computation used in the literature. All the presented methods use a Tabu list structure. Tabu tenure specification approaches are classified into four main method families: (a) approaches based on the running time, (b) approaches using a range of possible values, (c) approaches based on the current state of search, (d) approaches based on the historic of search.

2.1 Time Depending Tabu Tenure

In this approach, the Tabu tenure value is adjusted according to the spent time or to the current iteration number. The objective is then to progressively reduce the diversification level by decreasing the Tabu tenure value. This kind of approach is illustrated by Montemanni and Smith work [20] on the frequency assignment problem. The Tabu tenure is decreased during the run after each It_s iterations, independently of the search progression following the expression: $T = \beta \times T$, where β is a fixed value comprised in the interval $[0, 1[$. A value of β near to 1 allows to reduce slowly the Tabu tenure. The Tabu tenure T is equal to an initial value T_{init} at the beginning and to a minimum threshold T_{min} at the end. Parameters T_{min}, β et It_s are respectively fixed to 10, 0.96 and 5×10^4 for all tested FAP instances. Comparison made in [20] with three fixed Tabu tenure values shows that time depending Tabu tenure globally provides better results.

2.2 Random Bounded Tabu Tenure

In this second case, the Tabu tenure value is randomly chosen, at each iteration, inside a fixed interval. This principle is the most used in the literature and it is used as reference in this work. Intervals bounds are generally chosen according to some characteristics of the problem. For example, in Di Gaspero and Schaerf work [8], the Tabu tenure interval is calculated according to the number of variables N using the following expression: $[\frac{2}{3} \times N; N]$. Taillard [21] proposed a Tabu search method named *robust taboo search method* for quadratic assignment problem. In this method, Tabu tenure value is randomly chosen every $2 \times s_{max}$ iterations into the interval $[s_{min}; s_{max}]$, where $s_{min} = \lfloor 0.9 \times N \rfloor$ and $s_{max} = \lceil 1.1 \times N \rceil^1$. Another example is given by Bachelet and Talbi [1] where the Tabu tenure is randomly chosen within the interval $[\frac{N}{2}; \frac{3 \times N}{2}]$.

2.3 Reactive Tabu Tenure

In the third approach, the Tabu tenure variation is a reaction to current solution state and no historic of search past is used. At each iteration, several pieces

---

[1] $\lfloor \rfloor$ and $\lceil \rceil$ represent respectively the integer part and integer part plus 1.

of information are extracted from the current solution and used to define the value of the prohibition period. Typically, these data designate the number of conflicted variables or the number of neighbors of the current solution.

Galinier and Hao [11] have proposed to increase the Tabu tenure according to the evaluation $F(s)$ of the current solution s. In order to do that, two parameters L (randomly chosen into the interval $[0; 9]$) and λ (empirically fixed to 0.6) have been used as indicated in the following expression:

$$T = L + \lambda \times F(s) \tag{1}$$

In this work, Tabu tenure is maintained proportional to the cost function of the current solution. The given values for both L and λ parameters allow obtaining very good results. Although for other instances, parameter setting should be envisaged.

2.4   Adaptive Tabu Tenure

In the fourth approach, Tabu tenure is adjusted during the search according to the search history. We mention here three examples of adaptive Tabu tenure.

The first one is what Battiti et al. [2, 3] called *Reactive Tabu Search* method. The idea is to use a search memory composed by the previously visited solutions. After each move, the algorithm verifies if the current solution has already been found. If it is the case, the Tabu tenure T is increased. Otherwise the Tabu tenure value is decreased when no repetition is occurred during a sufficient long time. The Tabu tenure T is initialized to 1 at the beginning of search. When a solution is revisited, the Tabu tenure T is gradually increased using the following equation:

$$T = \min(\max(T \times 1.1, T + 1), L - 2) \tag{2}$$

where L represents the number of 0/1 variables. In the worst case, when the Tabu tenure is very high, two moves are possible. In [3], this method have been applied to the quadratic assignment problem. Comparison with Tabu search methods using static and interval based Tabu tenure shows very competitive results particularly for the highest size problems.

In the approach of Blöchliger [4] named *Approximated Cycle-Detection scheme (ACD)*, the method detects the cycles without storing the previously visited solutions. For that, a reference solution and a distance measurement are used. Every iteration, the distance between the reference solution and the current one is calculated. If this distance is equal to zero, a cycle is detected and the Tabu tenure is increased by an increment value η. Otherwise, it is slowly decreased. The parameter η varies during the search. Initialized to 5, it is incremented by 5 when a cycle is detected, and it is decremented by 1 every 15 000 iterations. The efficiency of the method depends on when the reference solution is updated and the relevance of used distance. The proposed method consists to update the reference solution when the current one is better in term of evaluation function and when the current solution is very far from the reference one.

In Consistent-Neighborhood Tabu (*CN-Tabu*) [10] method, the resolution is obtained by the exploration of partial solutions. Every iteration, a new assignment (x_i, v_i) is made, where x_i refers to an unassigned variable and v_i to the new assigned value. Each assignment (x_j, v_j) where x_j is an adjacent assigned variable is set Tabu and desinstantiated from the current solution. The Tabu tenure is dynamically calculated according to the number of times that the value v_j has been assigned to the variable x_j:

$$T(x_j, v_j) = nb_{mvt}(x_j, v_j) \tag{3}$$

where $T(x_j, v_j)$ is the prohibition period associated to the assignment (x_j, v_j), and $nb_{mvt}(x_j, v_j)$ is the number of times that the value v_j has been assigned to the variable x_j during the search.

Typically, the historic is not used to determine how much the Tabu tenure should be increased or decreased. Usually the historic provides only information on when the Tabu tenure needs to be adapted.

We propose here an adaptive Tabu tenure calculation where the value of Tabu tenure is adjusted separately for each decision variable. Memory structures and statistical data are used to determine when a decision variable is made Tabu and its prohibition period.

3   Adaptive Tabu Tenure in *ACL_TS* Method

In this section, we describe the process of Tabu tenure adaptation. In [6], we have compared dynamic and static Tabu tenure. In the dynamic case, the Tabu tenure value is randomly chosen into the interval $DT = [0.5 \times c(N); 1.5 \times c(N)]$ with N is the variables number and $c(N)$ the interval center. This variant is called $LD + DT$. Table 1 presents the results obtained by the $ACL\_TS : DT$ method combining loop detection and Tabu list for different Tabu tenure intervals on some Leighton instances (second DIMACS challenge instances[2]). These instances have been generated by Leighton [18]. All Leighton graphs used the same number of nodes (450) and the chromatic number is known. These instances are largely used in the literature. Three comparison criteria are retained: s the success rate over 10 runs, it the average number of iterations needed to resolve the instance and c the average conflicts number of the best solution found.

Tabu tenure interval is calculated according to the number of variables N. We observe that even if both instances presented in table 1 are composed of (450) nodes, the best results are not found by the same calculation function. All runs optimally solve the first problem with a Tabu tenure chosen in the first interval, whereas the second problem is never solved. Only the second settings allows to solve the second instance during 50% of runs. With the highest interval, both instances are never solved. Consequently, the instance size defined by the number of variables is not sufficient to determine the ideal value of Tabu tenure.

The dynamic adaptation of the Tabu tenure provides a serious alternative to statically determine the value or the set of values of the parameter. The used

---

[2] Available on `http://mat.gsia.cmu.edu/COLOR/instances.html`

Table 1. Basic ACL_TS (LD+DT): Influence of the interval of Tabu tenure

$f(N)$	$0.5\times\frac{\sqrt{N}}{8}; 1.5\times\frac{\sqrt{N}}{8}$			$0.5\times\frac{\sqrt{N}}{7}; 1.5\times\frac{\sqrt{N}}{7}$			$0.5\times\frac{\sqrt{N}}{4}; 1.5\times\frac{\sqrt{N}}{4}$		
DIMACS	s	it	c	s	it	c	s	it	c
le450_15a	10	429 118	0	5	147 775	2	0	-	4
le450_15b	0	-	1	5	296 529	2	0	-	4

method adjusts the Tabu tenure value according to the evolution of the search. The idea is to analyze the desired effect of the Tabu status. In fact, a good Tabu tenure value should prevent search cycles and should therefore be large enough to exclude variables provoking loops and orient the search toward new configurations.

We proposed to adapt Tabu tenure to each problem and to each variable of the problem by the use of variable search history. In previous work we have shown that the Tabu status is more efficient when it is applied only on variables provoking loops [6]. In addition, we aim to determine dynamically the Tabu tenure according to the number of loops provoked by each variable.

When a loop is detected the node *source* is made Tabu during a prohibition period calculated as follow:

$$AT(x_i) = \text{rand}(DT) + \frac{nbLoops(x_i)}{\sum_{x_j \in V} nbLoops(x_j)} \times N \qquad (4)$$

with $nbLoops(x_i)$ the number of loops provoked by the node x_i and DT the dynamic interval defined as follow:

$$DT = \left[0.5 \times \frac{\sqrt{N}}{2}; 1.5 \times \frac{\sqrt{N}}{2}\right] \qquad (5)$$

This variant is noted $LD+AT$ for Loop Detection and Adaptive Tabu tenure.

4   Study of the Tabu Tenure Repartition

This section presents the comparison between of the basic ACL_TS based on the LD+DT variant and the adaptive Tabu tenure (noted LD+AT) version. We observe here the Tabu tenure used for each node and the number of loops provoked per node. This study is made on some DIMACS instances: random graphs named DSJCN.p when N is the number of nodes and p the probability that two nodes are joined. DSJC instances are also largely used in the literature. Figures 2 and 4 refers to a single successful run on the instance DSJC125.1. Figures 3 and 5 concerns the same not successful run on DSJC500.1. The two first figures correspond to the basic method LD+DT and the two others to the adaptive one LD+AT. Spectrograms (a) show the use frequency of each Tabu

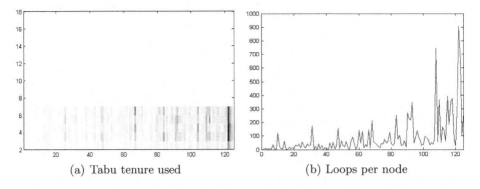

(a) Tabu tenure used   (b) Loops per node

Fig. 2. Method LD+TD instance DSJC125.1

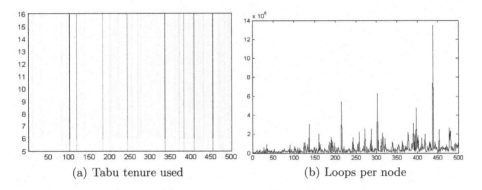

(a) Tabu tenure used   (b) Loops per node

Fig. 3. Method LD+TD instance DSJC500.1

tenure value (y-axis) per node. Darker points correspond to most used values. Curves (b) represent the number of loops provoked by each node. In all curves or spectrograms, the x-axis represents the nodes classified by ascending order of degree.

For the method LD+DT, Tabu tenure used per node are chosen randomly into an interval around $f(N) = \sqrt{N}/2$. Spectrograms 2(a) and 3(a) show uniform repartition over the values of the interval. The nodes provoking the highest number of loops (visible in 2(b) and 3(b) curves), appear with a dark color in spectrograms 2(a) and 3(a).

Unlike the first two figures, the spectrograms 4(a) and 5(a) are very different: Tabu tenure values are not uniformly used by the method LD+AT. We expect from this mechanism a better diversification of the search in favor of variables that are not involved in loops. Furthermore, according to the used equation 4, nodes causing most loops (curves 4(b) and 5(b)) use higher values of Tabu tenure. The penalty has increased significantly the values of used Tabu tenure, which corresponds to the expected behavior.

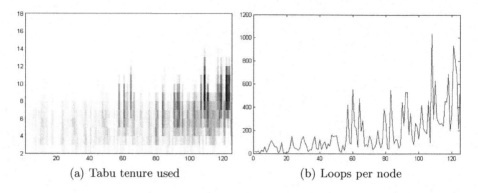

Fig. 4. Method LD+AT instance DSJC125.1

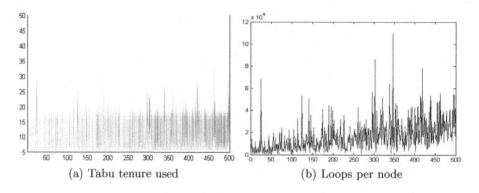

Fig. 5. Method LD+AT instance DSJC500.1

We have observed the difference of both methods in term of Tabu tenure used by each node during the search. Now, table 2 present comparatives results obtained with these both methods on Leighton instances. Two comparison criteria are used: the success rate s over 10 runs and the average iterations number it.

For all problems studied, we observed that adaptive Tabu tenure is generally better in term of success rate or in term of average iterations number for 7 out of 8 instances. Dynamic Tabu tenure is better only for the instance le450_5d. Furthermore, the method is robust; it finds a solution to all the problems, which demonstrates the effectiveness of the combination of adaptive Tabu tenure and loop detection mechanism.

These results are confirmed by DSJC instances in table 3. The success rate is better for the instance DSJC500.1 and the average number of unsatisfied constraints is lower for the adaptive method (the tests were carried out on five executions and graded from 0 to 10 according to their performance) for 5 out of 6 instances.

Table 2. Adaptation of the Tabu tenure on Leighton instances

problems	k	Dynamic Tabu tenure LD+DT		Adaptive Tabu tenure LD+AT	
		s	it	s	it
le450_5a	5	3	302 000	10	326 148
le450_5b	5	1	629 000	2	1 205 950
le450_5c	5	10	252 120	10	251 881
le450_5d	5	8	396 000	2	1 079 031
le450_15a	15	5	304 000	10	1 889 569
le450_15b	15	9	390 000	10	904 067
le450_15c	15	0	–	10	70.6×10^6
le450_15d	15	0	–	4	192.6×10^6

Table 3. Adaptation of the Tabu tenure on DSJC instances

problems	Dynamic Tabu tenure LD+DT		Adaptive Tabu tenure LD+AT	
	s	c	s	c
DSJC500.1	0	1	6	0.4
DSJC500.5	0	4.8	0	4.2
DSJC500.9	0	3.8	0	3.4
DSJC1000.1	0	15	0	13.4
DSJC1000.5	0	41.2	0	42.8
DSJC1000.9	0	4.4	0	4

5 Comparison with the Literature

In this section, we compare the results obtained by our method before and after
the adaptation of the Tabu tenure with four others well-known works published
in the literature. First column presents the studied instances and in the second
column, we give the chromatic number χ when it is known and the best number
of colors used to color each instance, noted k^*. Table 4 presents the minimum
number of colors used to solve each instance found by each method.

The first method is named HCA for Hybrid Coloring Algorithm published
by Galinier and Hao [11]. This hybrid method combines genetic algorithm and
Tabu search. The Tabu tenure is calculated using the mechanism explained in
the section 2.3. Notice that the algorithm has obtained the bests known results
for several DIMACS instances.

The second method published by Galinier et al. [12], is a population-based
method, named AMACOL (Adaptive Memory Algorithm for K-COLoring). This
algorithm is also one of the most efficient algorithms in the literature.

The third method is the Iterated Local Search (ILS) algorithm presented by
Chiarandini et al. in [5]. A Tabu search algorithm is run until the best solution
found does not change during a fixed number of iterations. A perturbation is
then applied on the best solution found so far and the Tabu search is run again.

The last algorithm is the Generic Tabu Search (GTS) published by Dorne and Hao [9]. This method uses reactive approach to compute the Tabu tenure (see section 2.3). The search is started from a greedy initial solution built by DSATUR procedure. The Tabu tenure depends on the number of conflicted nodes of the current solution.

Finally, we present our results in the last two columns: first, the basic version (LD+DT) and the adaptive Tabu tenure method (LD+AT) with DT interval equal to $\left[0.5 \times \frac{\sqrt{N}}{2} ; 1.5 \times \frac{\sqrt{N}}{2}\right]$ for both methods.

Table 4. Comparison with other methods on DSJC and Leighton instances

problems	(χ,k^*)	HCA	AMACOL	ILS	GTS	LD+DT	LD+AT
DSJC500.1	(-,12)	-	12	13	13	13	**12**
DSJC500.5	(-,48)	48	48	50	50	50	**49**
DSJC500.9	(-,126)	-	126	127	127	128	128
DSJC1000.1	(-,20)	20	20	21	21	21	21
DSJC1000.5	(-,83)	83	84	90	90	89	89
DSJC1000.9	(-,224)	224	224	227	226	230	**227**
le450_15c	(15,15)	15	15	15	-	16	**15**
le450_15d	(15,15)	-	15	15	-	16	**15**
le450_25c	(25,25)	26	26	26	-	26	26
le450_25d	(25,25)	-	26	26	-	26	26

Among the different methods presented here, the method HCA gets the best performance. Our method using adaptive Tabu tenure outperforms our dynamic Tabu tenure method. In particular, adaptive method allows to solve 5 instances (in bold) with less colors than dynamic method on the 10 instances presented in this table. Globally, our adaptive method obtained results of the same quality than HCA on 2 instances and than AMACOL on 5 instances. On random graphs, results obtained by LD+AT are very competitive compared to the others methods. Compared to ILS method, LD+AT obtained better results for 3 instances but it is worst for the instance DSJC500.9. For all others instances, performance are identical in term of minimum number of colors needed.

6   Conclusion and Perspectives

In conclusion, we have seen that the Tabu tenure calculation impacts on the global performance of Tabu list based methods. In the literature, several studies have been undertaken to determine the value of the parameter. We have proposed in this paper a new computation method of the Tabu tenure depending on the specific search history of each variable. The idea is to compute the Tabu tenure according to the number of loops provoked by each variable. We have used the k-coloring problem as a framework for test our method. The presented results show the effectiveness of our generic method comparing to competitive and specialized methods. The adaptive candidate list based Tabu Search was also applied to structured frequencies plan affectation problem [19]. The method has been ranked first among three proposed works (by other teams) in the frame of the ALGOPDF project. In ACL_TS method, only the variables provoking loops become Tabu. The loop detection is determined by a threshold value specifying the number of recent visits (during the M last iterations) after which a variable is considered in loop. The Tabu mechanism being strongly impacted by the loop detection, the study of this second parameter is critical.

References

[1] Bachelet, V., Talbi, E.-G.: COSEARCH: a co-evolutionary metaheuristic. In: Proceedings of the Congress on Evolutionary Computation CEC 2000, pp. 1550–1557 (2000)

[2] Battiti, R., Tecchiolli, G.: The reactive Tabu search. ORSA Journal on Computing 6(2), 126–140 (1994)

[3] Battiti, R.V.J., Rayward-Smith, I.O., Smith, G. (eds.): Reactive Search: Toward Self-Tuning Heuristics, pp. 61–83. John Wiley and Sons Ltd, Chichester (1996)

[4] Blöchliger, I.: Suboptimal colorings and solution of large chromatic scheduling problems, Ph.D. thesis, Mathematics Department, Ecole polytechnique federale de Lausanne, Lausanne, Suisse, 20056

[5] Chiarandini, M., Dumitrescu, I., Stutzle, T.: Stochastic Local Search for the Graph Colouring Problem, Technical Report AIDA-05-03 (2005)

[6] Devarenne, I., Mabed, H., Caminada, A.: Intelligent neighborhood exploration in local search heuristics. In: 18th IEEE International Conference on Tools with Artificial Intelligence, Washington D.C. USA (2006)

[7] Devarenne, I., Mabed, H., Caminada, A.: Self-adaptive Neighborhood Exploration Parameters in Local Search. In: 7th EU/MEeting on Adaptive, Self-Adaptive, and Multi-Level Metaheuristics, University of Málaga, Spain (2006)

[8] Di Gaspero, L., Schaerf, A.: A Tabu Search Approach to the Traveling Tournament Problem. In: MIC 2005: The 6th Metaheuristics International Conference, Vienna, Austria (2005)

[9] Dorne, R., Hao, J.K.: Tabu Search for graph coloring, T-coloring and Set T-colorings. In: Osman, I.H., et al. (eds.) Metaheuristics 1998: Theory and Applications, ch. 3, Kluver Academic Publishers (1998)

[10] Dupont, A.: Étude d'une métaheuristique hybride pour l'affectation de fréquences dans les réseaux tactiques évolutifs, PhD thesis, Univ. Montpellier II, France (2005)

[11] Galinier, P., Hao, J.-K.: Hybrid Evolutionary Algorithms for Graph Coloring. Journal of Combinatorial Optimization 3, 379–397 (1999)

[12] Galinier, P., Hertz, A., Zufferey, N.: An Adaptive Memory Algorithm for the k-colouring problem. Les Cahiers du GERAD G-2003-35 (2004)

[13] Glover, F.: Tabu Search - Part I. ORSA Journal on Computing 1(3), 190–206 (1989)

[14] Glover, F.: Tabu Search - Part II. ORSA Journal on Computing 2(1), 4–32 (1990)

[15] Hansen, P., Jaumard, B.: Algorithms for the maximum satisfiability problem. Computing 4(44), 279–303 (1990)

[16] Hao, J.-K., Galinier, P.: Tabu search for maximal constraint satisfaction problems. In: Smolka, G. (ed.) CP 1997. LNCS, vol. 1330, pp. 196–208. Springer, Heidelberg (1997)

[17] Hao, J.K., Dorne, R., Galinier, P.: Tabu search for frequency assignment in mobile radio networks. Journal of Heuristics 4(1), 47–62 (1998)

[18] Leighton, F.: A graph coloring algorithm for large scheduling problems. Journal of research of the national bureau of standards, 489–505 (1979)

[19] Mabed, H., Devarenne, I., Caminada, A., Defaix, T.: Frequency Planning for Military Slow Frequency Hopping System. In: International Network Optimization Conference 2007, Spa, Belgium (2007)

[20] Montemanni, R., Smith, D.H.: A Tabu search Algorithm with a dynamic Tabu list for the Frequency Assignment Problem, Technical Report, University of Glamorgan (2001)

[21] Taillard, E.: Robust taboo search for the quadratic assignment problem. Parallel Computing 17, 443–455 (1991)

[22] Neveu, B., Trombettoni, G., Glover, F.: A Candidate List Strategy with a Simple Diversification Device. In: Wallace, M. (ed.) CP 2004. LNCS, vol. 3258, pp. 423–437. Springer, Heidelberg (2004)

A Conflict Tabu Search Evolutionary Algorithm for Solving Constraint Satisfaction Problems

B.G.W. Craenen and B. Paechter

Napier University
10 Colinton Rd, Edinburgh, EH10 5DT
{b.craenen, b.paechter}@napier.ac.uk
http://www.soc.napier.ac.uk

Abstract. This paper introduces a hybrid Tabu Search - Evolutionary Algorithm for solving the binary constraint satisfaction problem, called CTLEA. A continuation of an earlier introduced algorithm, called the STLEA, the CTLEA replaces the earlier compound label tabu list with a conflict tabu list. Extensive experimental fine-tuning of parameters was performed to optimise the performance of the algorithm on a commonly used test-set. Compared to the performance of the earlier STLEA, and benchmark algorithms, the CTLEA outperforms the former, and approaches the later.

1 Introduction

Solving constraint satisfaction problems (CSP) with evolutionary algorithms (EAs) has been studied extensively over the years. This has resulted in the introduction of a large number of algorithms. A study of the performance of a representative sample of EAs, using a large randomly generated test-set, was carried out in [1]. A more comprehensive study, using an updated test-set, also including a large number of algorithm variants, can be found in [2]. There, it was found that one algorithm variant, the Stepwise-Adaptation-of-Weights EA with randomly initialised domain sets (rSAWEA), outperformed all other EAs. However, when comparing the effectiveness and efficiency of this algorithm with non-evolutionary algorithms, it was found that although the former could be approximated, the algorithm still fell short of achieving the later.

A reason for this lack of efficiency was identified to be the lack of preventing EAs from traversing already studied search-paths, something non-evolutionary algorithms are usually prevented from doing. In [3] therefore, the Simple Tabu List Evolutionary Algorithm (STLEA) was introduced, using a tabu list preventing it from wasting computational effort on already traversed search-paths. Tabu lists are a part of the Tabu Search (TS) meta-heuristic ([4]). They are used to ensure that an algorithm does not return to an already searched neighbourhood by making it tabu. The Tabu Search meta-heuristic has found its way into EAs before (e.g. [5,6,7,8]), especially for EAs handling constrained problems. An important feature of tabu lists is that they are only referenced, i.e., only insertion and look-up of elements is used. As such, they can be implemented as

J. van Hemert and C. Cotta (Eds.): EvoCOP 2008, LNCS 4972, pp. 13–24, 2008.

a hash-set, ensuring constant time cost when a suitable hash function is chosen. In [3] it was found that the STLEA, with a tabu list containing candidate solutions, increased efficiency enough to surpass the rSAWEA, with equal or better effectiveness. However, while comparing the performance of the STLEA to that of non-deterministic algorithms, it was still found to have equal effectiveness but inferior efficiency. Although, in [3], an important step was made in the right direction, it was still not good enough to beat deterministic algorithms.

This paper endeavours to set the next step by refining the tabu list used in the algorithm. Instead of focussing on candidate solutions for the tabu list, it focusses on storing conflicts instead. There are two advantages to this approach.

First, the number of conflicts in a CSP-instance is smaller than the number of possible candidate solutions, making the tabu list itself much easier to handle. This makes the tabu list approach more viable for large CSP-instances as well.

Second, a conflict tabu list can be used more directly to guide the search-path, than a candidate solution tabu list can. Whereas a candidate solution tabu list is only useful to exclude generated candidate solutions (individuals), the conflict tabu list can be used directly, for example by the crossover or mutation operator. This change of focus for the tabu list results in the introduction of the Conflict Tabu List Evolutionary Algorithm (CTLEA).

The paper is organised as follows: in section 2, a definition of constraint satisfaction problems is given. Section 3 defines the proposed algorithm. The experimental setup is explained in section 4. Section 5 discusses the results of the experiments, and finally in section 6, the conclusions that can be drawn from this paper are set forth.

2   Constraint Satisfaction Problems

The *Constraint Satisfaction Problem* (CSP) is a well-known NP-complete satisfiability problem ([9]). Defined informally as a set *variables* X and a set of *constraints* C between these variables, it only allows variables to be assigned values from their respective *domains*, denoted as $D_x, x \in X$. A *label* is then a variable-value pair, denoted: $\langle x, d \rangle, x \in X, d \in D_x$, and assigning a value to a variable is called *labelling* it. A *compound label* is a simultaneous assignment of several values to their respective variables, and a constraint is then a set of compound labels, with each compound label determining when the constraint is *violated*. A compound label not in a constraint is said to *satisfy* that constraint, while one that is, is called a *conflict*. A *solution* of a CSP is then defined as a compound label that contains all variables, but no conflicts from any constraint.

The number of distinct variables in the compound labels of a constraint is called the *arity* of that constraint, and these variables are said to be relevant to this constraint. The maximum arity of all constraints in a CSP is the arity of the CSP itself. In this paper, we only consider CSPs with an arity of two, meaning that all constraints in the CSP have arity two as well. Such CSPs are called *binary CSPs*. Restricting the arity of the studied CSPs is not a restriction in itself though, as [10] shows that every CSP can be transformed into an equivalent binary CSP.

This paper will use the same test-set constructed in [2], and [3]. It consists of model F generated solvable CSP-instances ([11]) with 10 variables and a uniform domain size of 10 values. Complexity of these instances is determined by using two complexity measures for the CSP: density (p_1) and average tightness $(\overline{p_2})$, both presented as a real number between 0.0 and 1.0 inclusive. *Density* is the ratio between the maximum number of constraints of a CSP $(\binom{|X|}{2}$ for a binary CSP) and the actual number of constraints $(|C|)$. The *tightness* of a constraint is defined as one minus the ratio between the maximum number of possible conflicts $(|D_{x_1} \times D_{x_2}|$ for a binary constraint relevant to variables x_1 and x_2) and the number of conflicts. The *average tightness* of a CSP is then the average tightness of all constraints in the CSP.

In the density-tightness parameter space of all randomly generated CSP-instances, the hard-to-solve instances can be found in what is called the *mushy region*. For the test-set used, the following density-tightness combinations lie within the mushy region $1 : (0.1, 0.9)$, $2 : (0.2, 0.9)$, $3 : (0.3, 0.8)$, $4 : (0.4, 0.7)$, $5 : (0.5, 0.7)$, $6 : (0.6, 0.6)$, $7 : (0.7, 0.5)$, $8 : (0.8, 0.5)$, and $9 : (0.9, 0.4)$. For each of these density-tightness combinations, 25 CSP-instances were selected from a population of 1000 randomly generated CSP-instances (see [2] for selection criteria). In total, the test-set includes $9 \cdot 25 = 225$ CSP-instances. The test-set can be downloaded at: http://www.emergentcomputing.org/csp/testset_mushy.zip.

3   The Algorithm

The *Conflict Tabu List Evolutionary Algorithm* (CTLEA) is an evolutionary algorithm using the Tabu Search meta-heuristic. In keeping with the simple definition of Tabu Search as "a meta-heuristic superimposed on another heuristic" ([4]), the CTLEA uses only the tabu list. In the STLEA, as described in [3], the tabu list was used to ensure that a compound label was not checked twice during a run. A major criticism of this type of tabu list is that the number of possible compound labels (candidate solutions), and therefore the amount of memory needed to maintain it, could become quite large, depending mostly on CSP parameters. The CTLEA therefore focusses on conflicts, and uses the tabu list to ensure that a *conflict* is not rechecked during a run. There are two advantages from using tabu lists in this way. First, the number of conflicts of a CSP-instance is smaller than the number of possible candidate solutions, addressing the criticism above. This not only makes the tabu list easier to maintain, but allows for the use of the tabu list for large CSP-instances as well. Second, a tabu list focussing on conflicts can be used by the algorithm to guide the search-path directly. Whereas a candidate solution tabu list is only useful to exclude whole candidate solutions (individuals), a conflict tabu list can be used by, for example, the crossover and mutation operators of the EA, to determine directly which variables and values can be labelled.

An example may provide more insight into the difference of size and subsequent cost of maintenance between a compound label and conflict tabu list. Let us consider the worst-case scenario for both tabu lists. The test-set used has 10 variables, and a uniform domain size of 10 elements. If the compound label tabu list were used, this means that in the worst-case, it should be able to store $10^{10} = 100,000,000,000$ compound labels, i.e., all possible candidate solutions. On the other hand, if the conflict tabu list were used, again in the worst-case, it should be able to store $\binom{10}{2} \times 10 \times 10 = 45 \times 10 \times 10 = 4500$ conflicts, i.e., all possible conflicts. This difference increases with scale as well. It must be noted however, that on average, only a fraction of the compound labels was stored by the STLEA in [3], although still substantially more than the number of conflicts stored by the CTLEA.

The basic structure of the CTLEA, shown in algorithm 1, was kept purposely close to that of the canonical EA. The biggest difference is that the CTLEA uses a single variance operator, called move-operator, instead of a separate crossover- and mutation operator.

The CTLEA works as follows. A population P of *popsize* individuals is initialised (line 2) and evaluated (line 3). The representation used by the individual, and initialisation is described in 3.1, the objective function used to evaluate them is described in section 3.2. The CTLEA then enters a while-loop wherein it iterates for a number of generations (line 4 to 9) until either a solution is found, or the maximum number of conflict checks allowed (*maxCC*) has been reached or exceeded (the *stop condition* in line 4). At the beginning of each iteration of the algorithm, parents are selected from P into population S using biased linear ranking selection ([12]) with bias *bias* (line 5). These parents are used by the move-operator to create a new offspring population (line 6), as described in section 3.4. The new offspring population is then evaluated by the objective function (line 7). Finally, at the end of each iteration, the survivor selection operator selects individuals from the offspring population (S) into a new population (P) to be used for the next iteration/generation (line 8). No 'elitism' is used by the CTLEA, i.e., no individuals from the previous iteration/generation are forcefully preserved for the next iteration/generation. The tabu list, described in section 3.3, is used by both the objective function and the move-operator.

Algorithm 1: CTLEA

```
1  funct CTLEA(popsize, maxCC, bias)  ≡
2      P := initialise(popsize);
3      evaluate(P);
4      while ¬solutionFound(P) ∨ CC < maxCC do
5              S := selectParents(P, bias);
6              S := moveOperator(S);
7              evaluate(S);
8              P := selectSurvivors(S);
9      od
```

3.1   Representation and Initialisation

A CTLEA individual consists of three parts: a compound label over all variables of the CSP used as a candidate solution; a subset of the constraints defined by the CSP, all violated by the candidate solution; and a field indicating which variable was altered previously, called the *changed variable field*. The variable stored by the changed variable field is called the changed variable.

A new individual is initialised by: labelling all variables in the compound label uniform randomly from the respective domains of each variable; initialising the subset of violated constraints to be empty (later to be used by the objective function); and setting the changed variable field to *unassigned* (later to be used by the move-operator).

Like the representation used in the STLEA in [3], the actual set of violated constraints is used, instead of the derivative *number* of violated constraints commonly used. This reduces the number of conflict checks needed by the objective function to determine the fitness of the individual. In exchange for this, more care has to be taken while maintaining this set during the run. Note that instead of actually storing the constraint itself, the index of the constraint in the set of constraints from the CSP is used. This index can be used to easily retrieve the actual constraint, reducing the memory needed were it to be stored in the individual itself.

The changed variable field is used by the objective function and set by the move-operator to quickly identify which variable has been changed, and consequently which relevant constraints need to be checked. Although limited here to a single variable, this mechanism can be extended in case more than one variable can be changed, although this is not necessary for the CTLEA.

3.2   Objective Function

The objective of the CTLEA is to minimise the number of violated constraints. A solution is found when a candidate solution violates no constraints. The objective function in the CTLEA then maintains the set of violated constraints of an individual. The number of conflict checks necessary for one fitness evaluation is reduced by only considering constraints relevant to the variable stored in the changed variable field. First, the constraints relevant to the changed variable are removed from the set of constraints stored by the individual. Then, all constraints in the CSP relevant to the changed variable are checked, and if violated by the candidate solution, added to the set of constraints stored by the individual. Eventually, the set of constraints stored by the individual will contain all constraints violated by the candidate solution stored by the individual.

For newly initialised individuals, all constraints are checked, and if violated by the candidate solution, added to the set of constraints stored by the individual.

The objective function of the CTLEA uses the tabu list by first checking if a conflict is in the tabu list before performing the conflict check on the CSP-instance. If the conflict is found to exist in the CSP-instance, it is added to the tabu list.

3.3   Conflict Tabu List

The CTLEA maintains a tabu list of conflicts implemented as a hash set, indexed over the constraint they are in. All conflicts found during the run of the CTLEA are added to the conflict tabu list. Only conflicts not already in the the tabu list are added, the list does not contain double entries.

The tabu list is used in only two ways, adding a conflict (insertion), and checking if a conflict is in the tabu list (look-up). Since there is no need to alter or remove a conflict once it has been added, both insertion and look-up can be done in constant time ($O(1)$) depending on the quality of the hash-function, and given adequate size of the hash table.

3.4   Move-Operator

The move-operator of the CTLEA takes a single individual (parent) to produce a single child (offspring). The basic premise of the move-operator is simple: select a variable to change, and change it in such a way that a child with fewer violated constraints is created. As such, there are two choices to be made: which variable to change; and what value to change the variable to.

The move-operator selects which variable to change by selecting one uniform randomly from a multi-set of variables created in the following way. First, all variables relevant to constraints in the set of constraints stored by the individual are added. Then, all variables transitively dependent to the variables already in the multi-set are added. A variable is *transitive dependent* to another variable, if it is relevant to a constraint which the other variable is relevant to. Take, for example, constraint c_1, with its two relevant variables x_1 and x_2. If there is another constraint c_2, with relevant variables x_1 and x_3, then x_3 is transitive dependent to x_1 and c_1. A multi-set is used so that variables that are relevant or transitive dependent to more than one constraint in the set stored by the individual have a higher probability of being selected.

Value selection follows the same idea as variable selection, in that a value is uniform randomly chosen from a set of values. The set of values is created by checking for each value in the domain if it violates a relevant constraint. If it does not, it is added to the set.

The move-operator uses the tabu list by first checking the tabu list if a value is tabu, before checking the CSP-instance. If a value violates a relevant constraint, the conflict is added to the tabu list as well. A simple example of this use of the tabu list goes as follows. Supposed variable x_1 is selected for change. A random value for x_1 is now chosen from the set of values V_1, say v_3. The tabu list is now used to check if this value is in the tabu list. Since the tabu-list stored value-pairs, all variables relevant through a constraint have are now selected. Say, only variable x_5 is relevant to x_1, and in the current individual it has value v_7. The tabu list is now checked for the occurance of value pair: $(\langle x_1, v_3 \rangle, \langle v_5, v_7 \rangle)$. If the value pair is on the tabu list, another value is selected for v_1, if not the move-operator ends. If in the object operator the tried value pair turns out to be a conflict, it is added to the tabu list.

The move-operator iteratively selects different variables and tries to select values to them. No variable is selected twice, and a new variable is selected only when no value can be found that does not violate its relevant constraints. Selecting a variable twice is prevented by not adding a variable to the variable multi-set if it has been selected earlier.

It is possible that all variables relevant, or transitively dependent have been selected by the move-operator already. At this point, the remaining variables are selected uniform randomly. If all possible variables have been selected, a new individual is initialised and inserted in the offspring population. At this point, the move-operator acts as a gradual restart strategy, by starting a new, randomly chosen, search-path for the CTLEA to explore.

4 Experimental Setup

The test-set introduced in [2] was used for experimentation with CTLEA (see section 2). Success rate (SR), and the average number of conflict checks to solution ($ACCS$), are used to measure the performance of the algorithm.

The SR measure is used to measure the effectiveness of an algorithm, and is calculated by dividing the number of successful runs performed by the algorithm by the total number of runs performed. A successful run is a run in which the algorithm solves the CSP-instance. Usually given as a real number between 0.0 and 1.0, the SR can also be expressed as a percentage. A SR of 1.0 or 100%, or perfect SR, means all runs solved their CSP-instance. Since the algorithm's primary task is to solve CSP-instances, the SR is perceived as the most important performance measure to compare algorithms on. Accuracy of the SR measure is affected by the total number of runs.

The $ACCS$ measure is used to measure the efficiency of our algorithm, and is calculated by averaging the number of conflict checks needed by an algorithm over several successful runs. A conflict check is defined as the check made to see if a conflict is in a constraint. Note that the $ACCS$ measure includes all conflict checks made by the algorithm, in the case of the CTLEA, this does also include those made in the move-operator. Conflict checks made during unsuccessful runs of an algorithm are discarded, and if all considered runs of an algorithm are unsuccessful, the $ACCS$ measure is undefined. Used as a secondary performance measure for comparing algorithms, the accuracy is $ACCS$ affected by the number of successful runs and the total number of runs of an algorithm (the ratio of which is the SR measure), i.e., $ACCS$ is more accurate when SR is higher.

Efficiency performance measures have to take into account the computation effort expended by an algorithm. The $ACCS$ uses the number of conflict checks as the atomic measure to quantify the expended computational effort however. The CTLEA also expends computational effort on maintaining the tabu list (see section 3). While comparing the effort spent on performing conflict checks and maintaining the tabu list, it was found that the latter was negligible in comparison to the former when the CSP-instance was sufficiently hard to solve. Given that the CSP-instances used in the test-set are all taken from the mushy region in the density-tightness parameter space (see section 2), complexity of the CSP-instances is sufficient to regard the computation effort of maintaining the tabu list as negligible compared to that of performing the conflict checks.

The CTLEA is blessed with relatively few parameters to fine-tune: the population size (*popsize*); the maximum number of conflict checks allowed (*maxCC*); the *bias* of the biased linear ranking parent selection operator; and the size of the parent population. Although it is possible to vary the size of the parent population, as in [3], we keep it equal to the number of individuals in the population (*popsize*), with no noticeable effects on performance in preliminary experiments. From [3], as well as other studies ([1,2]), we took 1.5 as bias for the biased linear ranking selection operator. This leaves us with just two parameters to fine-tune: *popsize*, and *maxCC*.

Although in [1] and [2] small population sizes were advocated, extensive experimentation in [3] shows that larger populations were more appropriate, mostly because of the beneficial effects on the population diversity. There is a trade-off to consider though. With small populations, more computational effort can be spend on increasing the fitness of the individuals over more generations. Although a relatively small number of search-paths can be followed in parallel, they can be followed to more depth. The drawback is that small populations have the tendency to lose population diversity, thus increasing the risk of getting the algorithm stuck in a local optimum from which it can not escape. On the other hand, larger populations allow for more search-paths to be followed in parallel, but to a lesser depth, while maintaining a higher population diversity. It is not possible to predict where in the *popsize-maxCC* parameter space the optimum parameter setting lies, and as such, we experimented with a large number of parameter combinations to find it. This also allows us to identify the optimum parameter settings for each density-tightness combination, in case this differs.

The experimental setup of the CTLEA is then as follows: for each CSP-instance in the test-set (of which there are 225), we run the algorithm 10 times. Varying combinations of population size (*popsize*) and maximum number of conflict checks allowed (*maxCC*) are used. The *popsize* parameter is taken from the following set: $\{10\} \cup \{50, 100, 150, \ldots, 2000\}$ (41 elements). The *maxCC* parameter is taken from the following set: $\{100000, 200000, \ldots, 2000000\}$ (20 elements). In total $225 \times 10 \times 41 \times 20 = 1,845,000$ runs were performed.

5   Results

Figure 1 summarises the results of the experiments described in the previous section. It consists of 9 graphs, each showing the result for one density-tightness combination in the test-set. The top row of graphs show the results for density-tightness combinations 1 to 3, the middle row the results for density-tightness combinations 4 to 6, and the bottom row the results for density-tightness combinations 7 to 9. Figure 1 shows the influence of different values of *maxCC* on the *SR* for different values of *popsize*. Along the *x*-axis of each graph in

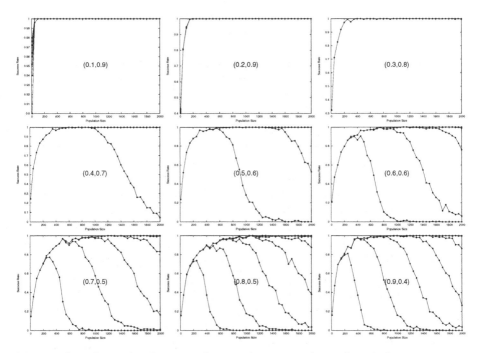

Fig. 1. The relationship between the population size (x-axis) and the success rate (y-axis) of the CTLEA for different maximum number of conflict checks allowed

figure 1 the *popsize* is shown, the y-axis shows the *SR*, while each curve in the graph was found for different values of *maxCC*.

The same trend is noticeable for *SR* in all graphs: the *SR* first increases when larger values for *popsize* and *maxCC* are used, but drops off sharply when the *popsize* gets too large relative to the available *maxCC*. At the point where the CTLEA solves all CSP-instances ($SR = 1.0$), just enough *maxCC* is available for the *popsize* but not more. Beyond this point, *SR* decreases for increased *popsize* but equal *maxCC*. Each curve therefore describes an arc with increasing *SR* for larger values of *popsize*, until *popsize* is increased to the maximum value able to be successfully maintained by the available *maxCC*, after which *SR* decreases again. Differences between the different graphs in figure 1 can partially be explained by differences in complexity between the different density-tightness combinations. CSP-instances in density-tightness combination 1, for example, are known to be easier to solve than those in density-tightness combination 9, and the number of conflict checks needed to sustain the population while reaching a perfect *SR* reflect that.

Table 1 shows, for each density-tightness combination, the first parameter combination for reaching a perfect *SR*, *popsize* minimised before *maxCC*, as well as the *ACCS* used to find the solutions. Note the increasing size of *popsize* and *maxCC* needed for reaching a perfect *SR* for the different density-tightness combinations. Because the CSP-instances for the different density-tightness combinations

Table 1. Success rate (SR) and average conflict checks to solution ($ACCS$) for the best population size ($popsize$) and maximum conflict checks allowed ($maxCC$) parameters.

	SR	ACCS	popsize	maxCC
1	1.0	1313	100	100000
2	1.0	4670	200	100000
3	1.0	20283	400	100000
4	1.0	50745	600	100000
5	1.0	94931	800	200000
6	1.0	167627	1100	300000
7	1.0	239106	1200	400000
8	1.0	254902	1050	700000
9	1.0	240046	950	500000

Table 2. Comparing the success rate and average conflict checks to solution of the CTLEA, the STLEA, Hill-climbing algorithm with Restart (HCAWR), Chronological Backtracking Algorithm (CBA), and Forward Checking with Conflict-Directed Backjumping Algorithm (FCCDBA)

	CTLEA		STLEA		HCAWR		CBA		FCCDBA	
	SR	ACCS	SR	ACCS	SR	ACCS	SR	ACCS	SR	ACCS
1	1.0	1313	1.0	2576	1.0	234242	1.0	3800605	1.0	930
2	1.0	4670	1.0	67443	1.0	1267015	1.0	335166	1.0	3913
3	1.0	20283	1.0	313431	1.0	2087947	1.0	33117	1.0	2186
4	1.0	50745	1.0	397636	1.0	2260634	1.0	42559	1.0	4772
5	1.0	94931	1.0	319212	1.0	2237419	1.0	23625	1.0	3503
6	1.0	167627	1.0	469876	1.0	2741567	1.0	44615	1.0	5287
7	1.0	239106	1.0	692888	1.0	3640630	1.0	35607	1.0	4822
8	1.0	254902	1.0	774929	1.0	2722763	1.0	28895	1.0	5121
9	1.0	240046	1.0	442323	1.0	2465975	1.0	15248	1.0	3439

(perhaps with the exception of density-tightness combination 1) were selected to minimise complexity variance, the increasing $popsize$ and $maxCC$ needed to solve the higher density-tightness combinations thus seems to reflect an aptitude of the algorithm to solve CSP-instances with a lower tightness, i.e., fewer average conflicts per constraint.

Table 2 shows a comparison of the performance of the CTLEA with the STLEA from [3], and benchmark algorithms from [2]. Table 2 shows that the CTLEA outperforms STLEA on all CSP-instances with density-tightness combinations. As the STLEA, the CTLEA compares favourably with the Hill-climbing with Restart Algorithm (HCAWR), with efficiency measured in $ACCS$ several magnitudes better. Compared with the Chronological Backtracking Algorithm (CBA), the CTLEA outperforms it on CSP-instances with density-tightness combinations 1, 2, and 3, but is outperformed on all others. This shows that in the area where CTLEA has shown good performance, CSP-instances with lower tightness, it can outperform a classical algorithm. Compared with the more sophisticated Forward Checking with Conflict-Directed Back-jumping Algorithm (FCCDBA) (a combination of forward-checking [13] and conflict-directed back-jumping [14]) however, the CTLEA can only approach performance on CSP-instances the first two density-tightness combinations, but is outperformed in all others. Overall, the CTLEA raised the performance bar a little higher for EAs, but remains unable to beat sophisticated deterministic algorithms on efficiency.

6   Conclusions

This paper introduced the Conflict Tabu Search Evolutionary Algorithm (CTLEA) for solving binary constraint satisfaction problems. The CTLEA is a hybrid algorithm, incorporating elements of an evolutionary and the tabu search meta-heuristic. The CTLEA is based on the Simple Tabu Search Evolutionary Algorithm (STLEA), introduced in [3], substituting its compound label tabu list with a tabu list limiting the search space by storing of conflicts. The rational behind choosing conflicts for the CTLEA tabu list is the comparatively limited number of conflicts and their usefulness in the new move-operator. Like the STLEA, the CTLEA maintains the basic structure of an evolutionary algorithm, but merges the crossover and mutation operator in one 'move-operator'. Further efficiency improvements were achieved by using the same representation as was used in the STLEA.

A large number of parameter tuning experiments were performed for different density-tightness combinations of a commonly used test-set. The performance of the CTLEA with the best parameter settings was found to outperform the STLEA, making it the best performing EA for solving the binary CSP found thus far.

Although comparable in performance to the Chronological Backtracking Algorithm on CSP-instances with lower tightness, the CTLEA continues to be outperformed by the more sophisticated Forward Checking with Conflict Directed Back-jumping Algorithm.

Future research will focus on comparing the relative behaviour of the CTLEA to other algorithms when size of the CSP-instances is increased and the effects of using different types of tabu lists on performance.

References

1. Craenen, B., Eiben, A., van Hemert, J.: Comparing evolutionary algorithms on binary constraint satisfaction problems. IEEE Transactions on Evolutionary Computing 7(5), 424–445 (2003)
2. Craenen, B.: Solving Constraint Satisfaction Problems with Evolutionary Algorithms. Doctoral dissertation, Vrije Universiteit Amsterdam, Amsterdam, The Netherlands (November 2005)

3. Craenen, B., Paechter, B.: A tabu search evolutionary algorithm for solving constraint satisfaction problems. In: Runarsson, T.P., et al. (eds.) Parallel Problem Solving from Nature – PPSN IX. Lecture Notes on Computer Science, vol. 4192, pp. 152–161. Springer, Berlin (2006)

4. Glover, F., Laguna, M.: Tabu search. In: Reeves, C. (ed.) Modern Heuristic Techniques for Combinatorial Problems, pp. 70–141. Blackwell Scientific Publishing, Oxford, England (1993)

5. Costa, D.: An evolutionary tabu search algorithm and the nhl scheduling problem. Orwp 92/11, Ecole Polytechnique Fédérale de Lausanne, Département de Mathématiques, Chaire de Recherche Opérationelle (1992)

6. Burke, E., Causmaecker, P.D.: VandenBerghe: A hybrid tabu search algorithm for the nurse rostering problem. In: Proceedings of the Second Asia-Pasific Conference on Simulated Evolution and Learning. Applications IV, vol. 1, pp. 187–194 (1998)

7. Greistorfer, P.: Hybrid genetic tabu search for a cyclic scheduling problem. In: Voß, S., Martello, S., Osman, I., Roucairol, C. (eds.) Meta-Heuristics: Advances and Trends in Local Search Paradigms for Optimization, pp. 213–229. Kluwer Academic Publishers, Boston, MA (1998)

8. Kratica, J.: Improving performances of the genetic algorithm by caching. Computer and Artificial Intelligence 18(3), 271–283 (1999)

9. Rossi, F., Petrie, C., Dhar, V.: On the equivalence of constrain satisfaction problems. In: Aiello, L. (ed.) Proceedings of the 9th European Conference on Artificial Intelligence (ECAI 1990), Stockholm, Pitman, pp. 550–556 (1990)

10. Tsang, E.: Foundations of Constraint Satisfaction. Academic Press, London (1993)

11. MacIntyre, E., Prosser, P., Smith, B., Walsh, T.: Random constraint satisfaction: theory meets practice. In: Maher, M.J., Puget, J.-F. (eds.) CP 1998. LNCS, vol. 1520, pp. 325–339. Springer, Heidelberg (1998)

12. Whitley, D.: The genitor algorithm and selection pressure: Why rank-based allocation of reproductive trials is best. In: Schaffer, J. (ed.) Proceedings of the 3rd International Conference on Genetic Algorithms, San Mateo, California, pp. 116–123. Morgan Kaufmann Publisher, Inc., San Francisco (1989)

13. Haralick, R., Elliot, G.: Increasing tree search efficiency for constraint-satisfaction problems. Artificial Intelligence 14(3), 263–313 (1980)

14. Prosser, P.: Hybrid algorithms for the constraint satisfaction problem. Computational Intelligence 9(3), 268–299 (1993)

Cooperative Particle Swarm Optimization for the Delay Constrained Least Cost Path Problem

Ammar W. Mohemmed[1,*], Mengjie Zhang[1], and Nirod Chandra Sahoo[2]

[1] School of Mathematics, Statistics and Computer Science
Victoria University of Wellington
P.O. 600, Wellington
`ammar.mohemmed@mcs.vuw.ac.nz`
[2] Department of Electrical Engineering
Indian Institute of Technology
Kharagpur-721302, India

Abstract. This paper presents a particle swarm optimization (PSO) algorithm for solving the delay constrained least cost (DCLC) path problem, i.e., shortest path problem (SPP) with a delay constraint on the total "cost" of the optimal path. The proposed algorithm uses the principle of Lagrange relaxation based aggregated cost, where PSO and noising metaheuristic are used for minimizing the modified cost function. It essentially consists of two PSOs. The main PSO is basically a hybrid PSO-Noising metaheuristic algorithm for efficient global search for the minimization part of the DCLC-Lagrangian relaxation by finding multiple shortest paths between a source and a destination. The second/auxiliary PSO is used to obtain the optimal Lagrangian multiplier for solving the maximization part of the Lagrangian relaxation of the DCLC path problem. For the main PSO, a new path encoding/decoding scheme based on heuristics has been devised for representing the paths as particles. The comparative simulation results on several networks with random topologies illustrate the efficiency of the proposed hybrid algorithm for constrained shortest path computation.

1 Introduction

Shortest path problem (SPP) is one of the most fundamental problems in graph theory. With the developments in communication, computer science, and transportation systems, more variants of the SPP have appeared. For example, in communication networks like IP, ATM, and optical networks, there is a need to find a path with minimum cost while maintaining a bound on delay to support Quality-of-Service applications. This problem is called the Delay Constrained Least Cost (DCLC) path problem and is known to be NP-hard [1]. This paper deals with the development of an evolutionary algorithm for this problem.

The Constrained Bellman-Ford (CBF) algorithm [2] solves DCLC by enumerating the Pareto set, therefore has exponential worst-case complexity. An early,

---

* Corresponding author.

J. van Hemert and C. Cotta (Eds.): EvoCOP 2008, LNCS 4972, pp. 25–35, 2008.

simple algorithm by Lee et al. [3] has tried each of the metrics and has checked the feasibility. Pornavalai et al. [4] improved upon this idea by combining paths calculated with different metrics. Another group of heuristic algorithms [5,6,7,8] referred to as Linear Aggregated Metric (LAM) first constructs a single metric from all metrics and then applies a shortest-path algorithm. Handler and Zang [9] proposed a Lagrangian method to solve the single constraint SPP. They first solve the Lagrangian dual of the problem and then close the (potential) duality gap with a k-shortest path algorithm. Jüttner et al. [10] applied the algorithm on practical networks with improving running time. In the Lagrangian method, when optimal multipliers are found, we have a lower bound on the problem. An upper bound could be achieved through the solution process of the relaxed problem. If the lower bound and the upper bound have equal value, then we have an optimum solution. If there is a duality gap, the lower bound is less than the upper bound then an extra step has to be performed to try to close the gap, if an optimal solution is wanted. The main issue is how to find the optimum Lagrangian multipliers. Handler and Zang [9] used the cutting plane method and Beasely [11] used subgradient optimization while Jüttner's method [10] is based on an algebraic approach. Although these methods find good near-optimum solutions in most of the cases, these solutions are not always the optimum path. Besides that, these methods are not extendable for more than one constraint.

In this paper, two cooperating PSOs have been used to solve the DCLC path problem being formulated as a Lagrange relaxation based aggregated cost problem. It essentially consists of two PSOs. The main PSO is basically a hybrid PSO-noising metaheuristic algorithm for efficient global search for the minimization part of the DCLC-Lagrangian relaxation by finding multiple shortest paths between a source and a destination. The noising metaheuristic [12] have been embedded into the main PSO for effective local search around any better particle found in every PSO iteration. The second/auxiliary PSO is a co-evolutionary PSO to obtain the optimal Lagrangian multiplier for solving maximization part of the Lagrangian relaxation of the DCLC path problem. For the main PSO, a new path encoding/decoding scheme based on heuristics has been devised for representing the network paths as particles.

The outline of this paper is as follows. Section 2 describes the mathematical formulation of the DCLC path problem. In Section 3, the basic algorithm of PSO is discussed along with a brief description of the noising metahueristic. Section 4 presents the path encoding/decoding technique used for particle representation of network paths for PSO. Section 5 describes how a cooperative PSO is used along with the main PSO for obtaining optimal Lagrange multiplier in solving the DCLC path problem. Section 6 discusses the experimental results followed by the conclusions in the final section.

2   Delay Constrained Least Cost (DCLC) Path Problem

Let $G = (V, E)$ be an undirected graph comprising a set of nodes $V = \{v_i\}$ and a set of edges $E = V \times V$ connecting nodes in V. Corresponding to each edge, there is a nonnegative number c_{ij} representing the cost of the edge/link and a nonnegative delay (transmission/propagation) d_{ij} from node v_i to node v_j. A path p from a source node s to a destination node t is a sequence of nodes in which no node is repeated. The total cost of this path is referred to by $c(p) = \sum_{(i,j)} c_{ij}$ and the total delay by $d(p) = \sum_{(i,j)} d_{ij}$. The delay constrained least cost (DCLC) path problem is to find the minimum cost path $p*$ such that the delay $d(p*)$ is under a given limit (constraint) Δ_{delay}:

$$\textbf{DCLC} : \min_{p \in P(s,t)} c(p) \text{ such that } d(p) \leq \Delta_{delay} \tag{1}$$

where $P(s, t)$ is the set of paths between node s and node t.

2.1   Lagrange Relaxation for the DCLC Problem

We use the same terminology as in [10] to describe the Lagrangian relaxation to the DCLC problem. For each link (i, j) in the network graph, an aggregated cost c_μ is defined as:

$$c_\mu = c_{ij} + \mu d_{ij} \ , \ \mu \geq 0 \tag{2}$$

For any specific μ, $c_\mu(p)$ denotes the aggregated cost of the path p. The Lagrangian relaxation of DCLC path problem, $L(\mu)$, is defined as:

$$L(\mu) = \min_{p \in P(s,t)} \{c(p) + \mu d(p)\} - \Delta_{delay} = \min_{p \in P(s,t)} c_\mu(p) - \mu \Delta_{delay} \tag{3}$$

where $L(\mu)$ is a lower bound to the DCLC problem. To obtain the best lower bound, $L(\mu)$ is maximized over:

$$L^* = \max_{\mu \geq 0} L(\mu) \tag{4}$$

For any fixed $\mu \geq 0$, solving Eq.(3) requires the solution of a shortest-path problem with Lagrangian-modified edge length. The main issue is to find the optimum μ that maximizes Eq.(4). Jüttner et al. [10] used an algebraic approach to solve Eq.(4). Their method (LARAC) is briefly described below.

If a path found at an iteration is minimum with respect to c_μ and its delay is larger than or equal to the threshold Δ_{delay}, this path is called p_c. If its delay is smaller than or equal to the threshold Δ_{delay}, this path is called p_d. If the aggregated cost c_μ of p_c is equal to that of p_d, a value $\mu \geq 0$ will maximize the function $L(\mu)$. In this case, $c_\mu(p_c) = c_\mu(p_d)$ and, hence, $\mu = \{c_\mu(p_c) - c(p_d)\}/\{d_\mu(p_d) - d_\mu(p_c)\}$. With this μ, a new c_μ-minimal path r is found. If $c_\mu(r) = c_\mu(p_d) = c_\mu(p_c)$, the optimal μ is found; otherwise, r is set as the new p_c or p_d according to whether r is infeasible ($d(r) > \Delta_{delay}$) or feasible ($d(r) \leq \Delta_{delay}$). Then, the same steps are repeated to find a new value of μ.

However, as noted in [10], this method does not always give the optimal path when there is a duality gap. In this study, a cooperative particle swarm technique is used to solve the DCLC path problem and compare its performance with LARAC algorithm.

3 Hybrid PSO and Noising Metahueristic for Shortest Path Problem

Particle swarm optimization (PSO) is a population based stochastic optimization tool inspired by social behavior of bird flock (and fish school etc.), as developed by Kennedy and Eberhart in 1995 [13]. The algorithmic flow in PSO starts with a population of particles whose positions, that represent the potential solutions for the studied problem, and velocities are randomly initialized in the search space. The search for optimal position (solution) is performed by updating the particle velocities, hence positions according to the following two equations:

$$PV_{id} = PV_{id} + \phi_1 r_1 (B_{id} - X_{id}) + \phi_2 r_2 (B_{id}^n - X_{id}) \tag{5}$$
$$i = 1, 2..N_s, \text{ and } d = 1, 2, .., D$$

$$X_{id} = X_{id} + PV_{id} \tag{6}$$

where ϕ_1 and ϕ_2 are positive constants, called *acceleration coefficients*, N_s is the total number of particles in the swarm, D is the dimension of problem search space, i.e., number of parameters of the function being optimized, r_1 and r_2 are two independently generated random numbers in the range [0, 1] and n represents the index of the best particle in the neighborhood of a particle. The other vectors are defined as: $X_i = [X_{i1}, X_{i2}...X_{iD}] \equiv$ Position of i-th particle; $PV_i = [PV_{i1}, PV_{i2}...PV_{iD}] \equiv$ Velocity of the i-th particle; $B_i \equiv$ Best position of the i-th particle, and $B_i^n \equiv$ Best position found by the neighborhood of the particle i. When the convergence criterion is satisfied, the best particle (with its position) found so far is taken as the solution to the problem.

However, in most cases, the velocities quickly attain very large values, especially for particles far from their global best. To control the increase in velocity, velocity clamping is used in Eq.(5). Thus, if the right side of Eq.(5) exceeds a specified maximum value $\pm PV_d^{max}$, then the velocity on that dimension is clamped to $\pm PV_d^{max}$. In [14], Maurice proposed the use of a constriction factor to prevent velocity from growing out of bound. The algorithm has been named the *constriction factor method*(CFM) where Eq.(5) is modified as:

$$PV_{id} = \chi[PV_{id} + \phi_1 r_1 (B_{id} - X_{id}) + \phi_2 r_2 (B_{id}^n - X_{id})] \tag{7}$$
$$i = 1, 2..N_s, \text{and} d = 1, 2, ..., D$$

$$\chi = 2(\left|2 - \phi - \sqrt{\phi^2 - 4\phi}\right|)^{-1} \tag{8}$$

To improve the search capability of PSO in terms of quality and speed, a hybrid PSO/Noising metaheuristic based algorithm is implemented to solve the shortest path problem. This hybrid algorithm constitutes the main PSO that is used to find multiple shortest paths. The basic idea of noising method is: for computation of the optimum of a combinatorial optimization problem, instead of taking the genuine data into account directly, they are perturbed by some progressively decreasing *"noise"* while applying local search [12]. The reason behind the addition of noise is to be able to escape any possible local optimum in the optimizing function landscape. A noise is a value taken by a certain random variable following a given probability distribution. More description of the hybrid PSO/Noising method can be found in [15].

4 Network Path Encoding for the Shortest Path Computation Using Hybrid PSO Algorithm

In order to exploit the global as well as the local search capability of the proposed hybrid PSO/Noising metaheuristic based algorithm for shortest path problems, an efficient path encoding/decoding is required for representing every possible path in the network as a particle in PSO. We have proposed in [16] a tree based encoding/decoding for the shortest path problem. Here, we give a brief description of this technique. The particle contains weights (real numbers) that are decoded to build a Shortest Paths Tree (SPT). This tree is represented by the predecessors vector and built progressively from iteration to iteration. In the end, the shortest path tree will contain the shortest path from the source to the destination. Each particle keeps two vectors, the $prev[v]$ representing the node previous to node v and the $C[v]$ recording the total cost of path from node v to the source. The $C[v]$ vector is initialized to ∞. In every iteration, the particle is decoded as follows: Initially, the tree consists only of the root (source node). The next node j from the nodes that have direct link (i, j) to the current one i is selected to be appended to the partial tree based on the following formula:

$$j = argmin\{c_{ij}w_j | (i, j) \in E\}, w_j \in [-1.0, 1.0] \tag{9}$$

where w_j is the weight of the node j in the particle and c_{ij} is the cost of the edge between node i and node j. Thus the role of the weights in the particle is to bias the edge costs in order to select the next node to be appended to the current partial tree. In a relaxation test, if the next node j to be appended to the current tree has a better cost (c_j) than previously recorded in $C[j]$, the $C[j]$ will be updated and so does $prev[v]$; otherwise c_j is set to the previous value of $C[j]$. Algorithm 1 lists the steps of this procedure.

4.1 Fitness Function

While this decoding technique being implemented in a PSO framework, one point to consider is: how to decide on the "badness" or the "goodness" of a particle.

Since $C[v]$ gets updated only if a less cost is found, it does not reflect the fitness of the particle. That is because a bad particle does not update the cost vector. Another case is that a path deemed as an invalid path does not end at the destination. Therefore, a fitness function reflecting these issues needs to be devised. Such a strategy is given below.

The cost c_j, cost of the path ending at node j, calculated from decoding the particle's weights, is taken as the fitness of the particle. If the path does not end at the destination, a penalty is added to the fitness of the particle as shown in Eq.(10) (ρ is a penalty value):

Algorithm 1. Pseudo codes for tree based decoding procedure.

ParticleDecoding $(PARTICLE)$

$\sigma \leftarrow 0, i \leftarrow s, w_s \leftarrow N_\infty, k = 0$ # N_∞ is a specific large number

                                               # n is a number of nodes

while $(i \neq t$ **&&** $k < n)$ **do**

    $k \leftarrow k + 1$

    $j \leftarrow argmin\{c_{ij}w_j | j \in A(i), w_j \neq N_\infty\}$ # $A(i)$ the set of nodes adjacent to i

    $c_j \leftarrow c_i + c_{ij}$

    if $(c_j < C[j])$ **then**

        $C[j] \leftarrow c_j$

        $prev[j] \leftarrow i$

    else

        $c_j \leftarrow C[j]$

        $i \leftarrow j$

        $p_i \leftarrow N_\infty$

    end if

    if $(\{j \in A(i), w_j \neq N_\infty\} = \phi)$ **then**

        $\sigma = 1$

        break

    end if

end while

return $c_j + \sigma\rho$

$$Fitness = c_j + \sigma.\rho \tag{10}$$

$$\text{where } \sigma = \begin{cases} 1 & j \neq t \text{ (invalid path)} \\ 0 & j = t \text{ (valid path)} \end{cases}$$

According to Eq.(10), if the decoded path does not end at destination node t, a penalty factor ρ will be added.

The best path found by the particle is simply decoded from the vector $prev[v]$. For the case, when the edges have more than one weight (say,delay), the decoding process is modified with Eq.(9) being re-formulated as:

$$j = argmin\{(c_{ij} + d_{ij})w_j | (i, j)E\}, w_j \in [-1.0, 1.0] \tag{11}$$

Thus, the next node j will be appended to the partial tree if the combined effect of the cost and delay of the edge connecting the current node i with node j is the minimum among all the adjacent nodes to the current node i. The cost ,c_j, of the path decoded from a particle which ends at node j will represent $(c_j + d_j)$.

5   Cooperative Hybrid PSO-Noising Method Based Algorithm for DCLC Path Problem

This section describes cooperative PSO-based hybrid search algorithm for the DCLC path problem. There is a single constraint: $d(p) \leq \Delta_{delay}$ where $p \in P(s, t)$. Eq.(4) is now re-written as:

$$L^*(\mu, p) = \max_{\mu \geq 0}\{ \min_{p \in P(s,t)} c_\mu(p) - \mu \Delta_{delay}\} \tag{12}$$

Two separate populations of PSOs are involved in this cooperative PSO. The main (first) PSO, $PSO1$, hybridized with the Noising metaheuristic focuses on evolving weights (position vector) of the particles with a *"frozen"* value for μ. Inside the iteration of this main PSO, when a particle experiences improvement, the noising method is called to fine search and to further improve the fitness of the particle. If the noising method is able to find a better path, the fitness of the particle will be updated. The first PSO is responsible for generating multiple shortest paths based on the combined cost $c_{ij} + \mu d_{ij}$. The set of paths represents an approximation of the $L(\mu, p)$ over $\mu \geq 0$. The second (auxiliary) PSO, $PSO2$, focuses on evolving Lagrangian multiplier μ using the set of the paths found by the first (main) PSO population. Only the multiplier μ is represented in the auxiliary PSO population. The two PSOs work together and exchange information to obtain the best value for Eq.(12). For the main PSO, the problem is a minimization problem and the objective function is:

$$f_{main}(p) = \min_{\mu \in PSO2} c_\mu(p) \tag{13}$$

For the auxiliary PSO, the problem is a maximization problem and the fitness of each individual μ is evaluated according to

$$f_{aux}(p) = \max_{p \in PSO1} L(\mu, p) \tag{14}$$

The $PSO1$ population is initialized randomly and the fitness is calculated with initial μ being set at some value (say, 100). The $PSO1$ runs for a number of iterations ($iteraPSO1$) with edges' costs set to the aggregated cost $c_{ij} + \mu d_{ij}$. At the end of these iterations, a set of paths (the number of paths equals the population size) will be generated. This set represents the best paths found with a specified Lagrangian multiplier during that stage of running the main PSO. This set of constructed paths is passed to the auxiliary PSO. The cost and delay of these paths are used in the fitness function of the auxiliary PSO defined in Eq.(14). The auxiliary PSO is initialized and runs for a number of iterations ($iteraPSO2$) searching for a new optimum value of μ that maximizes Eq.(14). This value is passed to the first PSO to run again for ($iteraPSO1$) and generates a new set of paths and so on. In each iteration of the main PSO, a particle is checked for a path with less cost and delay that does not exceed the delay constraint.

The criterion for the path decoding process is now modified. Thus, Eq.(9) is changed to include the Lagrangian multiplier μ:

$$j = argmin\{(c_{ij} + \mu.d_{ij})w_j | (i, j) \in E\}, w_j \in [-1.0, 1.0] \tag{15}$$

The two PSOs run for a maximum number of cycles ($maxcycles$). The total number of iterations will be $maxcycles * (iteraPSO1 + iteraPSO2)$.

6   Simulation Results and Discussion

The proposed PSO-based hybrid cooperative algorithm for DCLC path problem is evaluated on networks with random and varying topologies[1] through computer simulations. The edges' costs and delays are randomly and independently chosen from different ranges. To prevent the DCLC problem from being trivial, the delay constraint is chosen as:

$$\Delta_{delay} = 0.75 * d(p_{LD}) + 0.25 * d(p_{LC}) \tag{16}$$

where $d(p_{LD})$ is the delay of the least delay path and $d(p_{LC})$ is the delay of the least cost path. In all the simulation tests, the optimal solution obtained using Constraint Bellman Ford (CBF) algorithm [2] is used as reference for comparison purposes. Ref. [10] shows that Lagrange Relaxation based Aggre-gated Cost (LARAC) has the best performance compared with other algo-rithms, therefore the performance of the proposed algorithm is compared with it. The selection of parameter settings of the co-evolutionary PSOs are shown in Table 1.

Table 1. Parameters of Main PSO ($PSO1$) and Auxiliary PSO ($PSO2$)

	PSO_1	PSO_2
Population size	50	10
Maximum velocity	[-1,1]	[0,10]
Constriction factor	0.74	0.74
Maximum Iterations	500	10
Neighborhood Topology	Ring	Ring

The parameters for the noising metaheuristic are set as described in [15]. The maximum and minimum noise rates are 100 and 0 respectivily. he maxi-mum number of trials is set to 50. The maximum number of trials at a fixed noise rate is 10. The elementary transformation for local neighborhood search is taken as the swapping of node weight values at two random positions of a particle weight/position vector and two such swapping transformations are successively applied in each trial for generating a trial solution in the local search.

Two parameters are used to measure the performance assessment. They are: (1) average success rate (SR) and (2) average excess cost (EC). The average calculation is performed over 100 runs for each network topology. The success rate is defined as the percentage of runs that the algorithm finds the DCLC

---

[1] The random network topologies are generated using Waxman model [17] in which nodes are generated randomly on a two dimensional plane of size 100×100, and there is a link between two nodes u and v with probability $p(u,v) = \alpha.e^{\frac{-d(u,v)}{\beta L}}$, where $0 < \alpha, \beta \leq 1, d(u,v)$ is the Euclidean distance between u and v, and L is the maximum distance between any two nodes.

path according to CBF algorithm over the total number of runs (CBF gives the optimum path). The second parameter, the average excess cost (EC), is defined as:

$$EC = \frac{C(p_A) - C(p_{CBF})}{C(p_{CPF})} \tag{17}$$

where $C(p_A)$ is the cost of the path found by algorithm "A" and $C(p_{CBF})$ is the cost of the path found by the CBF algorithm. Table 2 shows a comparison of the average success rates (SR) and the average excess cost (EC) under two ranges for the cost and delay: range $[1, 20]$ and range $[1, 100]$. For example, for the network of 100 nodes and 281 edges and for the cost and delay in the range $[1, 20]$, the PSO algorithm is able to find the optimum path (that is the path with the minimum cost and delay less than Δ_{delay}) with a success rate of 93% compared with 54% for the LARAC algorithm. In addition, the excess cost of the non-optimum paths found by PSO is much less than that of the paths found by LARAC (0.12% in case of PSO compared with 2.5% in the case of LARAC). This suggests that the non-optimum paths found by PSO is very close to the optimum ones while those found by LARAC have a big gap with the optimum ones in terms of the cost. The same applies for the other cases.

Table 2. Performance comparison between LARAC [10] and PSO for the DCLC path problem

No. of Nodes	No. of Edges	PSO Cost $=[1,20]$ Delay$=[1,20]$		LARAC Cost $=[1,20]$ Delay$=[1,20]$		PSO Cost $=[1,100]$ Delay$=[1,100]$		LARAC Cost $=[1,100]$ Delay$=[1,100]$	
		% SR	% EC	% SR	% EC	% SR	% EC	% SR	% EC
100	281	93	0.12	54	2.5	91	0.2	51	2.7
100	255	100	0.0	59	1.7	95	0.1	51	1.6
90	249	98	0.06	52	2.5	94	0.18	50	2.5
90	227	97	0.02	64	1.2	92	0.11	59	1.5
80	231	97	0.03	58	2.2	99	0.01	50	2.4
80	187	98	0.08	57	1.6	92	0.01	60	1.3
70	321	98	0.09	61	3.2	97	0.07	56	4.0
70	211	96	0.1	54	4.1	97	0.03	52	2.7
60	232	98	0.04	56	3.2	97	0.07	60	3.1
50	159	99	0.04	73	2.7	97	0.07	67	3.4

Next, the effect of changing the delay constraint Δ_{delay} on the performance of the algorithms is investigated. The network topology of 100 nodes and 281 edges in Table 2 is taken as an example. First the cost and the delay are in the $[1, 100]$ range and the delay bound Δ_{delay} is changed from 300 to 700. The costs of the best feasible paths (that with the minimum cost and delay less than the bound) found by the two algorithms are compared with the optimum ones computed by CBF algorithm. Fig.1 (a) shows the results of this comparison. The costs of the feasible paths found by PSO coincide with those found by CBF irrespective of

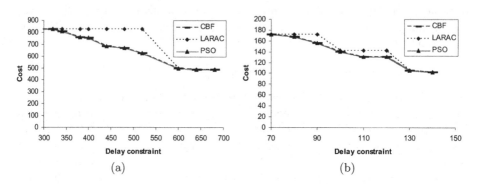

Fig. 1. Effect of the delay constraint on the performance, cost and delay taken in (a) 1 - 100 (b) 1 - 20

the values of Δ_{delay}. In the case of LARAC algorithm, the value of the constraint influences the performance, which can be seen in extra cost when the constraint is less than 600. The reason is that LARAC is not able to find the optimum path when Δ_{delay} is small. The same test has been conducted but with the cost and delay taken in the range of [1, 20] and the constraint changes from 70 to 150. Fig. 1 (b) shows the result of this experiment. Again the PSO algorithm has better performance. The two graphs show that the performance of the PSO based algorithm is more adaptable to the constraint's change and it is able to find the optimum path even when the constraint is small.

7   Conclusions

In this paper, a cooperative PSO method based algorithm is presented and tested for solving the delay constrained least cost (DCLC) path problem. A new tree based encoding/decoding scheme for the particles in PSO and network path construction from it has also been devised. The delay constraint is relaxed and added to the objective in Lagrangian fashion. Intensification of the refined search around potential regions is used to solve the minimization part of the Lagrangian relaxation problem in the main PSO while the auxiliary PSO is used to solve the maximization part of the problem by finding optimum Lagrangian multiplier. Comparison simulation with another heuristic algorithm (LARAC) show that the proposed algorithm produces good results in terms of higher success rates for getting the optimal path and less excess cost for the non-optimum paths. Although the PSO based algorithm can not compete yet in term of the time complexity but it has the advantage that it is easier to be extend for more than one constraint. This is can be done by including the extra constraints in the auxiliary PSO. For future work, we would like to extend this algorithm for solving multi-constrained SPP and enhance its performance for large-scale networks.

References

1. Garey, M.R., Johnson, D.S.: Computers and Intractability: A Guide to the Theory of NP-Completeness, vol. 1. W.H. Freeman, New York (1979)
2. R., W.: The design and evaluation of routing algorithms for real time channels. Technical Report TR-94-024, Tenet Group, Dept. EECS, Univ. California, Berkeley, CA (1994)
3. Lee, W.C., Hluchyj, M.G., Humblet, P.A.: Routing subject to quality of service constraints in integrated communication networks. IEEE Network 9, 14–162 (1995)
4. Pornavalai, C., Chakraborty, V., Shiratori, N.: Routing with multiple qos requirements for supporting multimedia applications. J. High Speed Networks (1998)
5. Jaffe, J.: Algorithms for finding path with multiple constraints. Networks 14, 95–116 (1984)
6. Blokh, D., Gutin, G.: An approximation algorithm for combinatorial optimization problems with two parameters (1995)
7. Neve, H.D., Mieghem, P.V.: A multiple quality of service routing algorithm for pnni. In: IEEE ATM 1998 Workshop, pp. 306–314 (1998)
8. Guoe, L., Matta, I.: Search space reduction in qos routing. Computer Networks 41, 73–88 (2003)
9. Handler, G., Zang, I.: A dual algorithm for the constrained shortest path problem. Networks 10, 293–310 (1980)
10. Juttner, A., Szviatovski, B., Mecs, I., Rajko, Z.: Lagrange relaxation based method for the qos routing problem. In: INFOCOM 2001, vol. 2, pp. 859–868 (2001)
11. Beasley, J., Christofides, N.: An algorithm for the resource constrained shortest path. Networks 19, 379–394 (1989)
12. Charon, I., Hurdy, O.: The noising method: a new method for combinatorial optimization. Operations Research Letters 14, 133–137 (1993)
13. Kennedy, J., Eberhart, R.C.: Particle swarm optimization. In: Neural Networks, pp. 1942–1948 (1995)
14. C.M.: The swarm and queen: Towards a deterministic and adaptive particle swarm optimization. In: IEEE Congress on Evolutionary Computation, vol. 2, pp. 1951–1957 (1999)
15. Mohemmed, A.W., Sahoo, N.: Efficient computation of shortest paths in networks using particle swarm optimization and noising metaheuristics. Discrete Dynamics in Nature and Society (2007)
16. Mohemmed, A.W., Sahoo, N.C., Geok, T.K.: A new particle swarm optimization based algorithm for solving shortest-paths tree problems. In: IEEE Congress on Evolutionary Computation, pp. 3221–3225 (2007)
17. Waxman, B.: Routing of multipoint connections. IEEE Journal of Selected Areas in Communications 6, 1617–1622 (1988)

Effective Neighborhood Structures for the Generalized Traveling Salesman Problem

Bin Hu and Günther R. Raidl

Institute of Computer Graphics and Algorithms
Vienna University of Technology
Favoritenstraße 9–11/1861, 1040 Vienna, Austria
{hu,raidl}@ads.tuwien.ac.at

Abstract. We consider the generalized traveling salesman problem in which a graph with nodes partitioned into clusters is given. The goal is to identify a minimum cost round trip visiting exactly one node from each cluster. For solving difficult instances of this problem heuristically, we present a new Variable Neighborhood Search (VNS) approach that utilizes two complementary, large neighborhood structures. One of them is the already known generalized 2-opt neighborhood for which we propose a new incremental evaluation technique to speed up the search significantly. The second structure is based on node exchanges and the application of the chained Lin-Kernighan heuristic. A comparison with other recently published metaheuristics on TSPlib instances with geographical clustering indicates that our VNS, though requiring more time than two genetic algorithms, is able to find substantially better solutions.

Keywords: Network Design, Generalized Traveling Salesman Problem, Variable Neighborhood Search.

1 Introduction

The Generalized Traveling Salesman Problem (GTSP) extends the classical Traveling Salesman Problem (TSP) and is defined as follows. We consider an undirected weighted complete graph $G = \langle V, E, c \rangle$ with node set V, edge set E, and edge cost function $c : E \rightarrow \mathbb{R}^+$. Node set V is partitioned into r pairwise disjoint clusters $V_1, V_2, \ldots, V_r$, $\bigcup_{i=1}^{r} V_i = V$, $V_i \cap V_j = \emptyset$, $i, j = 1, \ldots, r$, $i \neq j$.

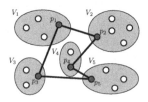

Fig. 1. Example for a GTSP solution

J. van Hemert and C. Cotta (Eds.): EvoCOP 2008, LNCS 4972, pp. 36–47, 2008.
© Springer-Verlag Berlin Heidelberg 2008

A solution to the GTSP defined on G is a subgraph $S = \langle P, T \rangle$ with $P = \{p_1, p_2, \ldots, p_r\} \subseteq V$ connecting exactly one node from each cluster, i.e. $p_i \in V_i$ for all $1 \leq i \leq r$, and $T \subseteq E$ being a round trip, see Fig. 1. The costs of such a round trip are its total edge costs, i.e. $C(T) = \sum_{(u,v) \in T} c(u, v)$, and the objective is to identify a solution with minimum costs. When edge costs satisfy the triangle inequality, even if we allow more than one node per cluster to be connected, an optimal solution of the GTSP always contains only one node from each cluster [11]. Obviously, the GTSP is NP-hard since it contains the classical TSP as the special case in which each cluster consists of a single node only.

The GTSP finds practical application particularly in many variants of routing problems, e.g. when some good can be delivered to multiple alternative addresses of customers. Occasionally, such applications can be directly modeled as the GTSP, but more often the GTSP appears as a subproblem [9].

In this paper, we present a general Variable Neighborhood Search (VNS) approach [6] for heuristically solving this problem. As local improvement within VNS, we use Variable Neighborhood Descent (VND) based on two different types of exponentially large neighborhoods, which can be seen as dual to each other. One neighborhood structure is the generalized 2-opt, which has been introduced in [19]; for it, we propose a new incremental evaluation scheme leading to a substantial speed-up. As second neighborhood structure we investigate a new approach: the nodes to be spanned from each cluster are fixed and TSP tours are derived via the chained Lin-Kernighan algorithm.

Section 2 gives an overview on research done on the GTSP so far. In section 3, we describe the initialization procedures, followed by section 4 explaining the neighborhood structures in detail. Section 5 shows the VNS settings, and experimental results are discussed section 6. Finally, we conclude in section 7.

2   Previous Work

The GTSP was introduced independently by Henry-Labordere [7], Srivastava et al. [22], and Saskena [20]. Laporte et al. [11,10] provided integer programming formulations for the symmetrical and asymmetrical GTSP, respectively. The formulation for the symmetrical case was later enhanced by Fischetti et al. [4] who proposed several classes of facet defining inequalities and corresponding separation procedures. Based on these, they developed a branch-and-cut algorithm [5] which could solve instances with up to 442 nodes to optimality.

Several approaches exist which transform the GTSP into the classical TSP. They have been studied by Noon and Bean [16], Lien et al. [13], Dimitrijevic and Saric [2], Laporte and Semet [12], and Behzad and Modarres [1]. Unfortunately, many transformations substantially increase the numbers of nodes and edges and are therefore of limited practical value. Furthermore, some transformations even require additional constraints, thus making general algorithms for the classical TSP inapplicable. Among the more efficient approaches, Dimitrijevic and Saric [2] proposed a transformation of the GTSP into the TSP on a digraph containing twice the number of nodes compared to the original graph. This technique was further improved by Behzad and Modarres [1] where the transformed graph has the same number of nodes as the original graph. However, the transformation increases edge costs significantly, what may lead to problems in some cases.

To approach larger GTSP instances, various metaheuristics have been suggested. Renaud et al. [19] developed a complex composite heuristic whose components can be used for other (meta-)heuristics as well. They introduced generalized k-opt heuristics which are derived from Lin's classical 2-opt and 3-opt local search for the TSP [14]. Snyder and Daskin [21] describe a Genetic Algorithm (GA) that achieves relatively good results in short running times. It uses random keys to encode solutions and a parameterized uniform crossover operator including local improvement based on the 2-opt heuristic to boost solution quality. Wu et al. [23] also proposed a GA using a direct representation in which the spanned nodes from each cluster and the sequence in which they are visited in the tour are stored. This approach has further been enhanced by Huang et al. [8] who apply a so-called hybrid chromosome encoding. However, reported results are on average inferior when compared to those of the GA from [21].

3   Solution Representation and Initialization

In our VNS, we represent a solution $S = \langle P, T \rangle$ in a direct way by storing the spanned nodes of each cluster $P = \{p_1, p_2, \ldots, p_r\}$ with $p_i \in V_i$, $i = 1, \ldots, r$, and additionally the visiting order in the round trip as circular permutation $\pi = \langle \pi_1, \ldots, \pi_r \rangle$ of the cluster indices $\{1, \ldots, r\}$.

Depending on the instance type we use two different procedures to compute feasible initial solutions for the VNS. Both are extensions of well-known greedy strategies for the classical TSP. The first algorithm is the (generalized) Nearest Neighbor Heuristic (NNH), and it can in principle be applied to all kinds of instances. The second procedure is specifically targeted to Euclidean instances where the clustering is based on geographical proximity. It exploits Euclidean coordinates of nodes and is called Generalized Insertion Heuristic (GIH). The following paragraphs describe these algorithms in detail.

3.1   Nearest Neighbor Heuristic for the GTSP (NNH)

Noon [17] suggested this approach, which computes a feasible solution as follows. We begin to construct a tour S_v from an arbitrarily chosen starting node $v \in V$. Iteratively, the algorithm always continues to the closest node belonging to a cluster that has not been visited yet and includes the corresponding edge. When nodes of all clusters have been reached, the tour is closed by including a final edge back to node v. This process is carried out once for each node in V as starting node, and the best tour is retained. See also Algorithm 1.

3.2   Generalized Insertion Heuristic for the GTSP (GIH)

This heuristic is inspired by the composite heuristic GI^3 from Renaud et al. [18]. In a first phase, it determines the set of spanned nodes P by calculating for

---

Algorithm 1. Nearest Neighbor Heuristic

---

for $v \in |V|$ **do**
  $S_v = \emptyset$
  $W = V$
  add v to S_v
  $v' = v$
  for $i = 1, \ldots, r - 1$ **do**
    remove from W all nodes belonging to the same cluster as v'
    $u =$ node in W nearest to v'
    add u to the partial tour S_v
    $v' = u$
return tour $S = S_{v^*}$ with $v^* = \operatorname{argmin}_{v \in V} C(S_v)$

---

each cluster V_i the node p_i having the lowest sum of distances to all other nodes in other clusters. After fixing these nodes, the CLOCK heuristic from [19] is performed to construct a tour containing many but not necessarily all nodes of P.

Recall that GIH only works on Euclidean instances where the nodes' coordinates are given. The CLOCK heuristic begins a partial tour S at the northernmost node from P. In case of a tie, the easternmost node among the northernmost nodes is chosen. This initial insertion is followed by four loops: In the first loop the procedure appends to S the northernmost node to the east of the last inserted node. In case of a tie, the easternmost node among these is chosen again. The process is repeated until there are no nodes to the east of the last appended node. The second, third, and forth loops work in the same way by appending to S the easternmost node to the south, the southernmost node to the west, and the westernmost node to the north of the last inserted node, respectively.

When the CLOCK heuristic terminates, there are in general some nodes from P left which are not yet included in the tour S. In contrast to the more complex GI^3 heuristic [18], we simply choose for each of these remaining nodes $p_j \in P \setminus H$ greedily the "cheapest" insertion position k so that $c(p_{\pi_{k-1}}, p_j) + c(p_j, p_{\pi_k}) - c(p_{\pi_{k-1}}, p_{\pi_k}) \le c(p_{\pi_{i-1}}, p_j) + c(p_j, p_{\pi_i}) - c(p_{\pi_{i-1}}, p_{\pi_i}) \ \forall i = 1, \ldots, |H|$ with $\pi_0 = \pi_{|H|}$.

As a final step, we try to improve the obtained feasible tour S by calling the shortest path algorithm, which will be introduced in Sect. 4.1. This procedure may replace nodes by other nodes of the same cluster, but it does not modify the visiting order π anymore. See Algorithm 2 for more details of the whole GIH.

This construction heuristic is much faster than the original GI^3, mainly because the latter uses a more sophisticated local improvement. Nevertheless, solutions obtained by GIH are typically only slightly inferior, and it usually takes just a few VNS iterations to catch up with or exceed the quality of GI^3's solutions.

While NNH has time complexity $\Theta(r \cdot |V|^2)$, GIH can be implemented in time $\Theta(|V|^2)$ and usually finds significantly better solutions to Euclidean instances with geographical clustering. However, GIH's applicability is far more limited.

Algorithm 2. Generalized Insertion Heuristic

for $i = 1, \ldots, r$ do
   ⌊ p_i = node in V_i with the least sum of costs to all other nodes in other clusters
partial tour $S = $ CLOCK heuristic($\{p_1, \ldots, p_r\}$)
for $j = 1, \ldots, r$ do
   if $p_j \notin S$ then
      ⌊ $k = $ minarg$_{i=1,\ldots,|S|}$ $(c(p_{\pi_{i-1}}, p_j) + c(p_j, p_{\pi_i}) - c(p_{\pi_{i-1}}, p_{\pi_i}))$, $\pi_0 = \pi_{|S|}$
      ⌊ insert p_j at position k in S

apply shortest path algorithm on S
return S

---

4   Neighborhood Structures

In our VNS, we use two complementary neighborhood structures. On the one hand, we approach the GTSP from the *global view* by first deciding in which order the clusters are to be visited and then computing an optimal selection of spanned nodes. On the other hand, we may start from the opposite direction and define a set of nodes for which we derive an appropriate tour.

4.1   Generalized 2-opt Neighborhood (G2-opt)

Renaud et al. [18] introduced the generalized 2-opt heuristic, which is based on the well known 2-opt heuristic for the classical TSP [14]. G2-opt is defined on a circular permutation $\pi = \langle \pi_1, \ldots, \pi_r \rangle$ indicating the visiting order of the clusters $\langle V_{\pi_1}, \ldots, V_{\pi_r} \rangle$, see Fig. 2(a). A particular permutation π thus represents the set of all feasible round trips $\langle p_{\pi_1}, p_{\pi_2}, \ldots, p_{\pi_r} \rangle$ with $p_{\pi_i} \in V_{\pi_i}$, $i = 1, \ldots, r$, and this set is in general exponentially large with respect to the number of nodes. However, the minimum cost round trip can be determined via a shortest path algorithm in polynomial time.

Given the visiting order of clusters, we can construct a graph containing edges only between nodes of consecutive clusters and a clone of the starting cluster attached to the last cluster, as it is shown in Fig. 2(b).

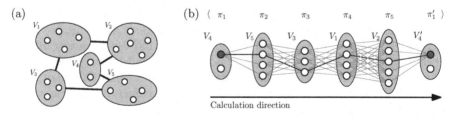

Fig. 2. (a) Visiting order of clusters characterized by permutation $\pi = \langle 4, 5, 3, 1, 2 \rangle$ and (b) corresponding graph on which the shortest path algorithm is applied, starting at the first node of cluster V_4 and ending at its clone in cluster V_4'

On this graph, we calculate shortest paths starting from each node of the starting cluster and ending at its clone. To ensure that at most one node is included from each cluster, we may simply assume the edges to be directed according to π. The overall cheapest path represents the optimal tour for this cluster order. Formally speaking, let L_{uv} denote the length of the shortest path from node $u \in V_{\pi_k}$ to node $v \in V_{\pi_l}$, $k < l$. Let L_u be the length of the shortest path containing r-edges, starting from $u \in V_{\pi_1}$, and ending at its clone in V'_{π_1}. The length L of the overall shortest tour respecting visiting order π is:

$$L = \min_{u \in V_{\pi_1}} L_u$$
$$L_u = \min_{v \in V_{\pi_r}}(L_{uv} + c(v, u)) \qquad \forall u \in V_{\pi_1}$$
$$L_{uv} = c(u, v) \qquad \forall u \in V_{\pi_1}, \ \forall v \in V_{\pi_2}$$
$$L_{uv} = \min_{w \in V_{\pi_{k-1}}}(L_{uw} + c(w, v)) \qquad \forall u \in V_{\pi_1}, \ \forall v \in V_{\pi_k}, \ k = 3, \ldots, r$$

To reduce the computational effort, we exploit the fact that π is rotation-invariant and choose V_{π_1} so that it is a cluster of smallest cardinality. The complexity of this dynamic programming algorithm is bounded by $O(|V_{\pi_1}| \cdot n^2/r)$.

Our generalized 2-opt neighborhood of a current solution S having cluster ordering π can now be defined as the set of all feasible round trips induced by any cluster ordering π' that differs from π by precisely one inversion I_{ij}, i.e. $\pi' = \langle \pi_1, \ldots, \pi_{i-1}, \pi_j, \ldots, \pi_i, \pi_{j+1}, \ldots, \pi_r \rangle$, $1 \leq i < j \leq r$.

Incremental bidirectional shortest path calculation. Instead of determining the shortest path L always from V_{π_1} (a cluster with the smallest number of nodes) to the cloned cluster V'_{π_1}, we can partition this task into three parts:

1. Perform shortest path calculations in forward direction from $u \in V_{\pi_1}$ to each node of a cluster V_{π_m} where m may be chosen arbitrarily.
2. Perform shortest path calculations in backward direction starting from $u' \in V'_{\pi_1}$ to each node of cluster $V_{\pi_{m+1}}$ where u' is the clone of node u.
3. Consider all edges in $E^m = \{(a, b) \in E \mid a \in V_{\pi_m} \wedge b \in V_{\pi_{m+1}}\}$ and the corresponding complete paths from u to u' including the above determined shortest paths to nodes in V_{π_m} and $V_{\pi_{m+1}}$, respectively. Take a $(a^*, b^*) \in E^m$ yielding an overall shortest path, i.e. $L_{ua^*} + c(a^*, b^*) + L_{b^*u'} \leq L_{ua} + c(a, b) + L_{bu'} \ \forall (a, b) \in E^m$.

This procedure, illustrated in Fig. 3, is in practice almost equally efficient as the simple one-way dynamic programming algorithm. When considering that we want to search the general 2-opt neighborhood, however, it provides the advantage of allowing for a substantially faster incremental evaluation scheme: If π' differs from π by an inversion I_{ij} with $i \leq m \leq j$, we do not have to recalculate the distances and predecessors of the nodes in clusters $V_{\pi_1}, \ldots, V_{\pi_{i-1}}$ and $V_{\pi_{j+1}}, \ldots, V'_{\pi_1}$, assuming we have stored these values in steps 1 and 2 before.

As a matter of course, m is always chosen to lie within the inversion interval. Clusters from V_{π_i} to V_{π_j} are marked "invalid" for both calculation directions

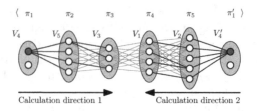

Fig. 3. Example for a bidirectional shortest path calculation with $m = 3$

after performing the inversion. Whenever we apply the shortest path algorithm in a particular direction, the evaluation is skipped for all clusters which are still valid, and the actual computation starts at the first invalid cluster. When processing these clusters, their "invalid" flags are removed.

To fully exploit this incremental evaluation, we further enumerate the possible inversions of π in a specific way: First, all inversions of pairs of two adjacent clusters are considered from left to right, then the inversions of all triplets in the reverse direction from right to left, next all 4-cluster inversions from left to right again, etc. Hereby, π_1 (and its clone in the corresponding graph for the shortest path calculation) remain fixed. See also Fig. 4. This strategy allows the largest data-reuse and minimizes the total number of clusters for which computations are necessary. It is in particular advantageous when we use a *next improvement* strategy in the local search, since we start with inversions of smallest size yielding the largest time savings; see Algorithm 3.

In the worst case, when we have to evaluate the whole neighborhood, $O(r^2)$ inversions must be considered. A naive complete enumeration would require time $O(r^2 \cdot |V_{\pi_1}| \cdot n^2/r) = O(|V_{\pi_1}| \cdot n^2 \cdot r)$. To be more precise, we have $(r - l + 1)$ possibilities for inversions of length l, $l = 2, \ldots, r - 2$. For each of them, the classical shortest path algorithm would have to consider all r clusters. However, with the incremental bidirectional shortest path calculation, we only have to consider $l + 1$ clusters after the first iteration. Hence, the classical algorithm evaluates $\sum_{l=2}^{r-2} r \cdot (r - l + 1) = \frac{r^3 - r^2 - 6r}{2}$ clusters while the incremental scheme

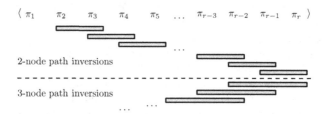

Fig. 4. Enumeration order of the inversions on π for making best use of the incremental bidirectional shortest path calculations

---

Algorithm 3. Search Generalized 2-opt Neighborhood (S)

---

for $l = 2, \ldots, r - 2$ do
   if l is even **then**
      for $i = 2, \ldots, r - l + 1$ do
         $\pi' = \langle \pi_1, \ldots, \pi_{i-1}, \pi_{i+l-1}, \ldots, \pi_i, \pi_{i+l}, \ldots, \pi_r \rangle$
         Apply incremental bidirectional shortest path calculation on π'
         if obtained solution S' is better than original solution S **then**
            ⌊ return solution S'

   else
      for $i = r - l + 1, \ldots, 2$ do
         $\pi' = \langle \pi_1, \ldots, \pi_{i-1}, \pi_{i+l-1}, \ldots, \pi_i, \pi_{i+l}, \ldots, \pi_r \rangle$
         Apply incremental bidirectional shortest path calculation on π'
         if obtained solution S' is better than original solution S **then**
            ⌊ return solution S'

return: no better solution found, i.e. S is a local optimum w.r.t. G2-opt

---

only processes $\sum_{l=2}^{r-2}(l + 1) \cdot (r - l + 1) = \frac{r^3 + 6r^2 - 25r - 6}{6}$ clusters for the whole neighborhood. Asymptotically, the latter is faster by factor 3.

4.2   Node Exchange Neighborhood (NEN)

With this new neighborhood structure, the search focuses on the set of spanned nodes $P = \{p_1, \ldots, p_r\}$. The node exchange neighborhood of a current solution S with spanned nodes P includes all feasible tours S' for each node set P' that can be derived from P by replacing one spanned node $p_i \in V_i$, $i \in \{1, \ldots, r\}$, by another node v of the same cluster V_i. This neighborhood therefore is induced by $\sum_{i=1}^{r}(|V_i| - 1) = O(|V|)$ different node sets resulting in a total of $O(|V| \cdot r!)$ round trips.

Unfortunately, determining the minimum cost round trip for a given node set P' is NP-hard since this subproblem corresponds to the classical TSP. Hence, instead of calculating the optimal round trip, we use the well known Chained Lin-Kernighan (CLK) algorithm [15] implemented in the Concorde library[1] to find a good but not necessarily optimal tour S' for a certain P'.

Though the size of this TSP is relatively small ($|P'| = r$), a complete evaluation of NEN is relatively time-demanding – even when using CLK – since we have to solve $O(|V|)$ different TSPs. To further speed up the neighborhood search, we restrict CLK to consider edges of the k-nearest-neighbor graph induced by P' only. For Euclidean instances and available point positions, this k-nearest-neighbor graph is efficiently derived using a KD-tree data structure. Tuning the parameter k, we can balance between speed and thoroughness of the search process. For the actual tests, we set k to 10. Algorithm 4 summarizes the steps of evaluating NEN.

---

[1] www.tsp.gatech.edu/concorde.html

Algorithm 4. Search Node Exchange Neighborhood (S)

---

for $i = 1, \ldots, r$ **do**
    forall $v \in V_i \setminus \{p_i\}$ **do**
        $P' = P \setminus \{p_i\} \cup \{v\}$
        determine k-nearest-neighbor graph G^k induced by P'
        apply CLK on G^k to obtain round trip S'
        if current solution S' better than so far best **then**
            └ save S' as so far best

restore and return best solution found

---

5 Variable Neighborhood Search Framework

We use the general variable neighborhood search (VNS) scheme with embedded variable neighborhood descent (VND) as proposed in [6].

Arrangement of the neighborhoods in VND: We alternate between G2-opt and NEN in this order. G2-opt is always considered first since its evaluation has a lower computational complexity.

Shaking in VNS: To perform shaking, we randomly exchange s spanned nodes by other nodes of the corresponding clusters and apply s random swap moves on the cluster ordering π. A swap move exchanges two positions in π. Parameter s depends on the number of clusters in the input graph and varies from 1 to $\frac{r}{7}$. We obtained the best results with these settings for s during our tests.

6 Computational Results

We tested the VNS on TSPlib[2] instances with geographical clustering which is done as follows [3]. First, r center nodes are chosen to be located as far as possible from each other. This is achieved by selecting the first center randomly, the second center as the farthest node from the first center, the third center as the farthest node from the set of the first two centers, and so on. Then, the clustering is done by assigning each of the remaining nodes to its nearest center node. We consider the largest of such TSPlib instances with up to 442 nodes, 97461 edges, and 89 clusters.

Our experiments were performed on a Pentium 4 2.6 GHz PC. In order to compute average values and standard deviations, we performed 30 runs for each instance. The VNS terminated after 10 consecutive outer iterations without finding a new best solution.

Table 1 presents results of our VNS and compares them to those of Fischetti et al's exact branch-and-cut algorithm (B&C) [5], the GI[3] heuristic [19], the random key GA (rk-GA) [21], and the hybrid chromosome GA (hc-GA) [8].

---

[2] http://elib.zib.de/pub/Packages/mp-testdata/tsp/tsplib/tsplib.html

Listed are for each instance its name, the numbers of nodes and clusters, the optimal solution value and run-time of B&C, and average percentage gaps of the heuristics' final objective values to the optimum solution value, as well as corresponding CPU-times. Best results are printed bold.

Since the B&C algorithm ran on a HP 9000/720, the GI^3 heuristic on a Sun Sparc Station LX, the rk-GA on a Pentium 4, 3.2 GHz PC, and the hc-GA on a 1.2 GHz PC, it is hard to compare the CPU-times directly. Nevertheless, it is obvious that in particular the rk-GA is very fast and computes high quality results within a few seconds. Comparing VNS with both GAs, VNS requires significantly more time, but it is often able to find superior solutions.

Especially on the larger instances with ≥ 299 nodes, VNS benefits from the sophisticated large neighborhood search and its average gaps of are consistently substantially smaller than those of all other considered heuristics. For 18 out of the 28 instances, VNS was even able to obtain optimal solutions in all of the 30 performed runs; its total average gap is only 0.05%, and no average gap exceeds

Table 1. Results on TSPlib instances with geographical clustering

Instance			B&C		GI^3		rk-GA		hc-GA		VNS		
Name	$|V|$	r	C_{opt}	*time*	*gap*	*time*	$\overline{gap}$	*time*	$\overline{gap}$	*time*	$\overline{gap}$	σ_{gap}	*time*
kroa100	100	20	9711	18.4s	**0.00%**	6.8s	**0.00%**	0.4s	–	–	**0.00%**	0.00	2.5s
krob100	100	20	10328	22.2s	**0.00%**	6.4s	**0.00%**	0.4s	–	–	**0.00%**	0.00	0.4s
rd100	100	20	3650	16.6s	**0.00%**	7.3s	**0.00%**	0.5s	–	–	**0.00%**	0.00	0.9s
eil101	101	21	249	25.6s	0.40%	5.2s	**0.00%**	0.4s	–	–	0.04%	0.12	16.3s
lin105	105	21	8213	16.4s	**0.00%**	14.4s	**0.00%**	0.5s	–	–	**0.00%**	0.00	0.6s
pr107	107	22	27898	7.4s	**0.00%**	8.7s	**0.00%**	0.4s	–	–	**0.00%**	0.00	0.5s
pr124	124	25	36605	25.9s	0.43%	12.2s	**0.00%**	0.6s	–	–	**0.00%**	0.00	26.6s
bier127	127	26	72418	23.6s	5.55%	36.1s	**0.00%**	0.4s	–	–	**0.00%**	0.00	1.4s
pr136	136	28	42570	43.0s	1.28%	12.5s	**0.00%**	0.5s	–	–	**0.00%**	0.00	48.1s
pr144	144	29	45886	8.2s	0.00%	16.3s	**0.00%**	1.0s	–	–	**0.00%**	0.00	4.0s
kroa150	150	30	11018	100.3s	0.00%	17.8s	**0.00%**	0.7s	0.00%	0.4s	**0.00%**	0.00	1.2s
krob150	150	30	12196	60.6s	0.00%	14.2s	**0.00%**	0.9s	0.00%	0.9s	**0.00%**	0.00	3.7s
pr152	152	31	51576	94.8s	0.47%	17.6s	**0.00%**	1.2s	0.00%	0.6s	**0.00%**	0.00	7.6s
u159	159	32	22664	146.4s	2.60%	18.5s	**0.00%**	0.8s	0.00%	1.0s	**0.00%**	0.00	22.6s
rat195	195	39	854	245.9s	**0.00%**	37.2s	**0.00%**	1.0s	–	–	0.01%	0.04	105.6s
d198	198	40	10557	763.1s	0.60%	60.4s	**0.00%**	1.6s	–	–	0.02%	0.05	141.3s
kroa200	200	40	13406	187.4s	**0.00%**	29.7s	**0.00%**	1.8s	0.01%	1.8s	**0.00%**	0.00	16.9s
krob200	200	40	13111	268.5s	**0.00%**	35.8s	**0.00%**	1.9s	0.06%	8.0s	**0.00%**	0.00	18.8s
ts225	225	45	68340	37875.9s	0.61%	89.0s	**0.02%**	2.1s	0.13%	19.0s	0.03%	0.07	274.4s
pr226	226	46	64007	106.9s	**0.00%**	25.5s	**0.00%**	1.5s	**0.00%**	0.6s	**0.00%**	0.00	1.7s
gil262	262	53	1013	6624.1s	5.03%	115.4s	0.79%	1.9s	**0.00%**	41.2s	0.05%	0.16	372.5s
pr264	264	53	29549	337.0s	0.36%	64.4s	**0.00%**	2.1s	**0.00%**	3.1s	0.01%	0.04	268.2s
pr299	299	60	22615	812.8s	2.23%	90.3s	0.11%	3.2s	0.10%	68.6s	**0.00%**	0.01	220.5s
lin318	318	64	20765	1671.9s	4.59%	206.8s	0.62%	3.5s	0.72%	18.3s	**0.30%**	0.61	320.1s
rd400	400	80	6361	7021.4s	1.23%	403.5s	1.18%	5.9s	2.15%	17.4s	**0.74%**	0.51	502.0s
fl417	417	84	9651	16719.4s	0.48%	427.1s	0.05%	5.3s	0.12%	19.4s	**0.00%**	0.00	92.4s
pr439	439	88	60099	5422.8s	3.52%	611.0s	0.26%	9.5s	0.76%	10.9s	**0.12%**	0.11	519.0s
pcb442	442	89	21657	58770.5s	5.91%	567.7s	1.70%	9.0s	0.94%	31.8s	**0.08%**	0.08	596.6s
Average gaps					1.26%		0.17%		0.30%		0.05%		

0.75%. From all considered heuristics, GI^3 was the weakest, with worst average results and running times in the same order of magnitude as our VNS.

In particular for the two large instances pr226 and fl417, already VND was able to directly identify the optimal solutions, i.e. merely alternating between G2-opt and NEN was sufficient to get to the global optima, and no VNS iterations were required. This documents how well these neighborhood structures complement each other.

7   Conclusions and Future Work

In this article, we proposed a variable neighborhood search approach for the generalized traveling salesman problem utilizing two large neighborhood structures. They can be seen as dual to each other: While G2-opt predefines the possible cluster orderings and uses a relatively sophisticated but efficient procedure for augmenting these partial solutions with appropriate selections of nodes, the situation is vice versa in the newly proposed NEN.

Considering in particular G2-opt, the described incremental evaluation scheme turned out to be a major speed-up factor in comparison to the previously used evaluation via independent standard shortest path calculations.

It further turned out that the VNS slightly benefits from a good starting solution. Therefore, we described the more generally applicable nearest neighbor heuristic and particularly for Euclidean instances with given point positions the generalized insertion heuristic. Both are reasonably fast and provide solutions of appropriate quality.

We tested the VNS on TSPlib instances with geographical clustering consisting of up to 442 nodes. Compared to two recent genetic algorithms, the VNS performs slower, but it is able to generate remarkably better solutions, in particular for larger instances.

Future work will in particular include tests on other types of instances, e.g. with non-Euclidean distances and incomplete graphs. An incremental evaluation scheme for NEN seems to be a challenging task but might further speed up the algorithm. Promising is also the combination of these neighborhood structures with others, and to investigate their application in other types of metaheuristics.

References

1. Behzad, A., Modarres, M.: A new efficient transformation of the generalized traveling salesman problem into traveling salesman problem. In: Proceedings of the 15th International Conference of Systems Engineering, pp. 6–8 (2002)
2. Dimitrijevic, V., Saric, Z.: An efficient transformation of the generalized traveling salesman problem into the traveling salesman problem on digraphs. Information Science 102(1-4), 105–110 (1997)
3. Feremans, C.: Generalized Spanning Trees and Extensions. PhD thesis, Universite Libre de Bruxelles (2001)
4. Fischetti, M., Salazar, J.J., Toth, P.: The symmetric generalized traveling salesman polytope. Networks 26, 113–123 (1995)

5. Fischetti, M., Salazar, J.J., Toth, P.: A branch-and-cut algorithm for the symmetric generalized traveling salesman problem. Operations Research 45, 378–394 (1997)
6. Hansen, P., Mladenovic, N.: An introduction to variable neighborhood search. In: Voss, S., et al. (eds.) Meta-heuristics, Advances and trends in local search paradigms for optimization, pp. 433–458. Kluwer Academic Publishers, Dordrecht (1999)
7. Henry-Labordere.: The record balancing problem: A dynamic programming solution of a generalized traveling salesman problem. RAIRO Operations Research B2, 43–49 (1969)
8. Huang, H., Yang, X., Hao, Z., Wu, C., Liang, Y., Zhao, X.: Hybrid chromosome genetic algorithm for generalized traveling salesman problems. Advances in Natural Computation 3612, 137–140 (2005)
9. Laporte, G., Asef-Vaziri, A., Sriskandarajah, C.: Some applications of the generalized travelling salesman problem. Journal of the Operational Research Society 47(12), 1461–1467 (1996)
10. Laporte, G., Mercure, H., Nobert, Y.: Generalized traveling salesman problem through n sets of nodes: The asymmetric case. Discrete Applied Mathematics 18, 185–197 (1987)
11. Laporte, G., Nobert, Y.: Generalized traveling salesman problem through n sets of nodes: An integer programming approach. INFOR 21(1), 61–75 (1983)
12. Laporte, G., Semet, F.: Computational evaluation of a transformation procedure for the symmetric generalized traveling salesman problem. INFOR 37(2), 114–120 (1999)
13. Lien, Y.N., Ma, E., Wah, B.W.S.: Transformation of the generalized traveling salesman problem into the standard traveling salesman problem. Information Sciences 74(1–2), 177–189 (1993)
14. Lin, S.: Computer solutions of the traveling salesman problem. Bell Systems Computer Journal 44, 2245–2269 (1965)
15. Martin, O., Otto, S.W., Felten, E.W.: Large-step Markov chains for the traveling salesman problem. Complex Systems 5, 299–326 (1991)
16. Noon, C., Bean, J.C.: An efficient transformation of the generalized traveling salesman problem. INFOR 31(1), 39–44 (1993)
17. Noon, C.E.: The Generalized Traveling Salesman Problem. PhD thesis, University of Michigan (1988)
18. Renaud, J., Boctor, F.F.: An efficient composite heuristic for the symmetric generalized traveling salesman problem. European Journal of Operational Research 108, 571–584 (1998)
19. Renaud, J., Boctor, F.F., Laporte, G.: A fast composite heuristic for the symmetric traveling salesman problem. INFORMS Journal on Computing 8(2), 134–143 (1996)
20. Saskena, J.P.: Mathematical model of scheduling clients through welfare agencies. Journal of the Canadian Operational Research Society 8, 185–200 (1970)
21. Snyder, L.V., Daskin, M.S.: A random-key genetic algorithm for the generalized traveling salesman problem. Technical Report 04T-018, Dept. of Industrial and Systems Engineering, Lehigh University, Bethlehem, PA, USA (2004)
22. Srivastava, Kumar, S.S.S., Garg, R.C., Sen, P.: Generalized traveling salesman problem through n sets of nodes. CORS Journal 7, 97–101 (1969)
23. Wu, C., Liang, Y., Lee, H.P., Lu, C.: Generalized chromosome genetic algorithm for generalized traveling salesman problems and its applications for machining. Physical Review E 70(1) (2004)

Efficient Local Search Limitation Strategies for Vehicle Routing Problems

Yuichi Nagata[1] and Olli Bräysy[2]

[1] Graduate School of Information Sciences,
Japan Advanced Institute of Science and Technology, Japan
nagatay@jaist.ac.jp
[2] Agora Innoroad Laboratory, Agora Center, P.O.Box 35, FI-40014
University of Jyväskylä, Finland
Olli.Braysy@jyu.fi

Abstract. In this paper we examine five different strategies for limiting the local search neighborhoods in the context of vehicle routing problems. The vehicle routing problem deals with the assignment of a set of transportation orders to a fleet of vehicles, and the sequencing of stops for each vehicle to minimize transportation costs. The examined strategies are applied to three standard neighborhoods and implemented in a recently suggested powerful memetic algorithm. Experimental results on 26 well-known benchmark problems indicate significant speedups of almost 80% without worsening the solution quality. On the contrary, in 12 cases new best solutions were obtained.

1   Introduction

In this paper we focus on the Capacitated Vehicle Routing Problem (CVRP). It can be defined as follows. Let $G = (V, E)$ be a complete undirected graph consisting of $n + 1$ nodes, and a set of edges E with non-negative weights and with associated travel times. Let one of the nodes be designated as the depot. With each node i, apart from the depot, is associated a demand q_i that can be a delivery from or a pickup to the depot. The problem is to minimize the total travel distance of a routing plan such that the total demand of any route does not exceed a vehicle capacity Q (the capacity constraint), the duration of any route does not exceed an upper limit L (the route duration limit), each route must start and end at the depot, and each customer must be served by exactly once by one vehicle. Note that the above described VRP with the route duration limit is often separated from CVRP and called distance-constrained VRP (DVRP). In this paper, the DVRP is not considered.

The VRP is a NP-hard problem which makes it difficult to solve it to optimality. In practice, heuristic or metaheuristic solution methods are often the only option. For more details, see e.g. Laporte and Semet [10], Gendreau et al. [5], and Cordeau et al. [4].

In broad terms, the actual search in almost all known heuristic and metaheuristic solution methods is based on various local search procedures and often

J. van Hemert and C. Cotta (Eds.): EvoCOP 2008, LNCS 4972, pp. 48–60, 2008.

the local search is clearly the most time consuming part of the algorithms. Because of this, it is crucial that strategies for efficient implementation and for speeding up the local searches are developed and tested. An obvious approach is to forbid moves that apparently seem to degenerate the current solution.

The main contribution of this paper is to examine five different approaches that are used to rule out non-improving local search moves and to demonstrate that significant speedups can be gained without worsening the solution quality. The five limitation approaches are combined with a composite of three standard local search neighborhoods and implemented in a recent powerful memetic algorithm by Nagata [16]. Moreover, 12 new best solutions are found on standard benchmark problems of Golden at el. [6] and Taillard [19] in reasonable computation time.

Memetic algorithms (MAs) [15] have recently received a lot of attention due to their good performance in solving various optimization problems. Briefly speaking, MAs are hybrid Evolutionary Algorithms (EAs) that combine the global and local search by using an EA to perform exploration while the local search method performs exploitation. More precisely, local searches are usually applied to solutions generated by recombination and mutation operators. Typically, the local search is the most time consuming part of MAs. For example, in the MA applied here, 80 – 90 % of the computation time is spent on the local search without the limitation approaches.

The remainder of this paper is organized as follows. In Section 2 we describe the main local search framework, the applied local search procedures and the suggested limitation strategies. Section 3 details the MA and experimental results are given in Section 4. Conclusions are presented in Section 5.

2   The Search Strategy

2.1   The Framework of the Local Search

A local search framework and strategies for limiting the local search are illustrated in Figure 1. Here these strategies are applied to a MA but they could be combined also e.g. with other metaheuristics.

In procedure Local-Search(s), $List$ represents a set of customers considered by a composite local search. The basic idea is to limit the local search to a set of customers in the $List$ during the whole search process. The $List$ is initialized by a user-defined limitation strategy (line 1).

For each customer $v \in List$, an improving move is searched in a neighborhood $\mathcal{N}(v, s)$ (line 5). Here, $\mathcal{N}(v, s)$ is defined as a set of moves around v (see Section 2.2). If an improving move is found, the current solution is updated (line 7) and the customers associated with the improving move are optionally added to the $List$ (line 8). Correspondingly, in case improvement is not found, customer v is optionally deleted from the $List$ (line 11). The updating of the $List$ depends on the applied limitation strategy.

Six different strategies for limiting the local search are summarized in Figure 1. The strategy FUL in the figure does not limit the local search and is used as a baseline for comparisons. The details are described in Section 2.3.

Procedure Local-Search(s)
begin
1 : Initialize *List* using a user-defined strategy;
2 : Set *List* in random order;
3 : **for** $i := 1$ **to** $|List|$ **do**
4 : $v := List[\ i\ ]$;
5 :   Search for an improving move in $\mathcal{N}(v,s)$;
6 :   **if** an improving move s' is found **then**
7 : $s := s'$;
8 :   Add the customers associated with the improving move to *List*; (Update)
9 :   **goto** line 2;
10:    **else**
11:       Remove v from *List*; (Update)
12:    **end if**
13: **end for**
14: **return** s;
end

Strategy	Initialize *List*	Update
Full search (FUL)	All customers	No
Don't-look bids (DLB)	All customers	Yes
Forbidding moves of common edges (FCE)	All customers	No
	Note: All moves changing any common edges are forbidden in $\mathcal{N}(v,s)$.	
Limiting moves to new edges (LNE)	A set of customers consisting of endpoints of the new edges ($E_{CB} \cup E_{CN}$)	No
Limiting moves to new routes (LNR)	A set of customers consisting of customers in the new routes (routes including the new edges)	No
LNR with DLB (LNRD)	A set of customers consisting of customers in the new routes (routes including the new edges)	Yes

Fig. 1. Local search framework and limitation strategies

2.2   The Local Search Neighborhood

The local search is based here on three standard local search neighborhoods: *2-opt*, *insertion*, and *swap*. These three local search procedures are always applied together with the first-accept strategy and in random order, forming a composite local search.

Moreover, the neighborhoods is restricted to the moves which are defined within geographically close customers only. To be more precise, given a current solution s and a customer node v, the three neighborhoods are redefined as described below. In the definition, w denotes other customer nodes which are close to v, meaning that w must be selected from a set of customers that are among the N_{near} closest customers with respect to v where N_{near} is a user-defined parameter. Note that w may or may not belong to the same route with v with two exceptions; v and w must belong to different routes in the second and third moves in 2-*opt* neighborhood. v^- (w^-) and v^+ (w^+) denote a predecessor and a successor of v (w), which may be the depot or a customer. (i, j) denotes an edge between two customers, i and j. The four neighborhoods listed below are defined in such a way that an edge (v, w) must be included after the moves.

We define $\mathcal{N}(v, s)$ as a composite of the following four neighborhoods.

2-opt(v, s)
- Remove (v, v^-) and (w, w^-), and add (v, w) and (v^-, w^-).
- Remove (v, v^-) and (w, w^+), and add (v, w) and (v^-, w^+).
- Remove (v, v^+) and (w, w^-), and add (v, w) and (v^+, w^-).
- Remove (v, v^+) and (w, w^+), and add (v, w) and (v^+, w^+).

(1,0)-Interchange(v, s)
- Insert v between w and w^-, and link v^- and v^+.
- Insert v between w and w^+, and link v^- and v^+.

(0,1)-Interchange(v, s)
- Insert w between v and v^-, and link w^- and w^+.
- Insert w between v and v^+, and link w^- and w^+.

(1,1)-Interchange(v, s)
- Insert v between w and $(w^-)^-$, and insert w^- between v^- and v^+.
- Insert v between w and $(w^+)^+$, and insert w^+ between v^- and v^+.

2.3   The Suggested Limitation Strategies

In this section we describe in detail the applied limitation strategies. The important idea is limiting the local search around the elements that are changed from one of parents by the crossover or any other procedures. This idea is similar to the technique used in the iterated local search algorithm [7] where the local search is limited around the elements that are changed by the perturbation, but it has never been applied to MAs.

First, we define some notations. In general, a crossover operator (sometimes together with a mutation (or any other)) generates offspring by combining fundamental elements from selected parents . When VRPs are considered, one possibility is to employ the edge as an element. Figure 2 illustrates an example of application of a crossover. Let A and B be a pair of parents to be combined by a crossover and mutation (or any other), and let C be a child. Let E_A, E_B and E_C be sets of edges consisting of A, B, and C, respectively. Now, E_C is separated into four disjoint sets: E_{CS}, E_{CA}, E_{CB}, and E_{CN}. E_{CS} is defined as a set of edges that exist in both of E_A and E_B. E_{CN} is defined as a set of edges that do not exist in neither E_A nor E_B. E_{CA} (E_{CB}) is defined as a set of edges that exist only in E_A (E_B).

Limiting moves to new edges. The basic idea of this strategy is limiting the composite local search to the neighborhood of the edges moved by the crossover

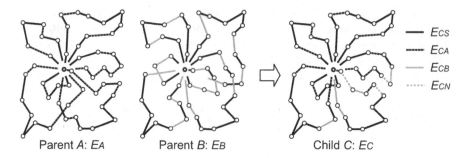

Fig. 2. Application of a crossover and mutation (or any other) in VRPs

and mutation (or other procedures). Key concept of this strategy is considering E_C to be generated from E_A by replacing some edges with $E_{CB} \cup E_{CN}$. More precisely, the *List* is initialized by a set of customers consisting of endpoints of the edges belonging to $E_{CB} \cup E_{CN}$ in the initialization step. See Figure 2 for example, the *List* is initialized by 12 customers.

Limiting moves to new routes. This strategy is similar to the previous one. The basic idea of this strategy is limiting the composite local search to the neighborhood of the new routes (routes including at least one edge belonging to $E_{CB} \cup E_{CN}$) generated by the crossover and mutation (or other procedures). More precisely, the *List* is initialized by a set of customers consisting of the customers in the new route in the initialization step. See Figure 2 for example, there are three new routes.

Forbidding moves of common edges. The basic idea is forbidding moves related to common edges among the parent solutions (or e.g. in a population or pool of solutions). The idea is motivated by the 'big valley' structure [9], i.e., solution features that appear frequently in the population have relatively high probability of belonging to the optimal solution. This strategy was applied to MAs for solving TSP [11], QAP [12], and CVRP [8]. However, Kubiak [8] reported that there was no significant improvement in the computation time on the CVRP. We test this strategy for comparison. This strategy is incorporated to the composite local search by directly forbidding the moves which change any common edges belonging to E_{CS}.

Don't-look bids strategy. The concept of *don't-look bits* was originally developed for speeding up local searches in the TSP [2]. This idea can be straightforwardly incorporated to the framework of the local search as follows: If an improving move is not found in $\mathcal{N}(s, v)$, v is removed from the *List*. Correspondingly, in case improvement is found, the customers associated with the improving move (endpoints of the moved edges) are added to the *List* (if they are not already included).

3   The Applied Memetic Algorithm

In this section we briefly outline the main ideas of the applied MA. For more details, we refer the reader to [16].

In short, the proposed MA is based on the EAX crossover, the above described standard local search neighborhoods and allowance of infeasible intermediate solutions during the search.

In the beginning, each customer is served by a separate route, as in the well-known savings heuristic. Then the composite local search, described in Section 2.2, is applied to routes to improve the solution. Here also moves that violate the capacity constraint are allowed during the local search. This procedure is repeated until a defined population size (N_{pop}) is obtained.

The selection of parents is done randomly by putting the population members to random order in the beginning of each generation and selecting the parents directly according to this order. Note that N_{pop} pair of parents are mated for each generation. For each pair of parents, A and B, a defined number (N_{ch}) of children (offspring) solutions are generated with the EAX crossover. If the generated offspring violates the vehicle capacity constraint, an attempt is made to eliminate the violation by a modification procedure. Briefly speaking, the modification procedure equals the composite local search where a penalty function is defined to deal with the overcapacity.

The feasible offspring solutions are further locally optimized using the composite local search. If the best feasible offspring found has a smaller total distance than A, it replaces A. No separate procedure is applied. The search is stopped when no more improvements have been found for the last N_g generations.

The important features of the EAX and modification procedure can be summarized as follows by using the notations defined in Section 2.3 : (i) Most of the offspring edges belong to $E_A \cup E_B$ (usually more than 95 %). (ii) Most of the common edges among the parent solutions ($E_A \cap E_B$) are inherited to offspring. (iii) New edges (E_{CN}) created for the offspring tend to be short. (iv) Offspring are generated from parent A by replacing a relatively small number of geographically close edges ($E_{CB} \cup E_{CN}$). This approach was motivated by the observation that geographically scattered edge exchanges often destroy important features of good solutions.

In our observation, 80 – 90 % of the computation time is spent on the composite local search applied after the EAX and modification procedures. The proposed limitation strategies are applied to this composite local search.

4   Experimental Results

A computational experiment has been conducted to analyze the performance of the proposed strategies. In this section, we describe the experimental setting and analyze the effect of the proposed approaches. Finally, we present comparative analysis to other recent metaheuristics that have shown the best performance.

4.1   Experimental Setting

The proposed limitation strategies are applied to the composite local search in the MA by Nagata [16] as described above. The optimization parameter values of the MA are same as those of the original MA; $N_{pop} = 100$, $N_{ch} = 30$, $N_g = 50$. N_{near} (see Section 2.2) is set to 10 if v^- (and v^+) is not the depot and otherwise 30. This strategy is motivated by the observation that edges incident to the depot are often longer than other edges in good routing plans.

The computational tests were carried out with the standard CVRP benchmarks. The 14 classical benchmark problems of Christofides et al. [3] (C1–C14) consist of 50 – 199 customers, the 20 large scale benchmark problems of Golden et al. [6] (G1–G20) consist of 200 – 483 customers, and the 13 benchmark problems of Taillard [19] (T75a–T385) consist of 75 – 385 customers. Problems, C6–C10, C13–C14, and G1–G8, including the route duration constraint, are not considered because this constraint is not supported by the MA used here.

For each problem, the MAs were executed 10 times. The MAs were implemented in C++ and executed with Xeon 3.2 GHz single processor computers.

4.2   Analysis of the Limitation Strategies

The analysis of the limitation strategies is executed on the benchmarks of Golden et al. Table 1 lists the detailed results. In the table, the results of the MA with the limitation configurations are compared with each other and also with the original MA. The first column lists the names of the instances, the number of customers, and the best-known solutions. Most of the best-known solutions (except for G20) were found by the previous MA. The first row lists the examined configurations. The successive rows show the best travel distance (top), the average travel distance (middle), and average computation time in seconds (bottom). New best solutions are marked with boldface. The last row presents the averaged values over the all instances. 'Dev. best' and 'Dev. aver.' describe the deviation of the best results and the average results, respectively, from the best-known solutions in percentage. 'Ave. CPU' gives the average CPU time in seconds.

The Table 2 summarizes the relative differences from configuration FUL (with no limitations). In the table, Rel. dev. shows averaged relative deviation of the solution qualities. For example, Rel. dev. for configuration 'A' is calculated by $\frac{1}{|Problems|} \cdot \sum_{i \in problems} \frac{average_i^A - average_i^{FUL}}{average_i^{FUL}} \times 100$, where $average_i^A$ refers to the average total distance of problem i obtained by configuration 'A'. Rel. time shows the average relative CPU time in percentage in a similar way.

As shown in the tables, the examined five limitation strategies, especially LNE, LNR, and LNRD, reduce the computation time of the MA significantly. Moreover, the deterioration of the solution quality by DLB, LNR, and LNRD is not observed or very small (≤ 0.007 %). FCE and LNE slightly deteriorate the solution quality (≥ 0.07 %). LNRD appears to be the best strategy, reducing 76 % of the computation time without worsening the solution quality.

We should also refer to the improvement of the MA with configuration FUL over the previous MA. The time reduction is caused by the idea that two values

Table 1. Experimental results for the benchmark problems of Golden et al. [6]

Problem best		Limitation strategies						Prev. MA
		FUL	DLB	FCE	LNE	LNR	LNRD	
G9	best	581.02	580.72	581.20	**580.02**	**580.40**	**580.48**	580.60
(255)	average	581.77	581.72	582.27	582.19	581.88	581.63	582.34
580.60	time (s)	1079	730	580	261	477	314	1552
G10	best	**738.44**	**738.81**	740.22	**738.56**	**738.80**	**738.73**	738.92
(323)	average	739.72	739.92	740.95	740.24	739.71	739.74	740.91
738.92	time (s)	1621	955	900	533	720	582	2508
G11	best	**914.82**	**914.24**	**915.15**	**915.63**	**914.03**	**914.75**	917.17
(399)	average	916.28	915.94	916.63	916.66	915.48	915.64	918.19
917.17	time (s)	2456	1733	1498	656	1098	794	3835
G12	best	**1105.21**	**1104.84**	**1107.09**	**1105.70**	**1105.90**	**1106.33**	1108.48
(483)	average	1107.89	1107.66	1109.67	1109.33	1107.80	1107.87	1110.71
1107.19	time (s)	3137	2552	1892	951	1340	1024	6801
G13	best	857.19	858.14	857.19	857.19	857.19	857.19	857.19
(252)	average	857.94	858.99	859.47	858.54	858.42	858.69	858.84
857.19	time (s)	978	695	532	268	342	252	1165
G14	best	1080.55	1080.55	1080.55	1080.55	1080.55	1080.55	1080.55
(320)	average	1080.83	1080.94	1081.07	1082.18	1081.19	1081.01	1080.93
1080.55	time (s)	1319	915	821	335	464	306	1620
G15	best	1341.40	1342.57	1343.46	1341.72	1341.55	1341.23	1340.24
(396)	average	1343.09	1343.74	1345.29	1343.29	1343.47	1343.24	1344.02
1340.24	time (s)	1987	1206	1103	502	590	510	1924
G16	best	**1616.55**	**1619.22**	1622.92	1621.19	**1619.52**	**1616.33**	1619.93
(480)	average	1621.84	1621.32	1624.95	1624.27	1621.82	1621.47	1625.07
1619.93	time (s)	3019	2049	1626	555	824	579	4295
G17	best	707.76	707.76	707.76	707.76	707.76	707.76	707.76
(240)	average	707.77	707.78	707.79	707.77	707.76	707.76	707.77
707.76	time (s)	661	447	344	105	145	118	718
G18	best	**995.13**	995.39	995.62	995.75	**995.13**	995.39	995.39
(300)	average	996.19	996.15	996.41	996.48	996.36	996.01	996.62
995.39	time (s)	1123	885	721	243	304	253	1261
G19	best	1366.16	**1366.13**	**1365.99**	1367.05	1366.21	1366.18	1366.14
(360)	average	1366.78	1366.51	1367.38	1367.49	1366.65	1366.76	1367.31
1366.14	time (s)	1774	1137	1101	243	299	234	2013
G20	best	1820.16	**1820.05**	1822.58	1820.88	1821.06	**1819.99**	1820.54
(420)	average	1822.33	1822.37	1824.62	1823.82	1822.68	1822.00	1822.97
1820.09	time (s)	2571	1642	1321	368	424	396	3169
Dev. of best. (%)		-0.045	-0.024	0.052	-0.010	-0.031	-0.045	0.012
Dev. of aver. (%)		0.081	0.088	0.181	0.151	0.086	0.078	0.183
Ave. CPU (sec)		1810	1245	1036	418	585	446	2571

Table 2. Relative differences from configuration FUL

	FUL	DLB	FCE	LNE	LNR	LNRD	Prev. MA
Rel. dev. (%)	0.000	0.007	0.100	0.070	0.005	-0.003	0.101
Rel. time (%)	100	69	57	23	32	24	134

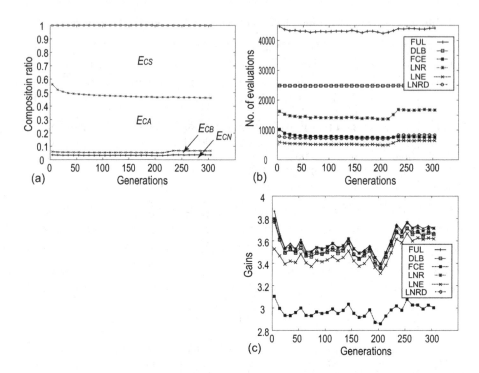

Fig. 3. Behaviors of EAX and the composite local search with different configurations

of N_{near} are used in the composite local search (see Section 4.1). The improvement in the solution quality is caused by varying the neighborhoods after each improvement instead of applying the same neighborhood until local minimum. A more detailed analysis is given in Figure 3 where the results are based on average output averaged every 10 generations in a run to problem **G9**.

Figure 3 (a) shows the composition ratio of E_{CS}, E_{CA}, E_{CB}, and E_{CN} in case of the MA with configuration FUL. As shown in the figure, children are generated from one parent (parent A) by replacing a relatively small number of edges; the ratio of $E_{CN} + E_{CB}$ is averagely less than 10 %.

Figure 3 (b) illustrates the number of evaluations per an application of the composite local search in the different limitation strategies. Here the starting point is made equal in each generation by applying each local search to the same solution obtained in the MA with configuration FUL. It appears that large reductions in the number of evaluations are possible with strategies LNE, LNR and LNRD because the ratio of $E_{CN} + E_{CB}$ is small, and these make it possible to reduce the computation time of the corresponding MAs. The computation time of the MA with configuration FCE seem to be larger in Table 2 than expected based on the results reported here. This is mainly because of the time needed for checking the common edges.

Table 3. Comparisons with other heuristics

Problem	Prev. best	Rochat & Taillard [18]	Tarantilis [20]	Mester & Bräysy [13]	Nagata aver. [16]	LNRD best	LNRD aver.	time (sec)
C3 (100)	826.14	826.14	826.14	826.14	826.14	826.14	826.14	12
C4 (150)	1028.42	1028.42	1028.42	1028.42	1028.51	1028.42	1028.42	31
C5 (199)	1291.29	1291.45	1311.48	1291.29	1293.93	1291.29	1294.67	156
C11 (120)	1042.11	1042.11	1042.11	1042.11	1042.11	1042.11	1042.11	6
C12 (100)	819.56	819.56	819.56	819.56	819.56	819.56	819.56	2
Dev. (%)		0.002	0.22	0.00	0.03	0.00	0.04	
Ave. CPU (min)		N/A	5.4	7.2	2.4	6.9	0.7	
Computer		SGI 100	Pentium 400	Penti. IV 2.8 GHz	Xeon 3.2 GHz	Xeon 3.2 GHz	Xeon 3.2GHz	

Problem	Prev. best	Tarantilis [20]	Pisinger aver. [17]	Mester & Bräysy [13]	Nagata aver. [16]	LNRD best	LNRD aver.	time (sec)
G9 (255)	580.60	585.43	590.33	583.39	582.34	**580.48**	581.63	314
G10 (323)	738.92	746.56	751.36	741.56	740.91	**738.73**	739.74	582
G11 (399)	917.17	923.17	926.57	918.45	918.19	**914.75**	915.64	794
G12 (483)	1107.19	1130.40	1125.22	1107.19	1110.71	**1106.33**	1107.87	1024
G13 (252)	857.19	865.01	874.24	859.11	858.84	857.19	858.69	252
G14 (320)	1080.55	1086.07	1103.53	1081.31	1080.93	1080.55	1081.01	306
G15 (396)	1340.24	1353.91	1366.23	1345.23	1344.02	1341.23	1343.24	510
G16 (480)	1619.93	1634.74	1645.67	1622.69	1625.07	**1616.33**	1621.47	579
G17 (240)	707.76	708.74	710.59	707.79	707.77	707.76	707.76	118
G18 (300)	995.39	1006.90	1007.84	998.73	996.62	995.39	996.01	253
G19 (360)	1366.14	1371.01	1377.88	1366.86	1367.31	1366.18	1366.76	234
G20 (420)	1820.09	1837.67	1834.70	1820.09	1822.97	**1819.99**	1822.00	396
Dev. (%)		1.51	2.42	0.32	0.31	-0.07	0.13	
Ave. CPU (min)		45.5	8.4	4.5	41.1	74.7	7.5	
Computer		Pentium 400	Penti. VI 3.0 GHz	Penti. VI 2.8 GHz	Xeon 3.2GHz	Xeon 3.2GHz	Xeon 3.2GHz	

Problem	Prev. best	Rochat & Taillard [18]	Mester & Bräysy [14]	Alba et al. [1]	LNRD best	LNRD aver.	time (sec)
T100a	2041.34	2047.90	2041.34	2047.90	2041.34	2041.34	18
T100b	1939.90	1940.61	1939.90	1940.36	1939.90	1940.37	14
T100c	1406.20	1407.44	1406.20	1411.66	1406.20	1406.20	14
T100d	1581.25	1581.25	1581.25	1584.20	**1580.46**	1580.46	27
T150a	3055.23	3070.91	3055.23	3056.41	3055.23	3055.23	19
T150b	2727.67	2733.60	2727.67	2732.75	**2727.20**	2727.35	61
T150c	2341.84	2364.31	2343.11	2364.08	2359.30	2359.61	116
T150d	2645.39	2663.20	2645.40	2654.69	2645.39	2645.46	36
T385	24431.44	24435.50	24855.32	25015.01	**24369.13**	24384.58	349
Dev. (%)		0.47	0.30	0.80	0.07	0.09	
Ave. CPU (min)		N/A	28.8	N/A	12.1	1.2	
Computer		SGI 100	Penti. IV 2.0GHz	Penti. IV 2.8 GHz	Xeon 3.2GHz	Xeon 3.2GHz	

Figure 3 (c) shows the gains, i.e., the amount of improvement in total travel distance by the composite local search with the different strategies. As shown in the figure, there is no difference between the results of FUL, DLB, LNR, and, LNRD whereas FCE and LNE are worse than the others. These results are consistent with the solution qualities in Table 2.

4.3   Comparisons with Other Heuristics

The MA with the best limitation strategy, LNRD, is compared with the other recent CVRP heuristics using three standard benchmark sets. The three tables in Table 3 show the comparative results on the benchmarks of Christofides et al., Golden et al., and Taillard, respectively where the heuristics that have shown the best performance are selected in each benchmark set. Please note that some "easy" benchmarks ($n < 100$) are omitted due to lack of space.

In the first row of the tables, the comparative heuristics are represented by the author names ('aver.' means that averaged results are listed in the column). Note that the previous MA is represented by 'Nagata' and was not applied to the benchmark of Taillard. The two first columns give the names of the instances together with the number of customers and the best-known solutions. The last three columns gives the best travel distance over ten runs, the average travel distance, and the average computation time in seconds, respectively, related to the proposed method. New best solutions are marked with boldface. The last three rows give the average deviation from the best-known solutions in percentage, type and speed in MHz of the computers used, and the average CPU time in minutes.

According to Table 3, the solution quality of the MA with configuration LNRD is better than those of the other heuristics. Mester and Bräsys's method [13] in benchmark of Christofides et al. is the only exception. Moreover, the proposed MA found 12 new best solutions and 14 previous best solutions in the total of 26 benchmark problems.

Even allowing for the differences between the computers, the computation time of 'LNRD aver.' seems to be shorter or not so longer than those of other heuristics. This is important because the MA with configuration LNRD does not suffer from heavy computation time contrary to the popular opinion that MAs needs heavy computation time.

5   Conclusion

In this paper we have examined five different strategies for limiting the local search neighborhood in the context of capacitated vehicle routing problem. The suggested limitation strategies were applied to a composite of standard VRP neighborhoods and implemented in a powerful memetic algorithm. The experimental tests were carried out with well-known benchmark problems and it was shown that the considered limitation approaches can significantly reduce the computation time and at the same time maintain or even improve the solution quality. 12 new best-known solutions to benchmarks of Golden et al. and Tillard are also reported.

References

1. Alba, E., Dorronsoro, B.: Computing nine new best-so-far solutions for capacitated VRP with a cellular genetic algorithm. Information Processing Letters 98, 225–230 (2006)
2. Bentley, J.L.: Experiments on Traveling Salesman Geuristics. In: Proc. of the First Annual ACM-SIAM Symposium on Discrete Algorithms, pp. 91–99 (1990)
3. Christofides, N., Mingozzi, A., Toth, P.: The vehicle routing problem. In: Christofides, N., Mingozzi, A., Toth, P., Sandi, C. (eds.) Combinatorial optimization, pp. 315–318. Wiley, Chichester (1979)
4. Cordeau, J.-F., Gendreau, M., Hertz, A., Laporte, G., Sormany, J.S.: New heuristics for the vehicle routing problem. In: Langevin, A., Riopel, D. (eds.) Logistics systems: design and optimization, pp. 279–297. Springer, New York (2005)
5. Gendreau, M., Laporte, G., Potvin, J.Y.: Metaheuristics for the capacitated VRP. In: Toth, P., Vigo, D. (eds.) The vehicle routing problem, pp. 129–154. SIAM, Philadelphia (2001)
6. Golden, B.L., Wasil, E.A., Kelly, J.P., Chao, I.M.: Metaheuristics in vehicle routing. In: Crainic, T.G., Laporte, G. (eds.) Fleet management and logistics, pp. 33–56. Kluwer, Boston (1998)
7. Johnson, D.S., McGeoch, L.A.: The traveling salesman problem: a case study. In: Aarts, E., Lenstra, J.K. (eds.) Local search in combinatorial optimization, Jhon Wiley & Sons, Chichester (1997)
8. Kubiak, M., Wesolek, P.: Accelerating Local Search in a Memetic Algorithm for the CVRP. In: Cotta, C., van Hemert, J.I. (eds.) EvoCOP 2007. LNCS, vol. 4446, pp. 142–153. Springer, Heidelberg (2007)
9. Kubiak, M.: Systematic construction of recombination operators for the vehicle routing problem. Foundations of Computing and Decision Science 29, 205–226 (2004)
10. Laporte, G., Semet, F.: Classical heuristics for the capacitated VRP. In: Toth, P., Vigo, D. (eds.) The vehicle routing problem, pp. 109–128. SIAM, Philadelphia (2001)
11. Merz, P., Freisleben, B.: Memetic Algorithms for the Traveling Salesman Problem. Complex Systems 13, 297–345 (2001)
12. Merz, P., Freisleben, B.: A genetic local search approach to the quadratic assignment problem. In: Proc. of the 7th Int. conf. on Genetic Algorithms, pp. 238–245 (1997)
13. Mester, D., Bräysy, O.: Active-guided evolution strategies for large-scale capacitated vehicle routing problems. Computers & Operations Research 34, 2964–2975 (2007)
14. Mester, D., Bräysy, O.: Active guided evolution strategies for large scale vehicle routing problems with time windows. Computers & Operations Research 32, 1593–1614 (2005)
15. Moscato, P.: On evolution, search, optimization, genetic algorithms and martial arts: towards memetic algorithms, C3P Report 826, California Inst. of Tech. (1989)
16. Nagata, Y.: Edge Assembly Crossover for the Capacitated Vehicle Routing Problem. In: Cotta, C., van Hemert, J.I. (eds.) EvoCOP 2007. LNCS, vol. 4446, pp. 142–153. Springer, Heidelberg (2007)

17. Pisinger, D., Ropke, S.: A general heuristic for vehicle routing problems. Computers & Operations Research 34, 2403–2435 (2007)
18. Rochat, Y., Taillard, E.: Probabilistic diversification and intensification in local search for vehicle routing. Journal of Heuristics 1, 147–167 (1995)
19. Taillard, E.: Parallel iterative search methods for vehicle routing problems. Networks 23, 661–673 (1993)
20. Tarantilis, C.D.: Solving the vehicle routing problem with adaptive memory programming methodology. Computers & Operations Research 32, 2309–2327 (2005)

Evolutionary Local Search for the Minimum Energy Broadcast Problem

Steffen Wolf and Peter Merz

Distributed Algorithms Group
University of Kaiserslautern, Germany
{wolf,pmerz}@informatik.uni-kl.de

Abstract. The problem of finding the broadcast scheme with minimum power consumption in a wireless ad-hoc network is NP-hard. This work presents a new hybrid algorithm to solve this problem by combining evolutionary approaches with local search. The algorithm is benchmarked by solving instances with 20 and 50 nodes where results are compared to either optimum or best results found by an IP solver. For these instances, the proposed algorithm was able to find optimal and near-optimal solutions and outperform previous heuristics.

1 Introduction

Wireless ad-hoc networks have become very popular, as they are easily set up and do not need a wired backbone structure [1]. Nodes in such networks usually carry their own power supply, which makes the wireless ad-hoc network a good choice for a first responders infrastructure, or even as the main communications infrastructure in regions where installing a wired infrastructure would be too expensive or time consuming.

The communication in such ad-hoc networks can be performed either directly between two nodes, or by relaying the messages via intermediate nodes. Each node is able to adjust its transmission power based on the distance to the receiver, which allows to keep the interference between different simultaneous communications low, and also helps saving energy. Using omnidirectional antennas further brings the advantage of simple local broadcasts, as all nodes within the transmission range can receive the message without additional cost at the sender. This property of the wireless transmission is often referred to as the *wireless multicast advantage*. Because of the limited battery power of each node, it is crucial to find communication topologies that minimize the energy consumption.

A special communication pattern is the one-to-all communication (broadcast). Here, one source node has to distribute information to all other nodes. Broadcast routing in wireless ad-hoc networks differs largely from routing in wired networks. In wireless settings such a broadcast can be achieved by simply adjusting the transmission power of the source to reach all nodes in the ad-hoc network in one hop. However, because of the physical laws describing the power consumption as a function of the distance, the total energy consumption can often be reduced

J. van Hemert and C. Cotta (Eds.): EvoCOP 2008, LNCS 4972, pp. 61–72, 2008.
© Springer-Verlag Berlin Heidelberg 2008

by using intermediate nodes [2]. E. g., if the power consumption is proportional to the squared distance (this is the case when there are no obstacles), it is twice as expensive to send a message directly to the destination, instead of sending it to a node half the way between sender and destination and have this node relay the message to the final node.

In this work, we are searching for the broadcast tree that minimizes the total energy consumption. This problem is known as the Minimum Energy Broadcast (MEB) problem [3]. We present a new Evolutionary Local Search heuristic for the MEB problem. We also use Mixed Integer Programming (MIP) to compute lower bounds or optimal solutions and compare the solution quality of the proposed heuristic against these bounds as well as other heuristics.

This paper is structured as follows. In the remainder of this section we give a formal definition of the MEB problem and summarize related work. In Section 2 we present our heuristic, and then give results of experiments carried out with this heuristic in Section 3. Section 4 summarizes our findings and gives an outline for future research.

1.1   Minimum Energy Broadcast

The Minimum Energy Broadcast (MEB) problem is an NP-hard optimization problem [4,5]. It is also known as Minimum Power Broadcast (MPB) or Minimum Energy Consumption Broadcast Subgraph (MECBS). The MEB problem can be defined as the problem of finding the broadcast tree $T = (V, E_T)$ (a directed spanning tree, defined by its parent function $p^T : V \rightarrow V$) rooted at a source node $s \in V$ in an ad-hoc wireless network $G = (V, E, d)$, that minimizes the necessary total transmission power $c(T)$ to reach all nodes of the network:

$$c(T) = \sum_{i \in V} \underbrace{\max_{j:p^T(j)=i} d(i,j)^\alpha}_{\text{transmission power of node } i}$$

Here, the distance function $d : E \rightarrow \mathbb{R}^+$ refers to the Euclidean distance and the constant α is the distance-power gradient which may vary from 1 to more than 6 depending on the environment [6]. Each node is required to send to its farthest child, all other children are then implicitly covered by this transmission. The leaves of the tree T do not send to other nodes and thus do not contribute to the total cost.

1.2   Related Work

One of the first approaches for the MEB problem is the Broadcast Incremental Power algorithm (BIP) by Wieselthier *et al.* [3]. This heuristic builds the broadcast tree in a way that resembles Prim's algorithm for building Minimum Spanning Trees (MST). While Prim's algorithm is an exact algorithm for the MST, BIP is an heuristic and does not necessarily find an optimal solution for the MEB problem. The MST itself can also be used as a heuristic solution for the

MEB problem, but BIP explicitly exploits the wireless multicast advantage and thus produces solutions with lower costs than the corresponding MST solutions. The approximation ratio of MST is known to be 6 for the case of $\alpha \geq 2$ [6], whereas for $\alpha < 2$ the MST does not provide a constant approximation ratio [4]. The approximation ratio of BIP for $\alpha = 2$ is shown to be between $13/3$ and 6 [7].

The BIP heuristic can be further improved by a local search, e. g. r-shrink [8]. Here, the transmission power for one node is reduced by r steps, cutting off r nodes. These nodes will be assigned to other nodes, which increases the latter nodes' transmission power. If the total cost is not reduced, this change is rejected, otherwise it is accepted and the local search is repeated. Experiments have shown that BIP solutions can be improved considerably. Another improving heuristic is called Embedded Wireless Multicast Advantage (EWMA) [5]. Here, the transmission power of a node is increased, such that other nodes can be switched off completely. This can be thought of as the opposite of the r-shrink heuristic.

Several Mixed Integer Programming formulations (MIP) have been presented to compute optimal solutions [9,10]. While both approaches are based on a network flow model, the MIP from [10] uses an incremental mechanism over the transmission power variables, and is said to give better linear relaxations.

A similar heuristic as ours can be found in [11]. The authors present an Iterated Local Search heuristic which is based on an edge exchange neighbourhood perturbation and the Largest Expanding Sweep Search (LESS, [12]), an improved local search based on EWMA. This heuristic differs from our heuristic as it uses a shrinking operation as mutation and an increasing operation as local search, whereas we propose the opposite. As a result, the broadcast tree in [11] is broken up and repaired in each step of their local search, whereas our heuristic maintains a feasible broadcast tree at all times. Although the idea of increasing the transmission power in the local search is counterintuitive, the authors achieved good results using this heuristic.

A more detailed survey covering these and other heuristics as well as exact algorithms can be found in [13]. This survey does not include a Nested Partitioning Algorithm, which has been presented recently, and which promises good results and time complexity [14]. This algorithm uses the r-shrink local search and a randomized BIP to improve the quality of the random samples for the Nested Partitioning. Since the test instances used in [14] were made publicly available, other heuristics can be compared directly.

2   Evolutionary Local Search

The MEB Heuristic presented here is based on *evolutionary algorithms* and *local search* [15]. It operates on a global view of the wireless ad-hoc network. The general outline of the algorithm is shown in Fig. 1. The heuristic is quite similar to *iterated local search* [16], but uses more than one offspring solution in each generation.

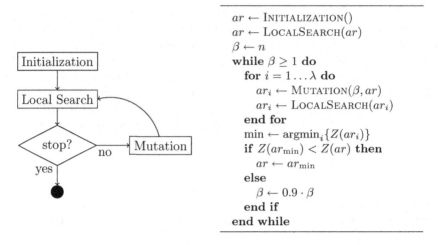

$ar \leftarrow$ INITIALIZATION()
$ar \leftarrow$ LOCALSEARCH(ar)
$\beta \leftarrow n$
while $\beta \geq 1$ **do**
   for $i = 1 \ldots \lambda$ **do**
      $ar_i \leftarrow$ MUTATION(β, ar)
      $ar_i \leftarrow$ LOCALSEARCH(ar_i)
   end for
   min $\leftarrow \mathrm{argmin}_i \{Z(ar_i)\}$
   if $Z(ar_\mathrm{min}) < Z(ar)$ **then**
      $ar \leftarrow ar_\mathrm{min}$
   else
      $\beta \leftarrow 0.9 \cdot \beta$
   end if
end while

Fig. 1. General overview of the Evolutionary Local Search

Representation. As can be seen from the survey [13] and also from the individual heuristics, most of the previous local search heuristics can be split in two groups based on their neighbourhood structure: Some are based on a tree representation, whereas others are based on the range assignments. A tree representation allows the use of simple tree operators, whereas the range assignment representation enables an easier calculation of the total cost. For our Evolutionary Local Search, we chose a combination of these two representations, combining their advantages.

A solution is represented by the assignment vector a and the range vector r. For each node i the value $a(i)$ gives the parent of i, while $r(i)$ holds the farthest possible child of i, or i itself if it does not transmit to another node. Both vectors are combined to a single vector ar.

Initialization. The initial solution is created by BIP [3]. Using BIP bears the advantage that there is an upper bound for the solutions of the Evolutionary Local Search heuristic for $\alpha = 2$, since in this case BIP already gives a 6-approximation to the MEB problem [6].

Local Search. After each step of the Evolutionary Algorithm a local search is applied to further improve the current solution. We use a modified r-shrink [8]. No levels were computed, instead we apply the local search step to all nodes in the order of their ID. Also, when decreasing the power level by more than one step, we do not restrict this decrease operation to one single node. Instead, a step of the modified 2-shrink may consist of two nodes decreasing their power level by one step each. This way, the local search becomes more complex but yields better results.

These local search steps are performed for each node i. The local search is restarted whenever an improving step was found and applied. The local search is thus repeated until no improvement for any node i can be found, i.e. a local

---

procedure LS(r)
if $r > 0$ **then**
  repeat
    for each transmitting node i **do**
      $x \leftarrow$ farthest child of i
      Reduce transmitting range of i by one step, cutting off x
      $j \leftarrow$ best foster parent for x
      Increase transmitting range of j to cover x
      if cost is higher **then**
        LS($r - 1$)
      end if
      if cost is still higher **then**
        Undo the move and all moves made by the recursive call
      end if
    end for
  until no improvement found
end if

Fig. 2. Modified r-shrink

optimum has been reached. The pseudocode for the modified r-shrink is shown in Fig. 2.

Mutation. Since local search alone will get stuck in local optima, we use mutation to continue the search. Mutation is done by increasing the transmission power of randomly chosen nodes to random levels. The 'gain' of such a move can be calculated as the sum of the power level changes. The mutation changes only the range vector, but not the assignment vector. In these intermediate solutions, the farthest possible children of some nodes are often not assigned to these nodes. The local search can later rearrange other nodes to such nodes and thus make use of the increased ranges.

Several mutation steps are applied in each round. The number of mutations is adapted to the success rate. The algorithm starts with $\beta = n$ mutations. If no better solution is found in one generation, the mutation rate β is reduced by 10 %. This way, the algorithm can adapt to the best mutation rate for the individual problem and for the phase of the search. It is our experience that it is favourable to search the whole search space in the beginning, but narrow the search over time, thus gradually shifting from exploration to exploitation.

Population. Our heuristic uses a population of only one individual. In this paper, we do not use recombination. Using mutation and local search, λ offspring solutions are created. The best solution is used as the next generation only if it yielded an improvement. This follows a $(1 + \lambda)$-ES selection paradigm. If there was no improvement in these λ children, the mutation rate β is reduced as described before.

Stopping criterion. The heuristic is stopped when the mutation rate drops below $\beta < 1$. This value ensures that the neighbourhood of the best solution found by the heuristic is searched especially thoroughly. However, in the smaller instances the heuristic often finds the optimum in the first or second generation.

3   Experiments

We used the same test sets as in [14]. Each set contains 30 instances, where n nodes are randomly located in an area of $1000 \times 1000\,\mathrm{m}^2$. The Euclidean distance was used and the distance-power gradient was set to $\alpha = 2$. The first set uses $n = 20$, the second $n = 50$.

The commercial MIP solver CPlex 10.1 [17] was used together with the MIP formulation from [10] to obtain optimal solutions for the problem instances. Since the MEB problem is NP-hard, CPlex was only able to provide optimal solutions for the 20 nodes problems and about half of the 50 nodes problems. Depending on the individual problem, CPlex took up to four days to find and prove the optimal solution. For the remaining problems, CPlex was stopped after it produced a gap of less than 10 %. Problem p50.07 was the hardest problem for CPlex, as it took about twenty days do arrive at this gap.

In the experiments, we set the number of offspring solutions to $\lambda = 500$. These settings have proven to be a good choice in preliminary experiments. Unless stated otherwise, we used $r = 1$ for the local search. Each experiment was repeated 30 times and average values are used for the following discussion. Calculation times refer to the CPU time on a 2.8 GHz Pentium IV running Linux 2.6; the algorithm was implemented in C.

As a comparison, we used our own implementation of the Iterated Local Search [11] (ILS). Here, we started from the BIP solution, used the suggested local search (LESS) and mutation operators (edge exchange), but relaxed the termination criterion. Instead of letting the ILS run for only 60 seconds, we allowed it to reach 20 000 iterations. We have not used any of the suggested speed-ups, since we are not interested in the CPU time of the ILS but only in the quality of the solutions.

We also compare our results to the results of the Nested Partitioning [14], and the results that can be achieved by applying r-shrink to BIP solutions.

3.1   Results

The results for the 20 nodes problems are shown in Table 1. The algorithm was able to find the optimum in almost every run. Also, the non-optimal solutions were very close to the optimum, with average excess of less than one percent.

Comparing the results of the algorithm against the results obtained by applying r-shrink to the BIP solution, which can be seen as the first generation of an Evolutionary Local Search, shows how much can be gained by the evolutionary approach. In some of the instances, BIP+r-shrink already found the optimal solution, but the average of BIP+r-shrink above the optimum was 11.63 %. The

Table 1. Results for the 20 nodes problems using $r = 1$. The average excess over the optimal solution is shown for the proposed ELS, the NP from [14], the ILS from [11], and BIP+r-shrink.

Instance	Optimum	ELS Excess	ELS found	ELS time	NP Excess	NP found	NP time	ILS Excess	ILS found	BIP+r-shrink Excess
p20.00	407 250.81	0.05 %	26/30	0.27 s	–	30/30	0.30 s	–	30/30	14.90 %
p20.01	446 905.52	–	30/30	0.42 s	–	30/30	0.36 s	7.09 %	5/30	–
p20.02	335 102.42	–	30/30	0.53 s	–	30/30	0.41 s	3.52 %	1/30	–
p20.03	488 344.90	–	30/30	0.67 s	0.16 %	27/30	0.46 s	–	30/30	4.79 %
p20.04	516 117.75	–	30/30	0.45 s	–	30/30	0.43 s	0.40 %	0/30	19.35 %
p20.05	300 869.14	–	30/30	0.41 s	–	30/30	0.35 s	3.19 %	0/30	31.06 %
p20.06	250 553.15	1.57 %	17/30	0.24 s	–	30/30	0.18 s	–	30/30	32.91 %
p20.07	347 454.08	–	30/30	0.49 s	–	30/30	0.31 s	–	30/30	7.25 %
p20.08	390 795.34	–	30/30	0.68 s	–	30/30	0.46 s	0.37 %	26/30	–
p20.09	447 659.11	0.004 %	27/30	0.60 s	–	30/30	0.41 s	0.64 %	0/30	14.82 %
p20.10	316 734.39	–	30/30	0.48 s	–	30/30	0.40 s	–	30/30	4.77 %
p20.11	289 200.92	–	30/30	0.30 s	–	30/30	0.24 s	1.93 %	0/30	6.35 %
p20.12	314 511.98	10.62 %	1/30	0.50 s	–	30/30	0.20 s	–	30/30	22.00 %
p20.13	486 234.51	0.25 %	28/30	0.36 s	–	30/30	0.26 s	–	30/30	26.85 %
p20.14	301 426.68	–	30/30	0.63 s	0.38 %	17/30	0.43 s	4.71 %	0/30	–
p20.15	457 467.93	–	30/30	0.25 s	–	30/30	0.24 s	1.74 %	0/30	20.57 %
p20.16	484 437.68	–	30/30	0.67 s	1.03 %	21/30	0.41 s	12.63 %	0/30	9.21 %
p20.17	380 175.41	–	30/30	0.46 s	–	30/30	0.39 s	3.72 %	5/30	2.80 %
p20.18	320 300.23	–	30/30	0.43 s	–	30/30	0.31 s	–	30/30	6.48 %
p20.19	461 267.52	–	30/30	0.63 s	0.21 %	18/30	0.34 s	–	30/30	3.06 %
p20.20	403 582.74	–	30/30	0.40 s	0.02 %	29/30	0.35 s	–	30/30	5.39 %
p20.21	271 958.28	–	30/30	0.23 s	–	30/30	0.19 s	–	30/30	–
p20.22	328 659.78	–	30/30	0.22 s	–	30/30	0.19 s	–	30/30	19.64 %
p20.23	326 654.08	–	30/30	0.41 s	–	30/30	0.28 s	–	30/30	20.61 %
p20.24	395 859.67	–	30/30	0.71 s	–	30/30	0.42 s	2.08 %	0/30	0.76 %
p20.25	453 517.28	–	30/30	0.22 s	–	30/30	0.21 s	–	30/30	19.17 %
p20.26	461 547.18	–	30/30	0.39 s	–	30/30	0.32 s	0.07 %	29/30	–
p20.27	389 057.00	0.23 %	24/30	0.40 s	–	30/30	0.38 s	1.13 %	0/30	29.81 %
p20.28	279 251.95	–	30/30	0.32 s	–	30/30	0.28 s	–	30/30	11.43 %
p20.29	299 586.76	–	30/30	0.26 s	–	30/30	0.31 s	–	30/30	14.86 %
avg		0.42 %	28.1/30	0.43 s	0.06 %	28.7/30	0.33 s	1.44 %	18.2/30	11.63 %

results of the proposed Evolutionary Local Search were also better than the results of the ILS, which found the optimum in only about two third of the runs, and showed an average excess of 1.44 %. It is worth mentioning, that the ILS sometimes failed to find the optimum when BIP+r-shrink was successful. Only the Nested Partitioning Algorithm [14] showed better results, it found the optimum more often for these instances.

Table 2 shows the results for the 50 nodes problems. Optimal solutions are known only for about half of these instances. For the remaining instances we use the best solution found by CPlex as a comparison. It can be seen that these instances, which are harder to solve for CPlex, are not harder for the Evolutionary Local Search. In many of the cases the Evolutionary Local Search still found the best known solutions, and in the remaining cases the solutions found by the heuristic were very close to the best known solutions. On average, every fourth run of the ELS found an optimum or a best known solution, and the average excess over the best known or optimal solutions was 2.50 %.

For problems p50.08 and p50.25, the algorithm found the optimum in all runs. The ELS showed its worst performance on p50.19 in terms of average

Table 2. Results for the 50 nodes problems using $r = 1$. The results of the ELS are compared to the optimal solution or the best known solution found by NP [14], ILS [11] and BIP+r-shrink. solutions found by NP [14], ILS [11] and BIP+r-shrink.

Table 3. Results for the 50 nodes problems for the ELS and BIP+r-shrink using $r = 2$.

Table 2.

Instance	Optimum	VI	ELS Excess	ELS found	ELS time	NP Excess	NP found	NP time	ILS Excess	ILS found	BIP+r-shrink Excess
p50.00	399 074.64		5.60 %	0/30	4.8 s	6.22 %	0/30	11.4 s	3.31 %	0/30	10.42 %
p50.01	373 565.15	VI	1.33 %	0/30	5.7 s	3.68 %	0/30	7.1 s	3.30 %	0/30	27.18 %
p50.02	393 641.09		3.21 %	3/30	5.8 s	10.11 %	0/30	10.3 s	12.06 %	0/30	22.19 %
p50.03	316 801.09		4.53 %	0/30	5.3 s	6.43 %	0/30	6.1 s	6.48 %	0/30	21.91 %
p50.04	325 774.22	VI	3.51 %	6/30	4.7 s	5.22 %	0/30	7.5 s	7.22 %	0/30	17.05 %
p50.05	382 235.90		2.25 %	4/30	5.1 s	3.28 %	1/30	10.9 s	2.57 %	0/30	10.61 %
p50.06	384 438.46	VI VI	0.99 %	8/30	6.5 s	1.19 %	5/30	10.2 s	0.14 %	18/30	18.83 %
p50.07	401 836.85	VI	5.55 %	2/30	6.5 s	6.70 %	0/30	8.9 s	7.80 %	0/30	14.80 %
p50.08	334 418.45		–	30/30	3.4 s	0.10 %	27/30	4.6 s	–	30/30	14.94 %
p50.09	346 732.05	VI	2.89 %	0/30	8.8 s	9.20 %	0/30	12.9 s	11.42 %	0/30	15.28 %
p50.10	416 783.45	VI VI	1.78 %	4/30	5.1 s	2.14 %	0/30	8.9 s	2.05 %	0/30	13.73 %
p50.11	369 869.41	VI	3.68 %	1/30	3.1 s	4.34 %	0/30	7.7 s	4.62 %	0/30	11.37 %
p50.12	392 326.01	VI	1.51 %	1/30	8.1 s	3.18 %	0/30	13.5 s	0.44 %	0/30	10.40 %
p50.13	400 563.83	VI	1.43 %	2/30	8.4 s	6.63 %	0/30	11.2 s	1.20 %	0/30	21.16 %
p50.14	388 714.91		0.37 %	1/30	2.4 s	0.08 %	22/30	6.6 s	–	30/30	37.11 %
p50.15	371 694.65		0.36 %	1/30	5.4 s	0.40 %	0/30	8.1 s	1.40 %	0/30	15.08 %
p50.16	414 587.42	VI	2.38 %	7/30	9.0 s	5.28 %	0/30	15.1 s	6.06 %	1/30	6.11 %
p50.17	355 937.07		5.49 %	0/30	3.6 s	2.17 %	1/30	11.9 s	4.94 %	2/30	9.00 %
p50.18	376 617.33	VI	1.34 %	0/30	8.2 s	5.96 %	5/30	11.3 s	0.98 %	12/30	7.55 %
p50.19	335 059.72		8.06 %	11/30	4.8 s	2.27 %	5/30	9.8 s	10.49 %	1/30	34.72 %
p50.20	414 768.96	VI VI	1.66 %	0/30	5.2 s	3.14 %	0/30	10.4 s	6.33 %	0/30	11.40 %
p50.21	361 354.27		1.52 %	23/30	5.7 s	2.93 %	2/30	10.4 s	–	30/30	22.61 %
p50.22	329 043.51		0.15 %	29/30	2.7 s	–	30/30	7.2 s	4.61 %	0/30	25.53 %
p50.23	383 321.04		1.89 %	11/30	8.7 s	6.39 %	0/30	11.1 s	2.47 %	0/30	10.94 %
p50.24	404 855.92		0.81 %	5/30	4.9 s	5.69 %	0/30	10.0 s	0.87 %	3/30	11.87 %
p50.25	363 200.32		–	30/30	1.3 s	–	30/30	3.2 s	–	30/30	29.72 %
p50.26	406 631.51		6.62 %	0/30	4.9 s	9.59 %	0/30	11.5 s	11.63 %	0/30	12.67 %
p50.27	451 059.62	VI VI	4.35 %	0/30	4.5 s	4.18 %	0/30	9.5 s	3.35 %	0/30	16.48 %
p50.28	415 832.44		1.52 %	5/30	9.0 s	4.48 %	0/30	11.6 s	0.28 %	0/30	8.64 %
p50.29	380 492.77		0.34 %	17/30	3.4 s	–	30/30	5.9 s	0.60 %	0/30	18.90 %
avg			2.50 %	7.53/30	5.5 s	4.03 %	5.1/30	9.5 s	3.89 %	5.23/30	16.94 %

Table 3.

Instance	ELS Excess	ELS found	ELS time	BIP+r-shrink Excess
p50.00	0.41 %	15/30	57 s	8.08 %
p50.01	0.16 %	5/30	47 s	14.96 %
p50.02	0.28 %	13/30	46 s	15.33 %
p50.03	1.71 %	11/30	57 s	18.65 %
p50.04	0.30 %	25/30	40 s	13.10 %
p50.05	0.83 %	16/30	31 s	10.06 %
p50.06	–	30/30	29 s	11.15 %
p50.07	0.54 %	24/30	64 s	11.52 %
p50.08	–	30/30	19 s	7.95 %
p50.09	3.29 %	0/30	102 s	9.40 %
p50.10	1.16 %	13/30	40 s	12.83 %
p50.11	2.87 %	1/30	28 s	8.31 %
p50.12	0.57 %	7/30	66 s	10.81 %
p50.13	0.04 %	29/30	74 s	9.87 %
p50.14	0.34 %	3/30	11 s	34.12 %
p50.15	0.20 %	5/30	35 s	7.10 %
p50.16	0.30 %	26/30	81 s	6.10 %
p50.17	1.88 %	17/30	33 s	9.00 %
p50.18	0.24 %	8/30	65 s	6.63 %
p50.19	–	30/30	28 s	28.13 %
p50.20	0.15 %	0/30	35 s	9.80 %
p50.21	–	30/30	41 s	7.36 %
p50.22	–	30/30	14 s	18.72 %
p50.23	–	30/30	109 s	3.99 %
p50.24	0.07 %	17/30	37 s	6.54 %
p50.25	–	30/30	7 s	27.55 %
p50.26	2.17 %	2/30	60 s	13.50 %
p50.27	0.18 %	22/30	40 s	12.45 %
p50.28	0.47 %	23/30	78 s	8.23 %
p50.29	0.08 %	27/30	18 s	15.94 %
avg	0.61 %	17.3/30	46 s	12.57 %

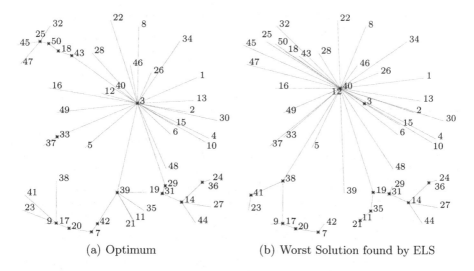

Fig. 3. Optimal and sub-optimal solutions for problem p50.19. The source is node 3, and transmitting nodes are highlighted.

excess above optimum, although it found the optimum in every third run. In the remaining runs, it often got stuck in an especially bad local optimum, as shown in Fig. 3. In this figure, we compare the optimal and the worst solution found by the ELS for problem p50.19. While in the optimal solution node 3 sends to the majority of the nodes, this part is given to node 40 in the worst solution.

The Evolutionary Local Search is again competitive as comparisons to other heuristics show. The BIP+r-shrink solutions are between 6 % and 37 % percent more expensive than the optimal or best known solutions. We have again used our implementation of the ILS [11] on these instances. Although the ILS gave better results on some of the instances, it did not find the optimal or best known solutions as often as the ELS, also the average excess was higher (3.89 % compared to 2.50 %). Although the mutation operator of the ILS is quite intriguing, the local search is counterintuitive and sometimes fails to make the right decisions. Comparing the results of our ELS to the Nested Partitioning [14] reveals that for the larger instances the ELS performed better, resulting in an average excess of 2.5 % compared to 4 % over the optimum or best known solutions. The ELS also found the optimal or best known solutions more often, i.e. in every fourth run as compared to in every sixth run. A stastistical significance test (95 % confidence) reveals that the ELS is superior to the ILS in 13 cases (against NP also in 13 cases), while it is inferior in 6 cases (against NP in 4 cases).

To further improve the quality of the ELS solutions, we increased the range of our local search to $r = 2$. This reduced the average excess to 0.61 %, but also increased the running times by a factor of 8, as Table 3 shows. Also, the solutions found by BIP+r-shrink improved slightly with $r = 2$. With the stronger local search, the ELS found the optimum in more than half of the runs, and clearly outperforms previous heuristics considering the quality of the results. The significance test (95 % confidence) reveals that the ELS with $r = 2$ is superior to the ILO in 22 cases (against NP in 24 cases), while it is inferior in two cases (against NP in one case).

3.2   Fitness Landscape Analysis

For a better understanding of the behaviour of the Evolutionary Local Search, we analysed the fitness distance correlation (FDC) of the local optima of the used r-shrink. We created 10 000 random feasible solutions for each problem instance and started the local search to find local optima. The fitness of the solution is defined by the cost of this solution, and the distance from the optimum is measured by the sum of the squared power range differences. The FDC using $r = 1$ for the 50 nodes problems lies between 0.70 and 0.86, with an average of 0.79. Using $r = 2$ it improves slightly to an average of 0.81 (0.68 to 0.87). Using the LESS local search, the FDC values are slightly worse (average 0.71, from 0.07 to 0.96), which indicates that using r-shrink as a local search operator is a good choice.

Since the FDC values alone do not always allow a meaningful deduction, we have also looked at the fitness distance scatter plots [18]. Figure 4 shows the scatter plots for both local search operators for p50.29. In these plots, the distance and the cost are normalized to the value of the optimum for the individual problem. Although the LESS local search gives a higher FDC value, it can be observed that it produces fewer different local optima, which are far-scattered in the search space. Most of the local optima for LESS are located some distance away from the global optimum, but show a low cost. An iterated or evolutionary local search may get stuck in these local optima. On the other hand, the r-shrink local search operator produces a large amount of different local optima. Again, most of these are located near the local optima from LESS, but there is a trail of local optima leading closer to the global optimum.

Still, in both instances there is a tendency for lower costs when the distance to the optimum decreases. This correlation can be exploited by the Evolutionary Local Search, which approaches the global optimum by jumping from one local optimum to the next one. Also, the results indicate that recombination may profit from the correlation and thus yield good results.

4   Conclusion

We have presented a new Evolutionary Local Search for the Minimum Energy Broadcast problem. As local search we used a modified r-shrink, which shows a higher complexity but also gives better results. The algorithm has shown to find optimal or near-optimal solutions in short time for the considered test instances. Also, comparisons show that the proposed heuristic outperforms previous heuristics.

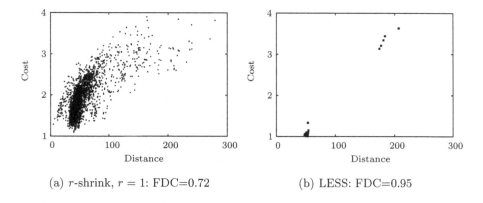

(a) r-shrink, $r = 1$: FDC=0.72   (b) LESS: FDC=0.95

Fig. 4. FDC scatter plots for p50.29. Distance and cost are normalized to the value of the optimum for the individual problem.

Future work focusses on the development of a powerful but fast recombination operator. We are also striving for a distributed algorithm that approximates the results of the heuristic proposed here.

Acknowledgements

This work was partially supported by the Rhineland-Palatinate cluster of excellence DASMOD.

References

1. Haas, Z.J., Tabrizi, S.: On some challenges and design choices in ad-hoc communications. In: Proceedings of the Military Communications Conference IEEE MILCOM 1998, Bedford, USA, pp. 187–192 (1998)
2. Rappaport, T.S.: Wireless Communications: Principles and Practices. Prentice-Hall, Englewood Cliffs (1996)
3. Wieselthier, J.E., Nguyen, G.D., Ephremides, A.: On the construction of energy-efficient broadcast and multicast trees in wireless networks. In: Proceedings of the 19th IEEE INFOCOM 2000, pp. 585–594 (2000)
4. Clementi, A.E.F., Crescenzi, P., Penna, P., Rossi, G., Vocca, P.: On the complexity of computing minimum energy consumption broadcast subgraphs. In: Ferreira, A., Reichel, H. (eds.) STACS 2001. LNCS, vol. 2010, pp. 121–131. Springer, Heidelberg (2001)
5. Čagalj, M., Hubaux, J.P., Enz, C.: Minimum-Energy Broadcast in All-Wireless Networks: NP-Completeness and Distribution Issues. In: MobiCom 2002: Proceedings of the 8th Annual International Conference on Mobile Computing and Networking, pp. 172–182. ACM Press, New York (2002)
6. Ambühl, C.: An optimal bound for the MST algorithm to compute energy efficient broadcast trees in wireless networks. In: Caires, L., Italiano, G.F., Monteiro, L., Palamidessi, C., Yung, M. (eds.) ICALP 2005. LNCS, vol. 3580, pp. 1139–1150. Springer, Heidelberg (2005)

7. Wan, P.J., Călinescu, G., Li, X.Y., Frieder, O.: Minimum-energy broadcasting in static ad hoc wireless networks. Wireless Networks 8(6), 607–617 (2002)

8. Das, A.K., Marks, R.J., El-Sharkawi, M., Arabshahi, P., Gray, A.: r-shrink: A heuristic for improving minimum power broadcast trees in wireless networks. In: Global Telecommunications Conference, GLOBECOM 2003, pp. 523–527. IEEE, Los Alamitos (2003)

9. Das, A.K., Marks, R.J., El-Sharkawi, M., Arabshahi, P., Gray, A.: Minimum power broadcast trees for wireless networks: Integer programming formulations. In: Proceedings of the 22nd IEEE INFOCOM 2003, pp. 1001–1010 (2003)

10. Montemanni, R., Gambardella, L.M., Das, A.: The minimum power broadcast problem in wireless networks: a simulated annealing approach. Wireless Communications and Networking Conference (WCNC) 4, 2057–2062 (2005)

11. Kang, I., Poovendran, R.: Iterated local optimization for minimum energy broadcast. In: 3rd International Symposium on Modeling and Optimization in Mobile, Ad-Hoc and Wireless Networks (WiOpt), pp. 332–341. IEEE Computer Society, Los Alamitos (2005)

12. Kang, I., Poovendran, R.: Broadcast with heterogeneous node capability. In: Global Telecommunications Conference, GLOBECOM 2004, pp. 4114–4119. IEEE, Los Alamitos (2004)

13. Guo, S., Yang, O.W.: Energy-aware multicasting in wireless ad hoc networks: A survey and discussion. Computer Communications 30(9), 2129–2148 (2007)

14. Al-Shihabi, S., Merz, P., Wolf, S.: Nested partitioning for the minimum energy broadcast problem. In: LION II: Learning and Intelligent Optimization Conference, Trento, Italy. LNCS, Springer, Heidelberg (to be published, 2008)

15. Hoos, H.H., Stützle, T.: Stochastic Local Search: Foundations and Applications. The Morgan Kaufmann Series in Artificial Intelligence. Morgan Kaufmann, San Francisco (2004)

16. Lourenço, H.R., Martin, O., Stützle, T.: Iterated Local Search. In: Glover, F., Kochenberger, G. (eds.) Handbook of Metaheuristics, pp. 321–353 (2002)

17. Ilog S.A.: ILOG CPLEX User's Manual, Gentilly, France, and Mountain View, USA (2006), http://www.cplex.com/

18. Merz, P., Freisleben, B.: Fitness Landscapes and Memetic Algorithm Design. In: Corne, D., Dorigo, M., Glover, F. (eds.) New Ideas in Optimization, pp. 245–260. McGraw–Hill, London (1999)

Exploring Multi-objective PSO and GRASP-PR for Rule Induction

Celso Y. Ishida[1], Andre B. de Carvalho[1], Aurora T.R. Pozo[1],
Elizabeth F.G. Goldbarg[2], and Marco C. Goldbarg[2]

[1] Federal University of Paraná
[2] Federal University of Rio Grande do Norte*
celsoishida@gmail.com, {andrebc,aurora}@inf.ufpr.br,
{beth,gold}@dimap.ufrn.br

Abstract. This paper presents a method of classification rule discovery based on two multiple objective metaheuristics: a Greedy Randomized Adaptive Search Procedure with path-relinking (GRASP-PR), and Multiple Objective Particle Swarm (MOPS). The rules are selected at the creation rule process following Pareto dominance concepts and forming unordered classifiers. We compare our results with other well known rule induction algorithms using the area under the ROC curve. The multi-objective metaheuristic algorithms results are comparable to the best known techniques. We are working on different parallel schemes to handle large databases, these aspects will be subject of future works.

1 Introduction

The current information age is characterized by an extraordinary expansion of data that is being generated and stored in all kinds of human endeavors. A significant need for techniques and tools, with the ability to intelligently assist humans in analyzing very large collections of data in a search for useful knowledge exists. In this sense, the area of data mining has received special attention. Our work deals with one of the main data mining task: classification. The classification task produces a model based on data. The model can be used to classify an unseen data item into one of the predefined classes based on its descriptor attributes. The model is learned in a supervised mode, which provides classified data (training set) as an entry to the learner.

One of the most used models to represent knowledge in data mining context is production rules. This is because of their simplicity, intuitive aspect, modularity and because they can be obtained directly from a dataset [7]. Therefore, rules induction has been established as a fundamental component of many data mining systems. Production rules, or rules induction, are formed by a set of rules, where each rule has the form: if the *antecedent* then *consequent*. The antecedent is a set of descriptor attributes and the consequent is the goal attribute. In this sense, the task of learning a rule is a combinatorial optimization problem where

---

* Our special thanks to PRH-22 project of ANP agency for the financial support.

J. van Hemert and C. Cotta (Eds.): EvoCOP 2008, LNCS 4972, pp. 73–84, 2008.

the complexity grows with the number of descriptor attributes and the number of examples on the database, furthermore, these examples can have noise (or disturbed values).

Numerous measures are used for performance evaluation of rules, but all of them are derived from the contingency table [16]. A contingency table for an arbitrary rule with antecedent B and consequent H is showed at Table 1. In Table 1, B denotes the set of instances for which the body of the rule is true, and $\overline{B}$ denotes its complement (the set of instances for which the body is false); similarly for H and $\overline{H}$. HB then denotes $H \cap B$, $\overline{H}B$ denotes $\overline{H} \cap B$, and so on.

Table 1. A contingency table

	B	$\overline{B}$	
H	$n(HB)$	$n(H\overline{B})$	$n(H)$
$\overline{H}$	$n(\overline{H}B)$	$n(\overline{H}\overline{B})$	$n(\overline{H})$
	$n(B)$	$n(\overline{B})$	N

Where $n(X)$ denote the cardinality of the set X, e.g., $n(\overline{H}B)$ is the number of instances for which H is false and B is true (i.e., the number of instances erroneously covered by the rule). N denotes the total number of instances in the dataset.

Traditional systems usually use a covered approach where a search procedure is iteratively called. On each iteration, the search algorithm finds the best rule and removes all the examples covered by the rule from the dataset. Then, the process is repeated with the remaining examples [17]. The process continues until all the examples are covered or some stop criterion is reached. In this way, on each iteration, a new rule is found. However, this approach has some problems; for example the remotion of the examples from the dataset at each new discovered rule causes the over-specialization of the rules after some iteration. This means that each rule covers few examples. Besides that, the classifier composed by the learned rules is an ordered list where the interpretation of one rule depends on the precedent rules. In this way, the interpretation becomes very difficult, mainly when the number of rules increases [2].

In this work, we propose a different approach based on Multi-objective Meta-heuristic (MOMH) techniques. MOMH techniques allow to conceive a novel approach where the properties of the rules can be expressed in different objectives and then, MOMH algorithm find these rules in a unique run. In this way, MOMH techniques allow creating classifiers composed by rules with specific properties exploring Pareto dominance concepts. These rules can be used as an unordered classifier. Two Metaheuristic algorithms are used and compared here: a Greedy Randomized Adaptive Search Procedure with path-relinking (GRASP-PR), and Multiple Objective Particle Swarm (MOPS). The preliminary ideas of the MOPS algorithm were presented in [4], here we explore more deeply the quality of the solutions compared to the GRASP-PR algorithm. Two Objectives are chosen for both systems: the sensitivity and specificity criteria, defined in equation 1 and 2. Both techniques are applied to different dataset and the results are compared to traditional systems. Furthermore, the analysis of the coverage of the Pareto Front of both systems was performed. Both techniques are applied to different dataset and the results are compared to traditional systems. Furthermore, the analysis of the coverage of the Pareto Front of both systems was performed.

$$sensitivity = \frac{n(HB)}{n(H)} \tag{1}$$

$$specificity = \frac{n(\overline{HB})}{n(\overline{H})} \tag{2}$$

This paper is organized as follows. The next section describes related works. Section 3 describes the GRASP-PR Rule Learning Algorithm and Section 4 describes Multiple Objective Particle Swarm for Rule Discovery Algorithm. Section 5 explains the methodology and results for the experiment made to evaluate the algorithms. Finally, Section 6 concludes the work.

2 Related Work

Recently, increasing interest has emerged in applying the concept of Pareto-optimality to machine learning inspired by the successful developments in evolutionary multiobjective optimization. These researches include multiobjective feature selection, multiobjective model selection in training multilayer perceptrons, radial-basis-function networks, support vector machines, decision trees and intelligent systems [12]. In the literature, few works deal with multiobjective evolutionary algorithms for rule learning among them [11], [6] and [10]. The first work focuses on the rule selection phase; it presents a genetic-based multiobjective rule selection algorithm to find a smaller rule subset with higher accuracy than the heuristically extracted rule sets. The algorithm has the objective to maximize the accuracy, and to minimize the number of rules. In [10], multiobjective association rules generation and selection with NSGA-II (Non-Dominated Sorting Genetic Algorithm) are discussed. In [5], a multiobjective optimization evolutionary algorithm with Pareto concepts is used to discover interesting classification rules for a target class. It presents an implementation of NSGA with positive confidence and sensitivity as objectives. This work is extended in [6] using multiobjective metaheuristics to produce sets of interesting classification rules. A measure of dissimilarity of rules was introduced to promote diversity on the population.

Our work focuses in the induction of an unordered classifier where the rule generation, and the rule selection happens at the same time, so we do not need to select best rules for the classifier after the generation process. The multiobjective metaheuristic approaches aim to induce classifiers composed by rules with specific properties. So, to tackle this purpose we choose GRASP-PR and MOPS techniques presented at next Sections.

3   The GRASP-PR Rule Learning Algorithm

Greedy Randomized Adaptive Search Procedure (GRASP) [8] is a meta-heuristic algorithm for combinatorial optimization. GRASP is a multi-start process where each iteration consists of two phases: the construction of initial solutions and the local search. The construction phase builds a feasible solution using a greedy randomized construction. From these initial solutions the local search phase explores the neighbourhood until finding a local minimum.

At each iteration of the construction phase, the initial solution is built using a set of candidate elements. A candidate element is a piece that can be incorporated to the partial solution without destroying the feasibility. The candidate elements are ordered using a function that measures the benefit of the element for the solution. The best n-candidate elements are selected to be at the Restricted Candidate List (RCL), where n is a parameter. The GRASP randomly chosen one element from the RCL to incorporate in the solution. This way, different solutions are obtained at each iteration, but does not necessarily compromise the power of the adaptive greedy component of the method [8]. After then, the candidate elements are updated and the functions are evaluated again.

The second phase, the local search, improves the constructed solutions exploring the neighborhood. These solutions are going to a new refined process through Path-relinking procedure. Path-relinking was proposed as a search strategy that explores trajectories connecting elite solutions obtained by tabu search [9]. The use of path-relinking within a GRASP was first proposed by Laguna and Marti [15]. Given two solutions, their common elements are kept constant and the space of solutions spanned by these elements is searched with the aim of finding a better solution. The path-relinking may be viewed as a strategy that seeks to incorporate attributes of high quality solutions, by the benefit of these attributes in the selected moves. The path-relinking can be applied as an intensification strategy, as a post-optimization step to all pairs of best solutions resulting of the local search [19].

We use these meta-heuristics to propose the GRASP-PR Rule Learning Algorithm. GRASP-PR Rule learning pseudo-code is presented at Algorithm 1. The algorithm has two phases: The GRASP phase with the construction of rules (*ConstructGreedyRandomizedSolution*) and the local search (*LocalSearch*) to improve the solutions. This phase creates a list of rules, called Elite Rules (*EliteRules*), that will be processed further in the algorithm with a path-relinking procedure (*path-relinking*). The goal of the path-relinking step is to explore intermediary solutions between each pair of Elite Rules in order to ensure a good Pareto Front coverage.

The construction of the rules is based on the RCL. A candidate element of RCL is one pair of attribute-value. For each possible pair, the algorithm computes the TP rate and the TN rate and chooses a $n\%$ of the best candidates according to TP rate (or a $n\%$ of the best pairs according to TN rate, shift from one to another). The constructed rule begins with one pair randomly choose from the RCL. After then, all possible candidates are evaluated and the RCL is again built. The algorithm randomly selects another pair from the RCL. If

Algorithm 1. GRASP-PR Rule Learning Algorithm

```
InputInstance()
procedure grasp()
  boolean TPbias = true // true=TP rate as bias; false=TN rate as bias
  for i = to maxLoop
    Solution = ConstructGreedyRandomizedSolution( TPbias, percentageElements)
    localBest = LocalSearch(Solution, TPbias)
    UpdateSolution( localBest, EliteRules)
    TPbias = not TPbias
  endfor
  EliteRules = filterGeneric( EliteRules)
  return( EliteRules)
end grasp
non_dom = InsertParetoFront ( EliteRules )
procedure path-relinking( EliteRules,non_dom )
  foreach rule₁ ∈ EliteRules
    foreach rule₂ ∈ EliteRules
      path (rule₁, rule₂, non_dom )
    endforeach
  endforeach
return(non_dom)

procedure path(rule₁, rule₂, non_dom)
  intermediate = rule₁
  while intermediate ≠ rule₂
    foreach attribute ∈ rule₂
      intermediate.add( attribute)
      evaluate( intermediate)
      non_dom = InsertParetoFront( intermediate )
    endforeach
  endwhile
  return(Pareto)
```

this additional pair improves the TP rate or the TN rate, the pair is included in the rule currently examined. Otherwise, the pair is removed from the RCL. The construction ends when the RCL is empty.

Each rule is improved with the local search procedure. The algorithm randomly selects one attribute or value to replace in the current solution at each iteration. If the replacement improves the TP rate (or the TN rate), then, the new rule replaces the current one and the process continues until there is no improvement. At the end of this phase, a procedure ($filterGeneric$) checks for rules more general or specific in the archive before a rule is added to the non-dominated solutions archive. A rule is more specific than other rule, if it has less attribute constraints and the same contingency table. Only the more general rules are kept in the archive.

The second phase applies the path-relinking between the Elite Rules and also initializes the non-dominated solutions archive ($InsertParetoFront$). For each two rules, the path-relinking creates intermediate rules between them (*path*). Considering an initial and a terminal rule, the idea is to replace one pair attribute-value of the initial rule with a pair attribute-value of the terminal rule, at each iteration, until the former reaches the latter. Each intermediate rule is analyzed (*evaluate*) and if it is non-dominated by any other solution in the Elite Archive, it is included in that archive. The procedure removes solutions that are dominated by the new rule from that archive. The non-dominated solutions archive is built selecting all the non-dominated rules according to sensitivity and specificity.

4   Multiple Objective Particle Swarm

Particle swarm optimization (PSO) is a population based stochastic optimization technique, inspired by social behavior of bird flocking or fish schooling [13]. PSO shares many similarities with evolutionary computation techniques such as Genetic Algorithms (GA). The system is initialized with a population solutions and searches for optima by updating generations. However, unlike GA, PSO has no evolution operators such as crossover and mutation. In this model, there is a set of particles, called swarm, that are possible solutions for the problem. These particles move through an n-dimensional space based on their neighbors' best positions and on their own best position. For that, on each generation, the position and velocity of particles are updated, considering the best position already obtained by the particle and the best global position obtained by all particles of the swarm. The best particles are found based on the fitness function, which is the problem's objective function. Each particle p, at some iteration t, has a position in R^n, $x(t)$, and a displacement velocity in this space, $v(t)$. It has also a little memory that contains its best position already fetched, $pbest(t)$, and the best position, $gbest(t)$, already fetched by particles that p knows, i.e., the best $p(t)$ of all particles attached to the neighborhood of $p(N(p))$. It is important to tell that $x(t)$, $v(t)$, $pbest(t)$ and $gbest(t)$ are n-dimensional vectors.

The algorithm works just like follows. The swarm is initiated at the time t=0, spreading the particles randomly in the space. After that, the iterative process initiates. The particle's velocity and position, on the next iteration, are calculated by the equations 3 and 4.

$$\vec{v}(t+1) = \varpi * \vec{v}(t) + \phi_1 * (\overrightarrow{pbest}(t) - \vec{x}(t)) + \phi_2 * (\overrightarrow{gbest}(t) - \vec{x}(t)) \quad (3)$$

$$\vec{x}(t+1) = \vec{x}(t) + \vec{v}(t+1) \quad (4)$$

where ϕ_1 and ϕ_2 , in equation 3, are coefficients that determine the influence of the particle's best (*pbest(t)*) and the particle global best (*gbest(t)*) on particle's velocity, respectively. The coefficient ϖ is the inertia of the particle, i.e., how much its previous velocity affects the current velocity. After the velocities and positions of all the particles have been updated, $pbest(t+1)$ and $gbest(t+1)$ are calculated and it continues to the next iteration or until the execution is ended. A particle swarm algorithm for the solution of multiobjective problems was presented in [3]. In MOPSO (Multiple Objective Particle Swarm Optimization), in contrast with PSO, there are many fitness functions. Differently from PSO, in MOPSO there is no global best, but a repository with the non-dominated solutions found. At each generation, the velocity of the particles is updated by the equation 5.

$$\vec{v}(t+1) = \varpi * \vec{v}(t) + \phi_1 * (\overrightarrow{pbest}(t) - \vec{x}(t)) + \phi_2 * (\overrightarrow{R_h}(t) - \vec{x}(t)) \qquad (5)$$

where R_h is a position of a particle of the repository, chosen as a guide. There are many forms to make this choice [3], in this work, we use the sigma distance method.

The rule learning algorithm using MOPSO works in the following manner. Each particle of the swarm represents a single classification-rule for an input dataset with n attributes. Each attribute, beyond the values that appear in the base, can accept a '?' value. That means, for that rule, this attribute does not matter for the classification. Thus, for example, for a base with the attributes: (Sky, Temperature, Humidity, Wind, Play), being the last one the class attribute, a particle with the position: (sun,?,?,yes,yes), for example, is equivalent to the rule: IF Sky = Sun AND Wind=Yes THEN Play = Yes.

The codification is conceived by integer numbers related with each attribute value of a dataset. For example, if we have the base and the position of the particle above mentioned, each of this values are going to be replaced by a corresponding integer number, sun=2, ?=0, ?=0, yes=1, yes=2, respectively. Thus, the vector: (2,0,0,1,2) represents its position on N^5 space.

First of all, the particles are spread randomly in the discrete search space (step 1). This is made by means of roulette where the most frequent values of the database have greater possibility to be chosen. The probability of the generic value of each attribute, '?', is calculated in function of the number of possible values for attribute. Thus, the more is the number of values, greater is the probability of the generic value in the roulette. This restriction avoids creating rules or particles that are not very specific, generating, for example, rules that cover few examples. After that, the particles are evaluated in all objectives (step 2). In our case, we used sensitivity and specificity. By means of the criterion of Pareto dominance, we put the particles with non-dominated solutions in the global repository (step 3). The solutions that already will be in the repository, but turned to be dominated by another solution, must be excluded. Analysing the repository, the more specific rules are removed and just the more generic rules are kept (step 4). From that point, we must divide the objective space between each particle of the repository (step 5). Basically, the particles of the swarm must choose one nondominated particle of the repository like their global optimum (in our case, we used a process known as sigma distance). Then, the velocity and the position of the particles in the space are updated through the equation 5, using the particles numeric position. After that, again, we evaluate the particles and divide the objective space between the particles of the repository, restarting the loop (step 6). The loop stops when it reaches the number of generations established by the user. We believe that, after some generations, the best classification-rules for the class, by means of used objectives, will be in the repository (step 7), the pseudo-code is presented at Algorithm 2.

Algorithm 2. Rule learning algorithm with MOPSO

1. For each particle i, do:
   a. Initialize $\vec{x_i}$ with a random solution to the problem
   b. Initialize $\vec{v_i}$ with a random velocity.
   c. Initialize $\overrightarrow{pbest_i} = \vec{x_i}$
2. Evaluate particles using fitness functions.
3. Find non-dominated solutions, storing them in the repository.
4. Filtering the repository, keeping the more generic rules.
5. Divide the search space between the solutions of the repository.
6. While not reach a stop criterion:
   a. For each particle i of the swarm, do:
      1) $\varpi = random[0, 0.8]$; $\phi_1 = random[0, 4]$; $\phi_2 = random[0, 4]$;
      2) $\vec{v}(t+1) = (\varpi * \vec{v}(t) + \phi_1 * (\overrightarrow{pbest}(t) - \vec{x}(t)) + \phi_2 * (\overrightarrow{R_h}(t) - \vec{x}(t))) mod \overrightarrow{N_i}$
      Note: $\overrightarrow{N_i}$ is a vector of the number of possible values to each attribute of the
      database. It restricts the particle inside the search space.
      3) $\vec{x}(t+1) = (\vec{x}(t) + \vec{v}(t+1)) mod \overrightarrow{N_i}$
      4) Evaluate particles. The particle will have one value
      for each objective of the problem.
      5) Update $\overrightarrow{pbest}(t)$
   b. Update the repository with non-dominated particles.
   c. Divide the search space, finding $\overrightarrow{R_h}(t)$ of the particles.
7. Return Repository

5   Experiments Results

In this work, two different empirical studies were made to evaluate the MOMH
techniques. The first one compares GRASP-PR Rule and MOPS algorithms with
other systems using the area under the ROC curve. The area under the ROC
curve (AUC) is considered a relevant criterion to deal with imbalanced data,
misclassification costs and noisy data. The second empirical study investigates
the quality of the solutions provided by GRASP-PR Rule and MOPS algorithms
based on the concept of Pareto dominance. This Section presents the methodology followed in these studies.

5.1   Methodology

GRASP-PR Rule and MOPS algorithms were compared with other traditional
systems with recognized good performance in terms of AUC, we use the results reported by [17]. There, the algorithms: ROCCER, C4.5, CN2, Ripper and
Slipper were compared. The experiments were done with datasets from UCL
Machine Learning Repository [9], using 10-fold stratified cross-validation and
for all inducers were given the same training and test files. The datasets with
more than two classes were reduced to two-class problems selecting the class
with the lower frequency as positive, and the remaining examples as negative.
Table 2 presents the description of datasets used in the experiments. It shows

Table 2. Description of datasets used in the experiments

#	Data set	Att	Exa	Cl	#	Data set	Att	Exa	Cl	#	Data set	Att	Exa	Cl
1	breast	10	683	2	7	haberman	4	306	2	12	new-thyroid	6	215	3
2	bupa	7	345	2	8	heart	14	270	2	13	nursery	9	12960	5
3	ecoli	8	336	8	9	ionosphere	34	351	2	14	pima	9	768	2
4	flag	29	174	6	10	kr-vs-kp	37	3196	2	15	satimage	37	6435	6
5	german	21	1000	2	11	lettera	17	20000	26	16	vehicle	19	846	4
6	glass	10	214	6										

the number of attribute ('Att' column), number of examples ('Exa' column) and number of classes ('Cl' column) of the sixteen datasets. The results of the AUC values were estimated using the trapezoidal rule and are showed at Table 3. The numbers between brackets indicate standard deviations.

The experiments with our algorithms were done with the same methodology and paired datasets. The parameters for GRASP-PR Rule algorithm were 50% for RCL and a maximum of 200 constructed solutions (100 for positive rules and 100 for negative rules). The MOPS algorithm was executed with 500 particles and 50 generations for each class. The rules obtained with each algorithm were used as a classifier using a weighted voted classification process (based on the confidence). In this process, there is a set of rules voting example by example, each one on its class. Thus, for each instance of the database, we obtain a numeric rank. This rank can be used as a threshold to produce a binary classifier. If the rank of the instance goes beyond the threshold, the classifier produces a "yes", otherwise a "no". Then, for a given threshold exists a point in the ROC plane, so, varying the threshold from -1 to $+1$ produces a curve on the ROC plane and we are able to calculate the AUC [18] using the trapezoidal rule.

5.2 Comparison with Other Systems

Both algorithms were executed 50 times for each dataset and the mean of all AUC and its standard deviation are presented at second column of Table 3. The distributions of the AUC values for each dataset were verified with Shapiro-Wilk test [20], all the distributions follow a normal distribution, with exceptions of datasets #1, #2, #3 and #16 for GRASP-PR, and datasets #5, #11, #12 and #15 for MOPS algorithm. Then, all the analysis were done using T-test with 95% confidence level. The GRASP-PR algorithm was taken as basis. The cells, at Table 3, without background colour indicate no statistical difference between algorithms; dark gray is used to represent cells statistically worse than GRASP-PR Rule algorithm and light gray indicates cells statistically better than our meta-heuristic algorithm. For space reasons, only summary results are reported here, the graphics of Pareto fronts can be on-line accessed at http://grasppr.googlepages.com/. Future work will explore these results in depth.

Considering the results of GRASP-PR and MOPS algorithms for each dataset, we can observe a great difference of AUC values between them. The first one

Table 3. Mean AUC

#	GRASP-PRMOPS	ROCCER	C45	C45NP	CN2	CN2OR	Ripper	Slipper
1	99.04(0.25)	98.75(0.16)	98.63(1.88)	97.76(1.51)	98.39(1.3)	99.26(0.81)	99.13(0.92)	98.72(1.38)	99.24(0.57)
2	68.07(0.64)	66.15(1.1)	65.3(7.93)	62.14(9.91)	57.44(11.92)	62.74(8.85)	62.21(8.11)	69.1(7.78)	59.84(6.44)
3	90.33(1.51)	82.69(2.22)	90.31(11.56)	50(0)	90.06(7.75)	90.17(6.9)	85.15(11.38)	61.86(25.49)	74.78(15.94)
4	70.66(1.76)	63.26(4.37)	61.83(24.14)	50(0)	68.68(17.22)	53.22(24.12)	42.78(24.43)	45.28(14.93)	52.35(7.44)
5	74.67(0.46)	73.39(0.83)	72.08(6.02)	71.43(5.89)	67.71(4.12)	75.25(5.38)	70.9(4.7)	64.02(13.62)	71.32(6.2)
6	82.13(1.31)	65(4.19)	79.45(12.98)	50(0)	81.5(12.65)	73.74(15.4)	79.64(13.24)	49.75(0.79)	50(2.30)
7	59.32(1.05)	59.96(1.23)	66.41(11.54)	55.84(6.14)	64.33(13.58)	59.83(9.87)	59.28(10.13)	57.45(3.85)	50.4(11.14)
8	88.59(0.33)	87.48(0.6)	85.78(8.43)	84.81(6.57)	81.11(7.91)	83.61(6.89)	82.25(6.59)	84.89(7.68)	84.03(6.36)
9	96.13(0.76)	81.19(2.84)	94.18(4.49)	86.09(9.97)	90.91(6.03)	96.23(2.97)	92.18(7.54)	92.06(5.94)	93.95(6.82)
10	98.91(0.09)	95.78(0.53)	99.35(0.36)	99.85(0.2)	99.86(0.2)	99.85(0.16)	99.91(0.17)	99.85(0.21)	99.91(0.09)
11	96.67(0.26)	93.8(0.72)	96.08(0.52)	95.49(1.96)	99.33(0.46)	99.34(0.28)	99.44(0.63)	97.27(1.86)	98.82(0.44)
12	98.35(0.39)	87.62(0.43)	98.4(1.7)	87.85(10.43)	97.5(3.39)	99.14(1.19)	98.43(2.58)	94.95(9.94)	99.12(1.25)
13	99.48(0.08)	85.91(0.61)	97.85(0.44)	99.42(0.14)	99.74(0.13)	100(0)	99.99(0.01)	99.43(0.26)	94.4(1.59)
14	71.02(0.65)	71.67(0.83)	70.68(5.09)	72.07(4.42)	72.6(6.5)	70.96(4.62)	71.97(5.44)	68.07(9.46)	70.02(5.97)
15	90.7(0.27)	65.39(9.11)	89.39(2.38)	90.15(1.7)	91.31(1.32)	91.48(1.45)	91.48(0.9)	86.83(3.94)	89.06(1.98)
16	96.83(0.51)	88.83(1.02)	96.42(1.47)	94.76(3)	96.99(1.44)	97.38(2.05)	96.49(2.41)	95.01(2.22)	93.99(3.13)
Avg	86.31	84.09	85.13	77.98	84.84	84.51	83.20	79.03	80.08

is better for most datasets, and it is worse only on dataset (#14), and there is no difference at one dataset. Despite of the number that our algorithm is better than other algorithms, the greatest values are against the C4.5 and Slipper with twelve and nine greater values, respectively. Against the Ripper, GRASP-PR Rule algorithm wins at 6 datasets. Against the C45NP and CN2OR, GRASP-PR Rule algorithm wins at 4 datasets. The worst performance is against the CN2, GRASP-PR Rule algorithm wins at 3 datasets and lose at 4 datasets. Against the ROCCER, there is no difference at most part of the datasets, with two wins for each one. The datasets #10, #11 and #13 are the datasets where the GRASP-PR Rule algorithm does not have good results.

5.3   Pareto Dominance

The Pareto front of the both algorithms were compared through the dominance ranking [14] using the PISA framework [1]. Each Pareto front receives a rank based on the domination relation by counting the number of fronts that it dominates. The one-tailed Mann-Whitney rank sum test [14] is applied to verify statistical difference between ranking of the two algorithms. The results for most datasets, including all folds, indicate that none of them, neither GRASP-PR Rule or MOPS algorithms, generates a better Pareto front according to dominance ranking with a significance level of 5%.

Another comparison was made using the hypervolume indicator [21]. For each fold, a ranking is generated with the Pareto fronts, this ranking is tested using the one-tailed Mann-Whitney rank sum test. The GRASP-PR Rule algorithm outperforms MOPS algorithm at 78% of the Pareto according to the hypervolume indicator. And, at just 18%, MOPS has front that outperforms the GRASP-PR Rule algorithm.

Both experiments conducted here with GRASP-PR Rule learning and MOPS algorithm show better results for GRASP-PR Rule learning. The AUC mean values presents good values if we compare with other algorithms. Furthermore, the Pareto dominance indicates that GRASP-PR Rule algorithm outperforms MOPS algorithm in most part of the datasets. However, considering the time processing, the GRASP-PR Rule learning algorithm spent almost 200 hours for the 50 executions of all datasets, against approximately 100 hours for the MOPS algorithm.

6   Conclusions

In this work, it was presented a different approach for classification-rule learning based on multiobjective metaheuristics. The approach works finding the best nondominated rules of a problem and forming unordered classifiers. Two techniques were evaluated using the sensitivity and specificity criterion as objectives: a Greedy Randomized Adaptive Search Procedure with path-relinking (GRASPPR), and Multiple Objective Particle Swarm (MOPS). Two different empirical evaluations were performed. The first one compares GRASP-PR Rule and MOPS algorithms to different well known algorithms into different data sets using the area under the ROC curve. The results showed that the algorithms are competitive with others of the literature. Besides this, GRASP-PR Rule algorithm outperforms MOPS algorithm in almost datasets. The second empirical study investigates the quality of the solutions provided by GRASP-PR Rule and MOPS algorithms based on the concept of Pareto dominance. Once again, GRASP-PR Rule algorithm outperforms MOPS algorithm in almost datasets. Concluding, rules with high sensitivity and specificity can be used to compose a classifier which presents good AUC performance. Furthermore, these rules can be created using a multi-objective metaheuristic approach. However, both metaheuristics can be improved. For the GRASP-PR Rule algorithm, a depth study must be done about the influence of: the size of the RCL list, the local search strategy and the neighbourhood. In the case of MOPS rule algorithm, a better exploration of the Pareto front is needed. Future works will include the above aspects and others like: parallel versions of the algorithms to deal with large databases; multi-class datasets and a depth study on the features of the classifiers produced by these algorithms.

References

1. Bleuler, S., Laumanns, M., Thiele, L., Zitzler, E.: PISA — a platform and programming language independent interface for search algorithms. In: Fonseca, C.M., Fleming, P.J., Zitzler, E., Deb, K., Thiele, L. (eds.) EMO 2003. LNCS, vol. 2632, pp. 494–508. Springer, Heidelberg (2003)
2. Clark, P., Niblett, T.: Rule induction with cn2: Some recent improvements. In: ECML: European Conference on Machine Learning, Springer, Berlin (1991)
3. Coello, C., Lechuga, M.: MOPSO: A proposal for multiple objective particle swarm optimization. In: IEEE World Congress on Computational Intelligence, pp. 1051–1056. IEEE Press, Los Alamitos (2002)
4. de Almeida Prado, A., Toracio, G., Pozo, A.T.R.: Multiple objective particle swarm for classification-rule discovery. In: 2007 IEEE Congress on Evolutionary Computation, September 25-28, 2007, pp. 684–691. IEEE Press, Los Alamitos (2007)
5. de la Iglesia, B., Philpott, M.S., Bagnall, A.J., Rayward-Smith, V.J.: Data mining rules using multi-objective evolutionary algorithms. In: Congress on Evolutionary Computation, pp. 1552–1559. IEEE Computer Society, Los Alamitos (2003)

6. de la Iglesia, B., Reynolds, A., Rayward-Smith, V.J.: Developments on a multi-objective metaheuristic (momh) algorithm for finding interesting sets of classification rules. In: Coello Coello, C.A., Hernández Aguirre, A., Zitzler, E. (eds.) EMO 2005. LNCS, vol. 3410, pp. 826–840. Springer, Heidelberg (2005)
7. Fawcett, T.: Using rule sets to maximize roc performance. In: ICDM, pp. 131–138. IEEE Computer Society, Los Alamitos (2001)
8. Feo, T.A., Resende, M.G.C.: Greedy randomized adaptive search procedures. Journal of Global Optimization 6, 109–133 (1995)
9. Tabu search and adaptive memory programming - advances, applications and challenges. In: Glover, F., Barr, R.S., Helgason, R.V., Kennington, J.L. (eds.) Interfaces in Computer Science and Operations Research, pp. 1–75. kluwer, Dordrecht (1996)
10. Ishibuchi, H.: Multiobjective association rule mining. In: PPSN Workshop on Multiobjective Problem Solving from Nature, pp. 39–48, Reykjavik, Iceland (2006)
11. Ishibuchi, H., Nojima, Y.: Accuracy-complexity tradeoff analysis by multiobjective rule selection. In: ICDM, pp. 39–48. IEEE Computer Society, Los Alamitos (2005)
12. Jin, Y.: Multi-Objective Machine Learning. Springer, Berlin, Boston, MA (2006)
13. Kennedy, J., Eberhart, R.: Particle swarm optimization. In: IEEE International Conference on Neural Networks, pp. 1492–1948. IEEE Press, Los Alamitos (1995)
14. Knowles, J., Thiele, L., Zitzler, E.: A tutorial on the performance assessment of stochastic multiobjective optimizers. 214, Computer Engineering and Networks Laboratory (TIK), Swiss Federal Institute of Technology (ETH) Zurich (July 2005)
15. Laguna, M., Marti, R.: Grasp and path relinking for 2-layer straight line crossing minimization. INFORMS J. on Computing 11(1), 44–52 (1999)
16. Lavrac, N., Flach, P., Zupan, B.: Rule evaluation measures: A unifying view. In: Džeroski, S., Flach, P.A. (eds.) ILP 1999. LNCS (LNAI), vol. 1634, pp. 174–185. Springer, Heidelberg (1999)
17. Prati, R.C., Flach, P.A.: ROCCER: An algorithm for rule learning based on ROC analysis. In: Kaelbling, L.P., Saffiotti, A. (eds.) IJCAI, pp. 823–828, Professional Book Center (2005)
18. Provost, F., Fawcett, T.: Robust classification for imprecise environments. Machine Learning 42(3), 203 (2001)
19. Resende, M., Ribeiro, C.: Greedy randomized adaptive search procedures. In: Glover, F., Kochenberger, G. (eds.) Handbook of Metaheuristics, pp. 219–249. Kluwer Academic Publishers, Dordrecht (2002)
20. Shapiro, S.S., Wilk, M.B.: An analysis of variance test for normality (complete samples). Biometrika 52(3-4), 591–611 (1965)
21. Zitzler, E., Thiele, L.: Multiobjective Evolutionary Algorithms: A Comparative Case Study and the Strength Pareto Approach. IEEE Transactions on Evolutionary Computation 3(4), 257–271 (1999)

An Extended Beam-ACO Approach to the Time and Space Constrained Simple Assembly Line Balancing Problem*

Christian Blum[1], Joaquín Bautista[2], and Jordi Pereira[3]

[1] ALBCOM, Dept. Llenguatges i Sistemes Informàtics
Universitat Politècnica de Catalunya, Barcelona, Spain
cblum@lsi.upc.edu
[2] ETSEIB, Nissan Chair
Universitat Politècnica de Catalunya, Barcelona, Spain
joaquin.bautista@upc.es
[3] ETSEIB, Dept. d'Organització d'Empreses
Universitat Politècnica de Catalunya, Barcelona, Spain
jorge.pereira@upc.edu

Abstract. Assembly line balancing problems are concerned with the distribution of work required to assemble a product in mass or series production among a set of work stations on an assembly line. The specific problem considered here is known as the time and space constrained simple assembly line balancing problem. Among several possible objectives we consider the one of minimizing the number of necessary work stations. This problem is denoted by TSALBP-1 in the literature. For tackling this problem we propose an extended version of our Beam-ACO approach published in [3]. Beam-ACO algorithms are hybrid techniques that result from combining ant colony optimization with beam search. The experimental results show that our algorithm is able to find 128 new best solutions in 269 possible cases.

1 Introduction

One of the most extensively studied assembly line balancing problems is known as the simple assembly line balancing problem (SALBP) [12]. It concerns the distribution of work required to assemble a product among a set of work stations on an assembly line, which is a sequence of work stations that are connected by a transport system moving the product to be manufactured along the line. Approaches for solving the SALBP include constructive heuristics based on priority rules (see [16]), complete techniques such as branch & bound approaches

---

* This work was supported by grants TIN-2005-08818-C04-01 and DPI2004-03475 of the Spanish government, and by the *Ramón y Cajal* program of the Spanish Ministry of Science and Technology of which Christian Blum is a research fellow. Moreover, we acknowledge Nissan Spain and the UPC Nissan Chair for partially funding this work.

J. van Hemert and C. Cotta (Eds.): EvoCOP 2008, LNCS 4972, pp. 85–96, 2008.

(see [8,7,13,15]), and several metaheuristics such as tabu search [14,9], simulated annealing [17], evolutionary computation [6], and ant colony optmization [1]. Inspired by the Nissan plant in Barcelona, Spain, Bautista and Pereira proposed an extension of the SALBP in [2] that—in addition to time constraints—also covers space constraints. In this work we tackle this problem with the objective of minimizing the necessary number of work stations. Henceforth we refer to the tackled problem as the TSALBP-1.

Ant colony optimziation (ACO) [5] is a metaheuristic based on the probabilistic construction of solutions. Instead of a standard ACO algorithm we implement in this paper a hybrid ACO algorithm—known as Beam-ACO—that results from combining ant colony optimization with a heuristic branch & bound derivative called beam search. While in standard ACO algorithms artificial ants construct solutions independently of each other, in Beam-ACO the solution constructions at each iteration are non-independent and guided by a lower bound function. In [3] we presented a first Beam-ACO approach to the TSALBP-1. In this work, we present an extended version that results in significant performance improvements. The extension basically consists in a more general handling of the beam width, and a heuristic form of avoiding that partial solutions occur more than once in the generated beam at each construction step.

The paper is organized as follows. In Sect. 2 we present a technical definition of the TSALBP-1. In Sect. 3 we outline Beam-ACO. Finally, in Sect. 4 we present the computational results, and in Sect. 5 we offer conclusions and an outlook on future work.

2   TSALBP-1

Each TSALBP-1 instance is characterized by a quadruple (T, G, c, a), where $T = \{1, \ldots, n\}$ is a set of n tasks. Each task $j \in T$ has a processing time $t_j > 0$ and a space requirement $a_j > 0$. Moreover, given is an acyclic, directed precedence graph G whose node set is the set of given tasks. A directed arc (i, j) in G indicates that task i must be processed before j. Given a task $j \in T$, we denote by $\text{Pre}_j \subset T$ the set of tasks that must be processed before j. Finally, c is the so-called processing time limit of a work station (also know as the cycle time), and a is the available space of a work station. Note that all work stations are equal with respect to c and a.

Solutions. A solution is obtained by assigning each task to exactly one work station. In this work we represent a solution s as an ordered set $\langle S_1, \ldots, S_m \rangle$ of $m \leq n$ work stations. Each work station $S_k \subseteq T$ is a set of tasks.. A solution s is valid if the following 4 conditions are fullfilled: **(1)** $\bigcup_{k=1}^{m} S_k = \{1, \ldots, n\}$ and $\bigcap_{k=1}^{m} S_k = \emptyset$. These conditions ensure that each task is assigned to exactly one work station; **(2)** $\sum_{j \in S_k} t_j \leq c$, for $k = 1, \ldots, m$. This ensures that no work station has too much load; **(3)** $\sum_{j \in S_k} a_j \leq a$, for $k = 1, \ldots, m$. Herewith is ensured that the space limits of the work stations are not exceeded; **(4)** for each

$j \in S_k$ it is required that $\bigcup_{l=1}^{k} S_l$ contains Pre_j, which ensures that the precedence constraints between the tasks are respected. Our algorithm exclusively generates valid solutions.

Objective Function. In this work we aim to find a solution s, where $m = |s|$ is minimal. This objective contains large plateaus, that is, many different solutions have the same number of work stations. Hence, the number of work stations is not operational for guiding a metaheuristic to promising parts of the search space. Therefore, we introduce a second criterion in order to distinguish between solutions with the same number of work stations. The second criterion concerns the remaining time and space in the last work station $S_m \in s$. We use the following notations: $t^{\mathrm{rem}}(s) := c - \sum_{j \in S_m} t_j$ and $a^{\mathrm{rem}}(s) := a - \sum_{j \in S_m} a_j$. Using these notations, the second criterion is defined as $g(s) := \frac{t^{\mathrm{rem}}(s)}{c} + \frac{a^{\mathrm{rem}}(s)}{a}$. Given $g(\cdot)$ we can define a comparison operator $f(\cdot)$ as follows. Given two solutions $s \neq s'$, $f(s) < f(s') \Leftrightarrow |s| < |s'|$ **OR** $|s| = |s'|$ and $g(s) < g(s')$. This means that—in the case of equality concerning the number of work stations—preference is given to the solution with more time and space remaining in the last work station. The idea is that such a solution is somehow *closer* to a solution with only $m - 1$ work stations. Finally, despite the fact that Beam-ACO uses the comparison operator $f(\cdot)$ for guiding the search process, we will present the results only in terms of the original objective, that is, the number of used work stations.

Reverse Problem Instances. Given a problem instance (T, G, c, a), the corresponding reverse problem instance (T, G^r, c, a) is obtained by inverting all the arcs in the precedence graph G. Each solution $s^r = \langle S_1, \ldots, S_m \rangle$ to the reverse problem instance (T, G^r, c, a) can be converted into a solution s to the original problem instance (T, G, c, a) by inverting the ordered list of tasks, that is, $s = \langle S_m, \ldots, S_1 \rangle$. It is known from the literature (see, for example, [12]) that the reverse problem instance may be easier to solve than the original one, or vice versa.

3   The Algorithm

In this section we describe our implementation of Beam-ACO for the TSALBP-1. In general, Beam-ACO works as any other ACO algorithm. At each iteration candidate solutions are probabilistically constructed on the basis of a so-called pheromone model $\mathcal{T}$, which is a set of numerical values that encode the algorithms' search experience. In contrast to standard ACO algorithms in which each solution construction is independent from the others, Beam-ACO employs a parallel and non-independent construction of the solutions of an iteration in the style of beam search. After the construction phase, some of the generated solutions are used to update the pheromone values in a way that aims at biasing

---

Algorithm 1. Priority-rule (PR) heuristic for the TSALBP-1

1: **input:** A TSALBP-1 instance (T, G, c, a)
2: $m := 0$
3: **while** $T \neq \emptyset$ **do**
4:      $m := m + 1$
5:      $S_m \leftarrow$ FillWorkStation(T, m) /* see Algorithm 2 */
6:      $T := T \setminus S_m$
7: **end while**
8: **output:** Solution $s = \langle S_1, \ldots, S_m \rangle$

Algorithm 2. Function FillWorkStation(T, m)

1: **input:** A set T of tasks, and the index k of the work station to be filled
2: $T' := T$
3: $c_{\mathrm{rem}} := c$, $a_{\mathrm{rem}} := a$
4: $T^{\mathrm{av}} := \{i \in T' \mid c_{\mathrm{rem}} \geq t_i, a_{\mathrm{rem}} \geq a_i, \mathrm{Pre}_i \cap T' = \emptyset\}$
5: $S_m := \emptyset$
6: **while** $T^{\mathrm{av}} \neq \emptyset$ **do**
7:      $j \leftarrow$ ChooseTask(T^{av})
8:      $T' := T' \setminus \{j\}$
9:      $c_{\mathrm{rem}} := c_{\mathrm{rem}} - t_j$, $a_{\mathrm{rem}} := a_{\mathrm{rem}} - a_j$
10:     $T^{\mathrm{av}} := \{i \in T' \mid c_{\mathrm{rem}} \geq t_i, a_{\mathrm{rem}} \geq a_i, \mathrm{Pre}_i \cap T' = \emptyset\}$
11:     $S_m := S_m \cup \{j\}$
12: **end while**
13: **output:** Filled work station S_m

---

future solution constructions towards good areas of the search space identified during the search process.

3.1 A Priority Rule Heuristic

As solution construction mechanism we utilize in the Beam-ACO algorithm the construction process of a so-called priority rule (PR) heuristic in which work stations are filled one after the other. At each step, one of the available tasks (that is, tasks whose predecessors are already assigned to work stations) is selected and assigned to the current work station. This is done until no available task can be added to the current work station without exceeding the cycle time or the space limit. When this situation occurs, the algorithm opens the next work station. This process is continued until all tasks are assigned to work stations. The PR heuristic is shown in Algorithm 1.

When filling a work station, the successive choice of tasks is performed in function ChooseTask(T^{av}) (see Algorithm 2), which works as follows. Given a partial solution s, first the set of *available tasks* is determined:

$$T^{\mathrm{av}} := \{i \in T' \mid c_{\mathrm{rem}} \geq t_i, a_{\mathrm{rem}} \geq a_i, \mathrm{Pre}_i \cap T' = \emptyset\} \quad , \tag{1}$$

where $c_{\rm rem}$ and $a_{\rm rem}$ are as defined in Algorithm 2. In words, set $T^{\rm av}$ consists of all tasks from T' that still fit into the work station under consideration, and whose set of predecessors is already assigned to work stations. Moreover, let $T^{\rm sat} \subseteq T^{\rm av}$ be the set of available tasks such that $c_{\rm rem} - t_j = 0$ or $a_{\rm rem} - a_j = 0$, $\forall j \in T^{\rm sat}$. Henceforth, we call $T^{\rm sat}$ the set of *saturating available tasks*, because they saturate—in terms of available processing time—the current work station. If $T^{\rm sat}$ is non-empty, a task is chosen from $T^{\rm sat}$, otherwise from $T^{\rm av}$. The actual choice of a task is done by using a so-called priority rule. In our work we employ a mixed rule that gives joint priority to the duration of a task, the space requirement of a task, and the total number of tasks succeeding it. The *priority rule value* of a task $j \in T^{\rm av}$ is computed as follows:

$$\eta_j := \frac{t_j}{c} + \frac{a_j}{a} + \frac{\left|{\rm Suc}_j^{\rm all}\right|}{\max_{1 \le i \le n} \left|{\rm Suc}_i^{\rm all}\right|} \tag{2}$$

Hereby, ${\rm Suc}_j^{\rm all}$ denotes the set of all tasks that can be reached from j in the precedence graph G via a directed path. Function ChooseTask($T^{\rm av}$) is implemented such that the task with the maximal priority value is chosen. The use of priority rule heuristics is quite popular, because they are fast in execution and achieve reasonably good results.

3.2   Beam-ACO for TSALBP-1

Our Beam-ACO approach—shown in Algorithm 3—works as follows. First, the PR heuristic outlined in the previous section is used in function GenerateFirstSolution() for generating a first solution. In fact, the PR heuristic is applied to the original as well as to the reverse problem instance. After converting the latter solution to be a solution to the original instance, the function returns the better of the two solutions. Then, the pheromone values are initialized. At each algorithm iteration, a probabilistic beam search is applied in order to construct solutions to the original problem instance as well as to the reverse problem instance. Note that a solution construction is aborted in case the current partial solution must lead to a final solution worse than the best one found so far. This is determined by means of a lower bound. Finally, the pheromone values are updated in function UpdatePheromoneTrail($\mathcal{T}$,*). The pheromone values are re-initialized in case of algorithm convergence. In Algorithm 3 we use the following notations: $\mathcal{T} = \{\tau_{j,k}\}_{j,k=1,\ldots,n}$ is the set of pheromone values. A pheromone value $\tau_{j,k}$ represents the desirability of assigning task j to work station k. Furthermore, $s_{\rm ib}$ is the best solution constructed at an iteration, and $s_{\rm bsf}$ is the best solution found since the start of the algorithm. The functions of our algorithm are outlined in more detail below.

ProbabilisticBeamSearch($\mathcal{T}$,$s_{\rm bsf}$): This function—shown in Algorithm 4—performs a probabilistic beam search based on the solution construction mechanism of the PR heuristic (see Algorithm 1). Beam search is a classical tree search method that was introduced in the context of scheduling [11]. The central idea behind beam

---

Algorithm 3. Beam-ACO for TSALBP-1

1: **input:** A TSALBP-1 instance (T, G, c, a)
2: $s_{\text{bsf}} \leftarrow$ GenerateFirstSolution()
3: **forall** $\tau_{j,k} \in T$ **do** $\tau_{j,k} := 0.5$ **end forall**
4: $cf := 0$
5: **while** termination conditions not satisfied **do**
6:      $I_{\text{ori}} \leftarrow$ ProbabilisticBeamSearch(T, s_{bsf}) /* see Algorithm 4 */
7:      $I_{\text{rev}} \leftarrow$ ProbabilisticBeamSearch(Reverse)(T, s_{bsf})
8:      $I := I_{\text{ori}} \cup I_{\text{rev}}$
9:      **if** $I = \emptyset$ **then**
10:         UpdatePheromoneTrail(T, s_{bsf})
11:     **else**
12:         $s_{\text{ib}} := \min\{f(s_i) \mid s_i \in I\}$
13:         UpdatePheromoneTrail(T, s_{ib})
14:         **if** $f(s_{\text{ib}}) < f(s_{\text{bsf}})$ **then** $s_{\text{bsf}} := s_{\text{ib}}$
15:     **end if**
16:     $cf \leftarrow$ ComputeConvergenceFactor(T)
17:     **if** $cf < 0.05$ **then**
18:         **forall** $\tau_{j,k} \in T$ **do** $\tau_{j,k} := 0.5$ **end forall**
19:     **end if**
20: **end while**
21: **output:** s_{bsf}

---

search is to allow the extension of partial solutions in several possible ways. At each step the algorithm extends each partial solution from a set B—called the *beam*—at most k_{ext} times. In standard beam search, each extension is chosen in a deterministic way with respect to a heuristic function that gives weights to the possible extensions. In our case extensions are done partly in a probabilistic way depending on the pheromone model of the underlying ACO algorithm (see below). Each newly obtained partial solution is either stored in set B_{compl} in case it is a complete solution, or in set B_{ext} otherwise. At the end of each construction step, the beam search algorithm creates a new beam B by selecting up to k_{bw} (called the *beam width*) solutions from the set of further extensible solutions B_{ext}. This is done in function SelectSolutions$(B_{\text{ext}}, k_{\text{bw}})$ by means of a lower bound LB$(\cdot)$.

The extension of a partial solution s consists in filling the next work station. This is done by function FillWorkStation(T_s, m) which is shown in Algorithm 2. Hereby, T_s denotes the set of tasks that—with respect to s—are not yet assigned to work stations. Function ChooseTask(T^{av}) of Algorithm 2 is implemented as follows. First, we flip a coin in order to decide if the extension is performed deterministically, or probabilistically. In case of a deterministic extension, the set of tasks from which to choose an operation, denoted by T^c, is determined as follows: If the set of available saturating tasks $T^{\text{sat}} \subseteq T^{\text{av}}$ is non-empty, we set $T^c := T^{\text{sat}}$, otherwise $T^c := T^{\text{av}}$ (see Sect. 3.1 for the definition of T^{sat} and T^{av}). Then, from T^c is chosen the task that maximizes

---

Algorithm 4. Function ProbabilisticBeamSearch($\mathcal{T}$,s_{bsf}) of Algorithm 3

1: **input:** The set of pheromone values $\mathcal{T}$, the best-so-far solution s_{bsf}
2: $m := 0$
3: $s := \langle \rangle$
4: $B := \{s\}$
5: $B_{\mathrm{compl}} := \emptyset$
6: **while** $B \neq \emptyset$ **do**
7:     $B_{\mathrm{ext}} := \emptyset$
8:     $m := m + 1$
9:     **for all** $s \in B$ **do**
10:        **for** $i = 1, \ldots, k_{\mathrm{ext}}$ **do**
11:            $S_m^i \leftarrow$ FillWorkStation(T_s, m) /* see Algorithm 2 */
12:            $s_{\mathrm{ext}} := s \cup S_m^i$
13:            **if** $|T_{s_{\mathrm{ext}}}| = 0$ **then**
14:                $B_{\mathrm{compl}} := B_{\mathrm{compl}} \cup \{s_{\mathrm{ext}}\}$
15:            **else**
16:                **if** $\mathrm{LB}(s_{\mathrm{ext}}) < |s_{\mathrm{bsf}}|$ **then**
17:                    **if** S_m^i is different to the last work station of all $s' \in B_{\mathrm{ext}}$ **then**
18:                        $B_{\mathrm{ext}} := B_{\mathrm{ext}} \cup \{s_{\mathrm{ext}}\}$
19:                    **end if**
20:                **end if**
21:            **end if**
22:        **end for**
23:    **end for**
24:    $B \leftarrow$ SelectSolutions($B_{\mathrm{ext}}, k_{\mathrm{bw}}$)
25: **end while**
26: **output:** A (possibly empty) set of solutions B_{compl}

$$\mathbf{p}_j = \frac{\left(\sum_{i=1}^{k} \tau_{j,i} \right) \cdot \eta_j^{\mathrm{aco}}}{\sum_{l \in T^{\mathrm{c}}} \left(\sum_{i=1}^{k} \tau_{l,i} \right) \cdot \eta_l^{\mathrm{aco}}} \ . \tag{3}$$

Note that Eqn. 3 uses the summation rule introduced in [10] for scheduling problems. The heuristic information η_*^{aco} is derived as follows. Let $\eta_{\min} := \min\{\eta_j \mid j \in T\}$ and $\eta_{\max} := \max\{\eta_j \mid j \in T\}$ be the minimum, respectively the maximum, of the priority rule values as defined in Eqn. 2. Then,

$$\eta_j^{\mathrm{aco}} := \frac{\eta_j - \eta_{\min} + 1}{\eta_{\max}} \quad \forall j \in T \ . \tag{4}$$

In case of a probabilistic extension, T^{c} is set to T^{av}, and a task is chosen by roulette-wheel-selection with respect to the probabilities shown in Eqn. 3.

After the extension of a partial solution s, we first check if the result s_{ext} is a complete solution (see line 13 of Algorithm 4). If this is the case, it is added to the set B_{compl} of already completed solutions. Otherwise, two conditions must be satisfied for considering s_{ext} for further extension, that is, for adding s_{ext} to set B_{ext}. First, the lower bound value $\text{LB}(s_{\text{ext}})$ must be smaller or equal to the objective function value of the best-so-far solution s_{bsf}. And second, the last work station of s_{ext} is required to be different to the last work station of all solutions that are already in B_{ext}. This is a heuristic way of avoiding that the same solution appears more than once in B_{ext}.

In this work we use a relatively simple lower bound—denoted by $\text{LB}(\cdot)$—for evaluating partial solutions: Given a partial solution s and the set of tasks T_s (that is, tasks that are not yet assigned to work stations), the lower bound is defined as follows:

$$\text{LB}(s) = \max\left\{\left\lceil\frac{\sum\limits_{j\in T_s} t_j}{c}\right\rceil, \left\lceil\frac{\sum\limits_{j\in T_s} a_j}{a}\right\rceil\right\} \tag{5}$$

Note that this lower bound is a simple adaptation of the LM1 bound (see, for example, [12]) for the SALBP-1 to the TSALBP-1. Finally, the beam B of the next iteration of the construction procedure is selected from set B_{ext} in function SelectSolutions($B_{\text{ext}}, k_{\text{bw}}$) of Algorithm 4. First, the solutions in B_{ext} are ranked with respect to increasing lower bound values. In case of ties, we use the remaining time in the last work station as tie breaker, that is, we consider a work station with less remaining time as better. Further ties are randomly broken. Then, we select the $\min\{k_{\text{bw}}, |B_{\text{ext}}|\}$ highest ranked partial solutions from B_{ext}.

Note that function ProbabilisticBeamSearch(Reverse)(T, s_{bsf}) works in the same way as function ProbabilisticBeamSearch(T, s_{bsf}), just that it is applied to the reverse problem instance.

ComputeConvergenceFactor(T): Given the current pheromone values, this function computes a value cf to indicate the state of convergence of the algorithm:

$$cf := 2 \cdot \left(\frac{\sum\limits_{j=1}^{n}\sum\limits_{k=1}^{|s_{\text{bsf}}|}\min\{\tau_{\max} - \tau_{j,k}, \tau_{j,k} - \tau_{\min}\}}{n \cdot |s_{\text{bsf}}| \cdot (\tau_{\max} - \tau_{\min})}\right) \tag{6}$$

When the pheromone values are initialized, cf is 1; on the other side, when all pheromone values are either equal to $\tau_{\max}$ or to $\tau_{\min}$, cf is 0. We have set $\tau_{\max}$ to 0.99, and $\tau_{\min}$ to 0.01. Note that these value settings are motivated by the work presented in [4].

UpdatePheromoneTrail($T, *$): This function either uses solution s_{ib} or solution s_{bsf} for updating the pheromone values. s_{bsf} is only used in case no iteration best solution exists, which might occur due to solution construction abortions. Let us denote the updating solution by s_{upd}. Then, for $j = 1, \ldots, n$ and $k = 1, \ldots, |s_{\text{upd}}|$

the corresponding pheromone value $\tau_{j,k}$ is updated as follows:

$$\tau_{j,k} = \min \left\{ \max \left\{ \tau_{\min}, \tau_{j,k} + \rho \cdot (\delta_{j,k} - \tau_{j,k}) \right\}, \tau_{\max} \right\} \ , \tag{7}$$

where $\rho \in (0,1]$ is a learning rate (which we have set to 0.1 for all the experiments). Moreover, $\delta_{j,k}$ is 1, if task j is assigned to work station k in solution s_{upd}, and 0 otherwise. This concludes the description of the Beam-ACO algorithm. The experimental results are outlined in the following section.

4   Computational Results

We implemented the Beam-ACO algorithm in ANSI C++ using GCC 3.2.2 for compiling the software. Our experimental results were obtained on a PC with Intel Pentium 4 processor (3.06 GHz) and 1 Gb of memory, running Debian Linux.

We applied our algorithm to the 269 TSALBP-1 instances that were generated by Bautista and Pereira (see [2]) from the 269 SALBP-1 instances existing in the literature. This was done by setting $a_j := t_{n-j+1}$ for all $j \in T$, and $a := c$. Beam-ACO was applied 10 times to all 269 instances for 360 seconds per run. These are the same settings that were chosen for the first Beam-ACO approach published in [3]; henceforth referred to as Beam-ACO-1.

First, we show a comparison of Beam-ACO with the first algorithm developed to tackle the TSALBP-1—called ANTS [2]—and Beam-ACO-1. While ANTS was applied to all 269 problem instances, Beam-ACO-1 was only applied to the 26 instances based

Table 1. Comparison of Beam-ACO with ANTS and Beam-ACO-1 on the 26 instances based on the SCHOLL precedence graph

c, a	ANTS	Beam-ACO-1 [3]	Beam-ACO		
			best	average (std)	average time (std)
1394	60	59	56	56.90 (0.32)	29.31 (60.78)
1422	58	59	55	55.70 (0.48)	55.64 (101.65)
1452	58	57	54	54.90 (0.32)	18.90 (39.23)
1483	56	55	53	53.00 (0.00)	28.62 (17.73)
1515	54	54	50	51.00 (0.47)	125.62 (108.64)
1548	53	53	49	49.80 (0.42)	44.18 (79.60)
1584	53	51	48	48.00 (0.00)	116.88 (28.51)
1620	50	49	46	46.00 (0.00)	67.60 (13.98)
1659	49	48	45	45.00 (0.00)	66.56 (24.91)
1699	47	46	44	44.00 (0.00)	6.10 (6.03)
1742	46	45	42	42.90 (0.32)	26.10 (64.68)
1787	45	44	41	41.60 (0.52)	95.18 (138.50)
1834	43	43	40	40.00 (0.00)	133.68 (37.01)
1883	42	42	39	39.00 (0.00)	89.41 (15.98)
1935	41	41	38	38.00 (0.00)	120.25 (18.07)
1991	40	40	37	37.00 (0.00)	126.43 (18.33)
2049	39	38	36	36.00 (0.00)	124.99 (20.37)
2111	37	37	35	35.00 (0.00)	84.91 (26.08)
2177	36	36	34	34.00 (0.00)	44.09 (23.54)
2247	35	34	33	33.00 (0.00)	63.31 (11.82)
2322	34	33	32	32.00 (0.00)	32.85 (12.80)
2402	33	32	31	31.00 (0.00)	36.63 (19.73)
2488	32	31	30	30.00 (0.00)	4.80 (4.26)
2580	30	30	29	29.00 (0.00)	2.19 (0.05)
2680	29	29	28	28.00 (0.00)	2.17 (0.05)
2787	28	28	27	27.00 (0.00)	2.39 (0.65)

on the precedence graph called SCHOLL. Table 1 presents a comparison of the 3 algorithms on these 26 instances. Cycle time, respectively space limit, are indicated in the first table column. The second column contains the values of the best solutions found by ANTS, while the third column presents the values of the best solutions found by Beam-ACO-1. Finally, the last 3 table columns provide the results of Beam-ACO, concerning the best solution found in 10 runs (**best**), the average and standard deviation of the results (**average (std)**), and the times including the standard deviation at which the best solutions were found (**average time (std)**). While Beam-ACO-1 was only able to improve over ANTS in

14 out of 26 cases, our extended Beam-ACO approach can find new best solutions for all 26 instances, again finding better solutions even for the 14 cases in which Beam-ACO-1 was able to improve over ANTS. Note that the improvements are up to 4 work stations, which is impressive in the context of assembly line balancing problems.

Table 2. Results of Beam-ACO—in comparison to ANTS—when applied to 243 TSALBP-1 problem instances (part A)

Data set (size) c	ANTS	Beam-ACO best	avg. (std)	avg. time (std)	Data set (size) c	ANTS	Beam-ACO best	avg. (std)	avg. time (std)
Mukherje (94) 248	20	19	19.00 (0.00)	0.51 (0.02)	Warnecke (58) 58	36	35	35.00 (0.00)	26.91 (24.09)
263	18	18	18.00 (0.00)	0.48 (0.02)	60	33	34	34.00 (0.00)	1.94 (1.17)
281	17	16	16.00 (0.00)	0.47 (0.02)	62	32	31	31.00 (0.00)	2.45 (1.60)
301	16	15	15.00 (0.00)	1.49 (1.67)	65	30	29	29.00 (0.00)	0.84 (0.29)
324	15	14	14.00 (0.00)	0.44 (0.02)	68	28	27	27.00 (0.00)	0.55 (0.16)
351	13	13	13.00 (0.00)	0.44 (0.02)	71	27	26	26.00 (0.00)	0.51 (0.18)
Roszieg (X) 14	12	12	12.00 (0.00)	0.04 (0.01)	74	25	24	24.20 (0.42)	130.83 (118.83)
16	9	9	9.00 (0.00)	0.00 (0.00)	78	24	23	23.00 (0.00)	0.57 (0.02)
18	8	8	8.00 (0.00)	0.00 (0.00)	82	22	22	22.00 (0.00)	0.27 (0.02)
21	7	7	7.00 (0.00)	0.00 (0.00)	86	21	20	20.00 (0.00)	0.49 (0.24)
25	6	6	6.00 (0.00)	0.00 (0.00)	92	20	19	19.00 (0.00)	0.35 (0.12)
32	5	5	5.00 (0.00)	0.00 (0.00)	97	19	17	17.70 (0.48)	41.59 (70.78)
Sawyer (30) 25	16	16	16.00 (0.00)	0.10 (0.03)	104	17	16	16.00 (0.00)	0.54 (0.22)
27	14	14	14.00 (0.00)	0.10 (0.03)	111	16	16	16.00 (0.00)	0.00 (0.00)
30	13	13	13.00 (0.00)	0.10 (0.03)	Wee-Mag (75) 28	67	67	67.00 (0.00)	12.14 (7.73)
33	11	11	11.00 (0.00)	0.09 (0.03)	29	65	65	65.20 (0.42)	73.54 (95.21)
36	10	10	10.00 (0.00)	0.09 (0.04)	30	64	64	64.00 (0.00)	0.00 (0.00)
41	9	9	9.00 (0.00)	0.07 (0.01)	31	63	63	63.00 (0.00)	1.16 (0.04)
47	8	8	8.00 (0.00)	0.00 (0.00)	32	62	62	62.00 (0.00)	0.00 (0.00)
54	7	7	7.00 (0.00)	0.00 (0.00)	33	62	62	62.00 (0.00)	0.00 (0.00)
75	5	5	5.00 (0.00)	0.00 (0.00)	34	61	61	61.00 (0.00)	0.00 (0.00)
Tonge (70) 160	28	27	27.00 (0.00)	4.48 (3.35)	35	61	61	61.00 (0.00)	0.00 (0.00)
168	26	26	26.00 (0.00)	2.13 (1.42)	36	60	60	60.00 (0.00)	0.00 (0.00)
176	25	24	24.20 (0.42)	119.45 (103.50)	37	60	60	60.00 (0.00)	0.00 (0.00)
185	23	23	23.00 (0.00)	1.25 (1.39)	38	60	60	60.00 (0.00)	0.00 (0.00)
195	22	21	21.30 (0.48)	123.42 (121.85)	39	60	60	60.00 (0.00)	0.00 (0.00)
207	20	20	20.00 (0.00)	2.35 (0.93)	40	60	60	60.00 (0.00)	0.00 (0.00)
220	19	19	19.00 (0.00)	0.33 (0.11)	41	59	59	59.00 (0.00)	0.00 (0.00)
234	18	17	17.30 (0.48)	99.70 (115.25)	42	55	55	55.00 (0.00)	7.26 (3.95)
251	17	17	17.00 (0.00)	0.29 (0.01)	43	51	50	50.00 (0.00)	13.48 (17.21)
270	15	15	15.00 (0.00)	8.64 (7.43)	45	43	41	41.70 (0.48)	66.63 (101.47)
293	14	14	14.00 (0.00)	0.50 (0.09)	46	38	37	37.00 (0.00)	53.66 (57.22)
320	13	13	13.00 (0.00)	0.00 (0.00)	47	36	34	34.30 (0.48)	129.99 (128.18)
364	12	11	11.00 (0.00)	0.30 (0.20)	49	34	33	33.00 (0.00)	0.73 (0.23)
410	10	10	10.00 (0.00)	0.00 (0.00)	50	33	32	32.00 (0.00)	1.55 (1.07)
468	9	9	9.00 (0.00)	0.00 (0.00)	52	32	32	32.00 (0.00)	0.65 (0.02)
527	8	8	8.00 (0.00)	0.00 (0.00)	54	31	31	31.00 (0.00)	0.66 (0.02)
Warnecke (58) 54	40	39	39.00 (0.00)	5.10 (3.25)	56	30	30	30.00 (0.00)	6.09 (6.06)
56	37	36	36.00 (0.00)	1.47 (0.86)					

The results for the 243 remaining problem instances are shown in Tables 2 and 3. The format of these tables is the same as the format of Table 1, except for the fact that the column for Beam-ACO-1 is missing. The results show that Beam-ACO is able to improve in 102 cases (in addition to the 26 SCHOLL instances) over the results of ANTS. Only in one case—Warnecke, $c = a = 60$—the best solution found by Beam-ACO is 1 work station worse than the best solution found by ANTS. In the remaining cases Beam-ACO and ANTS found best solutions of the same quality. The obtained results show that Beam-ACO is clearly a current state-of-the-art algorithm for the TSALBP-1 problem.

Table 3. Results of Beam-ACO—in comparison to ANTS—when applied to 243 TSALBP-1 problem instances (part B)

Data set (size)	c	ANTS	Beam-ACO best	avg. (std)	avg. time (std)	Data set (size)	c	ANTS	Beam-ACO best	avg. (std)	avg. time (std)
Arcus1 (83)	5755	30	27	27.00 (0.00)	0.00 (0.00)	Hahn (X)	2004	9	9	9.00 (0.00)	0.00 (0.00)
	5785	30	27	27.00 (0.00)	0.00 (0.00)		2338	8	8	8.00 (0.00)	0.00 (0.00)
	6016	28	26	26.00 (0.00)	0.00 (0.00)		2806	6	6	6.00 (0.00)	0.00 (0.00)
	6267	26	25	25.00 (0.00)	0.00 (0.00)		3507	5	5	5.00 (0.00)	0.00 (0.00)
	6540	25	24	24.00 (0.00)	0.00 (0.00)		4676	4	4	4.00 (0.00)	0.00 (0.00)
	6837	24	23	23.00 (0.00)	0.00 (0.00)	Heskiaoff (28)	138	8	8	8.00 (0.00)	0.07 (0.01)
	7162	23	22	22.00 (0.00)	0.00 (0.00)		205	6	6	6.00 (0.00)	0.00 (0.00)
	7520	22	21	21.00 (0.00)	0.00 (0.00)		216	5	5	5.00 (0.00)	0.05 (0.00)
	7916	21	20	20.00 (0.00)	0.00 (0.00)		256	5	5	5.00 (0.00)	0.00 (0.00)
	8356	20	19	19.00 (0.00)	0.00 (0.00)		324	4	4	4.00 (0.00)	0.00 (0.00)
	8847	19	18	18.00 (0.00)	0.00 (0.00)		342	4	3	3.00 (0.00)	0.24 (0.29)
	9400	18	17	17.00 (0.00)	0.00 (0.00)	Jackson (11)	7	8	8	8.00 (0.00)	0.01 (0.00)
	10027	16	16	16.00 (0.00)	0.00 (0.00)		9	6	6	6.00 (0.00)	0.00 (0.00)
	10743	15	15	15.00 (0.00)	0.00 (0.00)		10	6	6	6.00 (0.00)	0.00 (0.00)
	11378	14	14	14.00 (0.00)	0.00 (0.00)		13	4	4	4.00 (0.00)	0.00 (0.00)
	11570	14	13	13.00 (0.00)	8.61 (15.40)		14	4	4	4.00 (0.00)	0.00 (0.00)
	17067	9	9	9.00 (0.00)	0.00 (0.00)		21	3	3	3.00 (0.00)	0.00 (0.00)
Arcus2 (111)	3786	25	24	24.00 (0.00)	7.70 (6.89)	Jaeschke (X)	6	9	9	9.00 (0.00)	0.00 (0.00)
	3985	23	22	22.00 (0.00)	0.83 (0.31)		7	7	7	7.00 (0.00)	0.00 (0.00)
	4206	22	21	21.00 (0.00)	0.40 (0.13)		8	7	7	7.00 (0.00)	0.00 (0.00)
	4454	21	20	20.00 (0.00)	0.61 (0.11)		10	4	4	4.00 (0.00)	0.00 (0.00)
	4732	19	19	19.00 (0.00)	0.31 (0.01)		18	3	3	3.00 (0.00)	0.00 (0.00)
	5048	18	18	18.00 (0.00)	0.29 (0.01)	Kilbridge (45)	56	11	11	11.00 (0.00)	0.21 (0.14)
	5408	17	16	16.00 (0.00)	2.21 (2.36)		57	11	11	11.00 (0.00)	0.12 (0.05)
	5824	15	15	15.00 (0.00)	0.00 (0.00)		62	10	10	10.00 (0.00)	0.00 (0.00)
	5853	15	15	15.00 (0.00)	0.00 (0.00)		69	9	9	9.00 (0.00)	0.00 (0.00)
	6309	14	14	14.00 (0.00)	0.00 (0.00)		79	8	8	8.00 (0.00)	0.00 (0.00)
	6842	13	13	13.00 (0.00)	0.00 (0.00)		92	7	7	7.00 (0.00)	0.00 (0.00)
	6883	13	13	13.00 (0.00)	0.00 (0.00)		110	6	6	6.00 (0.00)	0.00 (0.00)
	7571	12	11	11.00 (0.00)	0.49 (0.02)		111	6	5	5.00 (0.00)	1.35 (1.42)
	8412	10	10	10.00 (0.00)	0.00 (0.00)		138	5	5	5.00 (0.00)	0.00 (0.00)
	8898	10	9	9.30 (0.48)	95.67 (100.97)		184	4	3	3.00 (0.00)	3.87 (3.86)
	10816	8	8	8.00 (0.00)	0.00 (0.00)	Lutz1 (89)	1414	13	13	13.00 (0.00)	0.00 (0.00)
Barthold (148)	403	16	15	15.00 (0.00)	3.99 (2.38)		1572	12	12	12.00 (0.00)	0.06 (0.01)
	434	15	14	14.00 (0.00)	0.77 (0.03)		1768	10	10	10.00 (0.00)	0.05 (0.01)
	470	14	13	13.00 (0.00)	0.76 (0.04)		2020	9	9	9.00 (0.00)	0.00 (0.00)
	513	12	11	11.90 (0.32)	15.84 (47.71)		2357	8	8	8.00 (0.00)	0.00 (0.00)
	564	11	10	10.90 (0.32)	29.58 (93.52)		2828	6	6	6.00 (0.00)	0.00 (0.00)
	626	10	10	10.00 (0.00)	0.00 (0.00)	Lutz2 (89)	11	57	54	54.00 (0.00)	21.35 (23.71)
	705	9	8	8.30 (0.48)	97.10 (101.97)		12	54	52	52.00 (0.00)	4.68 (7.40)
	805	8	8	8.00 (0.00)	0.00 (0.00)		13	51	51	51.00 (0.00)	95.21 (85.48)
Barthol2 (148)	84	59	57	57.90 (0.32)	92.91 (79.46)		14	51	51	51.00 (0.00)	0.00 (0.00)
	85	58	56	56.00 (0.00)	66.67 (57.14)		15	42	40	40.00 (0.00)	2.26 (1.58)
	87	56	52	52.70 (0.48)	62.84 (110.83)		16	38	36	36.00 (0.00)	1.07 (0.68)
	89	54	51	51.00 (0.00)	8.56 (5.61)		17	35	34	34.00 (0.00)	0.69 (0.23)
	91	53	49	49.90 (0.32)	15.84 (28.29)		18	34	33	33.00 (0.00)	1.09 (0.04)
	93	52	48	48.00 (0.00)	32.21 (20.50)		19	33	33	33.00 (0.00)	0.60 (0.18)
	95	51	47	47.00 (0.00)	12.36 (11.89)		20	29	27	27.00 (0.00)	2.15 (2.08)
	97	49	46	46.00 (0.00)	4.72 (3.83)		21	27	26	26.00 (0.00)	1.43 (0.71)
	99	49	45	45.00 (0.00)	2.20 (1.20)	Lutz3 (89)	75	27	27	27.00 (0.00)	0.71 (0.29)
	101	47	44	44.00 (0.00)	3.12 (1.49)		79	26	24	24.00 (0.00)	6.73 (4.65)
	104	46	42	42.00 (0.00)	60.60 (33.18)		83	24	23	23.00 (0.00)	0.78 (0.81)
	106	45	41	41.40 (0.52)	118.76 (122.13)		87	23	22	22.00 (0.00)	0.83 (0.83)
	109	44	40	40.00 (0.00)	90.53 (50.49)		92	21	20	20.00 (0.00)	0.73 (0.53)
	112	43	39	39.00 (0.00)	22.43 (16.81)		97	20	19	19.00 (0.00)	5.68 (3.06)
	115	42	38	38.00 (0.00)	4.03 (1.91)		103	18	18	18.00 (0.00)	0.30 (0.03)
	118	40	37	37.00 (0.00)	7.59 (5.29)		110	17	17	17.00 (0.00)	0.31 (0.02)
	121	39	36	36.00 (0.00)	10.09 (10.27)		118	16	16	16.00 (0.00)	0.27 (0.01)
	125	38	35	35.00 (0.00)	2.89 (1.21)		127	15	15	15.00 (0.00)	0.23 (0.01)
	129	36	34	34.00 (0.00)	1.78 (0.73)		137	14	14	14.00 (0.00)	0.28 (0.01)
	133	36	33	33.00 (0.00)	1.43 (0.41)		150	12	12	12.00 (0.00)	0.25 (0.01)
	137	35	32	32.00 (0.00)	1.55 (0.57)	Mansoor (11)	48	5	5	5.00 (0.00)	0.00 (0.00)
	142	33	31	31.00 (0.00)	1.21 (0.03)		62	4	4	4.00 (0.00)	0.00 (0.00)
	146	32	30	30.00 (0.00)	1.21 (0.02)		94	3	3	3.00 (0.00)	0.00 (0.00)
	152	31	29	29.00 (0.00)	1.20 (0.01)	Mertens (X)	6	7	7	7.00 (0.00)	0.00 (0.00)
	157	30	28	28.00 (0.00)	1.17 (0.03)		7	7	7	7.00 (0.00)	0.00 (0.00)
	163	29	27	27.00 (0.00)	1.14 (0.02)		8	6	6	6.00 (0.00)	0.00 (0.00)
	170	28	26	26.00 (0.00)	1.14 (0.02)		10	3	3	3.00 (0.00)	0.00 (0.00)
Bowman (8)	20	6	6	6.00 (0.00)	0.00 (0.00)		15	2	2	2.00 (0.00)	0.00 (0.00)
Buxey (29)	27	14	14	14.00 (0.00)	0.07 (0.01)		18	2	2	2.00 (0.00)	0.00 (0.00)
	30	13	13	13.00 (0.00)	0.00 (0.00)	Mitchell (X)	14	10	10	10.00 (0.00)	0.00 (0.00)
	33	12	12	12.00 (0.00)	0.00 (0.00)		15	9	9	9.00 (0.00)	0.01 (0.00)
	36	11	11	11.00 (0.00)	0.15 (0.04)		21	6	6	6.00 (0.00)	0.00 (0.00)
	41	9	9	9.00 (0.00)	0.06 (0.00)		26	5	5	5.00 (0.00)	0.00 (0.00)
	47	8	8	8.00 (0.00)	0.00 (0.00)		35	4	4	4.00 (0.00)	0.00 (0.00)
	54	7	7	7.00 (0.00)	0.00 (0.00)		39	3	3	3.00 (0.00)	0.00 (0.00)
Gunther (35)	41	18	18	18.00 (0.00)	0.00 (0.00)	Mukherje (94)	176	27	27	27.00 (0.00)	0.74 (0.42)
	44	14	14	14.00 (0.00)	0.14 (0.03)		183	27	25	25.00 (0.00)	48.13 (39.36)
	49	12	12	12.00 (0.00)	0.08 (0.01)		192	26	24	24.00 (0.00)	4.32 (5.36)
	54	10	10	10.00 (0.00)	0.08 (0.01)		201	25	23	23.00 (0.00)	5.09 (4.91)
	61	9	9	9.00 (0.00)	0.00 (0.00)		211	23	22	22.30 (0.48)	83.74 (104.67)
	69	8	8	8.00 (0.00)	0.11 (0.03)		222	23	22	22.00 (0.00)	0.00 (0.00)
	81	7	7	7.00 (0.00)	0.00 (0.00)		234	21	20	20.00 (0.00)	0.50 (0.02)

5   Conclusions

In this work we have proposed an extended Beam-ACO algorithm—being a hybrid between ant colony optimization and beam search—for the TSALBP-1 problem. The results showed that our algorithm was able to improve the best known solutions in 128 out of 269 cases.

References

1. Bautista, J., Pereira, J.: Ant algorithms for assembly line balancing. In: Dorigo, M., Di Caro, G.A., Sampels, M. (eds.) Ant Algorithms 2002. LNCS, vol. 2463, pp. 65–75. Springer, Heidelberg (2002)
2. Bautista, J., Pereira, J.: Ant algorithms for a time and space constrained assembly line balancing problem. European Journal of Operational Research (2007)
3. Blum, C., Bautista, J., Pereira, J.: Beam-ACO applied to assembly line balancing. In: Dorigo, M., Gambardella, L.M., Birattari, M., Martinoli, A., Poli, R., Stützle, T. (eds.) ANTS 2006. LNCS, vol. 4150, pp. 96–107. Springer, Heidelberg (2006)
4. Blum, C., Dorigo, M.: The hyper-cube framework for ant colony optimization. IEEE Trans. on Systems, Man, and Cybernetics – Part B 34(2), 1161–1172 (2004)
5. Dorigo, M., Stützle, T.: Ant Colony Optimization. MIT Press, Cambridge (2004)
6. Gonçalves, J.F., de Almeida, J.R.: A hybrid genetic algorithm for assembly line balancing. Journal of Heuristics 8, 629–642 (2002)
7. Hoffmann, T.R.: EUREKA: A hybrid system for assembly line balancing. Management Science 38, 39–47 (1992)
8. Johnson, R.V.: Optimally balancing large assembly lines with "FABLE". Management Science 34, 240–253 (1988)
9. Lapierre, D.L., Ruiz, A., Soriano, P.: Balancing assembly lines with tabu search. European Journal of Operational Research 168, 826–837 (2006)
10. Merkle, D., Middendorf, M.: An ant algorithm with a new pheromone evaluation rule for total tardiness problems. In: Oates, M.J., Lanzi, P.L., Li, Y., Cagnoni, S., Corne, D.W., Fogarty, T.C., Poli, R., Smith, G.D. (eds.) EvoIASP 2000, EvoWorkshops 2000, EvoFlight 2000, EvoSCONDI 2000, EvoSTIM 2000, EvoTEL 2000, and EvoROB/EvoRobot 2000. LNCS, vol. 1803, pp. 287–296. Springer, Heidelberg (2000)
11. Ow, P.S., Morton, T.E.: Filtered beam search in scheduling. International Journal of Production Research 26, 297–307 (1988)
12. Scholl, A., Becker, C.: State-of-the-art exact and heuristic solution procedures for simple assembly line balancing. European Journal of Operational Research 168(3), 666–693 (2006)
13. Scholl, A., Klein, R.: SALOME: A bidirectional branch and bound procedure for assembly line balancing. INFORMS Journal on Computing 9, 319–334 (1997)
14. Scholl, A., Voss, S.: Simple assembly line balancing—Heuristic approaches. Journal of Heuristics 2, 217–244 (1996)
15. Sprecher, A.: Dynamic search tree decomposition for balancing assembly lines by parallel search. International Journal of Production Research 41, 1423–1430 (2003)
16. Talbot, F.B., Patterson, J.H., Gehrlein, J.H.: A comparative evaluation of heuristic line balancing techniques. Management Science 32, 430–454 (1986)
17. Vilarinho, P.M., Simaria, A.: A two-stage heuristic method for balancing mixed-model assembly lines with parallel workstations. International Journal of Production Research, 1405–1420 (2002)

Graph Colouring Heuristics Guided by Higher Order Graph Properties

István Juhos[1] and Jano van Hemert[2]

[1] Department of Computer Algorithms and Artificial Intelligence,
University of Szeged, Hungary
paper@juhos.info
[2] National e-Science Institute, University of Edinburgh, United Kingdom
jano@vanhemert.co.uk

Abstract. Graph vertex colouring can be defined in such a way where colour assignments are substituted by vertex contractions. We present various hyper-graph representations for the graph colouring problem all based on the approach where vertices are merged into groups. In this paper, we show this provides a uniform and compact way to define algorithms, both of a complete or a heuristic nature. Moreover, the representation provides information useful to guide algorithms during their search. In this paper we focus on the quality of solutions obtained by graph colouring heuristics that make use of higher order properties derived during the search. An evolutionary algorithm is used to search permutations of possible merge orderings.

1 Introduction

The *graph colouring problem* is an important problem from the class of non-deterministic polynomial problems. It has many applications in the real world such as scheduling, register allocation in compilers, frequency assignment and pattern matching. To allow these applications to handle larger problems, it is important fast algorithms are developed. Especially as high-performance computing is becoming more readily available, it is worthwhile to develop algorithms that can make use of parallelism. However, several exact end heuristic algorithms are developed to solve the graph vertex colouring problem, there is no best, they performance always depend on the problem. Here, we introduce some representation of the graph colouring and heuristics based on these representations which are suitable to handle larger graphs and support parallelism.

A graph $G = (V, E)$ consists of a set of vertices V and a set of edges $E \subseteq V \times V$ defines a relation over the set of vertices. We let $n = |V|$, $m = |E|$, then $d(v)$ denotes the degree of a vertex $v \in V$ and A is the adjacency matrix of G. A colouring c of a graph is a map of colours to vertices ($c : V \to C$) where C is the set of colours used. There can be several such mappings, which can be denoted if necessary (e.g. $c_1, c_2, \dots$). A colouring c is a k-colouring iff $|C| = k$. It is a proper or valid colouring if for every $(v, w) \in E : c(v) \neq c(w)$. In general, when we refer to a colouring we mean a valid colouring, unless the context indicates

J. van Hemert and C. Cotta (Eds.): EvoCOP 2008, LNCS 4972, pp. 97–109, 2008.

otherwise. The chromatic number $\chi(G)$ of a graph G is the smallest k for which there exists a k-colouring of G. A colouring algorithm makes colouring steps, i.e., it progressively chooses an uncoloured vertex and then assignes it a colour. Let $t \in \{1, \ldots, n\}$ be the number of steps made.

The graph colouring approaches discussed here will construct a colouring for a graph by progressively contracting the graph. This condensed graph is then coloured. Moreover, they allow us to extract heuristic information to better guide the search. We implement a number of heuristics and combine these with particular sequential contraction algorithms. The sequence in which nodes and multiple-nodes are merged defines a search space in terms of permutations. Using an evolutionary algorithm we search that space.

2   Representing Solutions to Graph Colouring

We define merge operations to perform contraction of the original graph and subsequent contractions. A merge operation takes two unconnected vertices from a graph $G = (V, E)$ and produces a new graph $G' = (V', E')$ where these vertices become one hyper-vertex. If edges exist between another vertex and both the original vertices, then these become one multiple-edge. If $v_1, v_2 \in V$ are merged to $\{v_1, v_2\} \in V'$ and both $(v_1, u), (v_2, u) \in E$ then $(\{v_1, v_2\}, u) \in E'$ is called a multiple-edge. If only one of (v_1, u) or $(v_2, u) \in E$ then we keep calling that one an edge. Examples of merge operations are shown in Figure 1. The merge operation is applied similarly to hyper-vertices.

By repeating merge operations we will end up with a complete graph. If during each merge operation we ensure only hyper-vertices and vertices are merged that have no edges between them, we then can assign all the vertices from the original graph that are merged into the same hyper-vertex one unique colour. The number of (hyper-)vertices in the final contracted graph corresponds to the number of colours in a valid colouring of the original graph. As Figure 1 shows, the order in which one attempts to merge vertices will determine the final colouring. Different colourings may use a different number of colours. Important to note is, if two vertices need to be coloured differently in every optimal colouring, merging them will prevent reaching an optimal colouring. We will investigate a number of different strategies for making choices about which vertices to merge. Note, the colouring in the example is only to follow the process. In reality, the actual assigning of colours happens after the whole merge process finishes.

Graph colouring solvers make use of constraint checks during colouring. The adjacency checks to verify if assigned colourings are valid play a key role in the overall performance (see [1,2]). The number of checks depends on the representation of the solution, the algorithm using the representation and the intrinsic difficulty of the problem. Graph colouring problem are known to exhibit an increase in difficulty in the so-called phase transition area [3]. In our experiments we will test several algorithms on problems in this area. The area is defined by the edge density of the graph where by increasing the density we arrive at the phase transition when problems, given a certain k, become unsolvable. It is

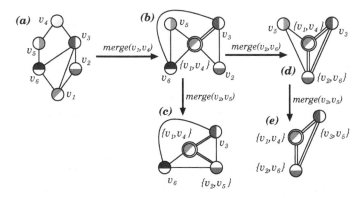

Fig. 1. Demonstration of the result of two different merge orders of rows: $P_1 = v_1, v_4, v_2, v_5, v_3, v_6$ and $P_2 = v_1, v_4, v_2, v_6, v_3, v_5$. The double-lined edges are multiple-edges and double-lined nodes are multiple-nodes. The P_1 yields a 4-colouring (c) while P_2 achieves a 3-colouring (e).

known, this point coincides with a significant increase in effort required to solve these problem instances.

To support the merge operations introduced above we define several data structures, each of which has its benefits when combined with different approaches to deciding in which order to perform the merge operations. We call these data structures the first order structures. Beside supporting the merge operations, these first order structures allow information to be derived that can aid in guiding search algorithms. That information will be stored in secondary order structures. Third order structures can be derived by summarising the second order structures.

First order structures are the cells of the representation matrices. They define the neighbourhood relation of the hyper-vertices for the binary and weighted relation for the integer models. Figure 2 shows four different representations in relation to an example graph and its adjacency matrix. An r_i refer to an appropriate row of the representation matrices describe relations of the normal or merged vertices. The *Binary Merge Table* (BMT) simply keeps track of the edges after merges by collapsing rows and keeping 1s from all the rows. The *Integer Merge Table* (IMT) also counts the number of edges that were collapsed. The *Binary Merge Square* (BMS) and *Integer Merge Square* (IMS) are similar to their table variants, but the merge operations become more efficient over time as columns belonging to the collapsed edges are discarded.

Secondary order structures are the summary of the first order structures, i.e., the rows and columns in the representation matrices respectively. They form four vectors. Since, sequential colouring algorithms take steps, and the coloured and uncoloured part of the graphs are changing step-by-step it is worth to define these structures separately to the coloured/merged and uncoloured/non-merged sub-graphs. To identify these partial sums we use ^{col} and ^{unc} superscripts. We can obtain the sum of the rows and columns of binary merge matrices from

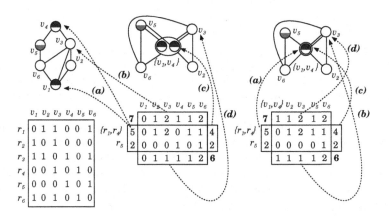

(a) Example graph

	$\{v_1, v_4\}$	v_2	v_3	v_5	v_6
$\{r_1, r_4\}$	0	1	1	1	1
r_2	1	0	1	0	0
r_3	1	1	0	0	1
r_5	1	0	0	0	1
r_6	1	0	1	1	0

(b) Binary Merge Square

	v_1	v_2	v_3	v_4	v_5	v_6
$\{r_1, r_4\}$	0	1	1	0	1	1
r_2	1	0	1	0	0	0
r_3	1	1	0	1	0	1
r_5	0	0	0	1	0	1
r_6	1	0	1	0	1	0

(c) Binary Merge Table

	v_1	v_2	v_3	v_4	v_5	v_6
r_1	0	1	1	0	0	1
r_2	1	0	1	0	0	0
r_3	1	1	0	1	0	1
r_4	0	0	1	0	1	0
r_5	0	0	0	1	0	1
r_6	1	0	1	0	1	0

(d) Adjacency matrix

	$\{v_1, v_4\}$	v_2	v_3	v_5	v_6
$\{r_1, r_4\}$	0	1	2	1	1
r_2	1	0	1	0	0
r_3	2	1	0	0	1
r_5	1	0	0	0	1
r_6	1	0	1	1	0

(e) Integer Merge Square

	v_1	v_2	v_3	v_4	v_5	v_6
$\{r_1, r_4\}$	0	1	2	0	1	1
r_2	1	0	1	0	0	0
r_3	1	1	0	1	0	1
r_5	0	0	0	1	0	1
r_6	1	0	1	0	1	0

(f) Integer Merge Table

Fig. 2. An example graph (a) with its adjacency representation (d) and four different representations shown after perform the same merge (rows r_1 and r_4 are merged)

their integer pairs by counting their non-zero elements. Hence, the example in Figure 3 shows only the left side of the sub-matrices sum the columns and the right side sum the non-zero elements to get the relevant summary in the binary matrix. This is the same for the columns, where the top vector is the sum of the rows and the bottom is the number of non-zeros. The second order structures

Fig. 3. The original graph, its induced sub-IMT and then its induced sub-IMS when colouring is in progress. (a) gives the sum of the degree of the nodes, (b) gives the number adjacent vertices already assigned a colour, (c) gives the degree of the hyper-vertex, and (d) gives the number of coloured multiple-edges.

are denoted by τ, using $_{t,b,l,r}$ indices as subscript to refer to the top, bottom, left and right vector.

Third order structures are formed by summarising the secondary order structures. These can be divided into two parts similar to the second order structures according to the coloured and uncoloured sub-graphs. These structures are denoted by ζ. In this study, they will be used in the fitness function of the evolutionary algorithm. The top-left sums the top vector (or the left vector) and the bottom-right sums the bottom vector (or the right vector). These are shown in bold font in Figure 3.

3   Combinations of Contraction Algorithms and Heuristics

Before we introduce heuristics, we first define three abstract sequential algorithms, which allow for a concise defining of existing or novel solvers.

Definition 1 (Sequential contraction algorithm 1 (SCA1))

1. *Choose a non-merged/uncoloured vertex v*
2. *Choose a non-neighbour hyper-vertex/colour-set r to v*
3. *Merge v to r*
4. *If there exists a non-merged vertex then continue with Step 1.*

Definition 2 (Sequential contraction algorithm 2 (SCA2))

1. *Choose a hyper-vertex/colour-set r*
2. *Choose a non-neighbor non-merged/uncoloured vertex v to r*
3. *Merge v to r*
4. *If there exists a non-merged vertex then continue with Step 1.*

Definition 3 (Sequential contraction algorithm 3 (SCA3))

1. *Choose two vertices/colour-sets (hyper or normal) that can be merged without violating any constraints*
2. *Merge them*
3. *If there exist more hyper-vertices or vertices to merge then continue with Step 1.*

We describe four heuristics; the well known ERDŐS and DSATUR heuristics make use of the secondary order structures, whereas our other two, DOTPROD and COS, make use of the first order information. The ERDŐS heuristic uses similar assumptions as DSATUR but in the opposite direction.

DSatur is of type SCA1. Here the choice of colouring the next uncoloured vertex v is determined by colour saturation degree. As colour saturation is calculated by observing the colours of neighbouring vertices, it requires $\mathcal{O}(n^2)$ constraint checks. However, if we merge the already coloured vertices and use a second

order structure, we have this information by observing at most the number of
hyper-vertices/colours for v. Hence, $\mathcal{O}(nk_t)$ constraint checks are required, where
k_t is the number of colours used in step t. The bottom second order structure
provide the saturation degree, which gives $\mathcal{O}(n)$ computational effort to find the
largest one. Hence, IMT is an appropriate structure for the definition. Here, the
choice for hyper-vertex/colour-set is done in a greedy manner.

Definition 4 (Non-merged/uncoloured vertex choice of the DSatur$_{imt}$)

1. *Find those non-merged/uncoloured vertices which have the highest saturated
   degree:* $S = \arg\max_u \tau_b^{col}(u) : u \in V^{unc}$
2. *Choose those vertices from S that have the highest non-merged/uncoloured-
   degree:* $N = \arg\max_v \tau_b^{unc}(v)$
3. *Choose the first vertex from the set N.*

DotProd is based on a novel technique, which is shown to be useful in [4]. Two
vertices of a contracted graph are compared by consulting the corresponding
BMS rows. The dot product of these rows gives a valuable measurement for the
common edges in the graph. Application of the DOTPROD heuristics to the BMS
representation using SCA2 colouring scheme provides the *Recursive Largest First*
(RLF) algorithm of Dutton and Birgham [5]. Unfortunately the name RLF is
somewhat misleading, since "largest first" does not say where it is relating to.
The meaning of it differs throughout the literature, but we shall define it exactly
and then generalise this heuristics. The identification schemes introduced provide a way for the classification of other heuristics with the possibility for well
described comparisons. Here, we introduce a DOTPROD heuristic that is combined with the BMS representation and the SCA3 type algorithm. We explore
here only this combination, but other combinations are possible.

**Definition 5 (Non-merged/uncoloured vertex choice of the
DotProd$_{bms}$)**

1. *Find those two vertices (hyper or normal) which have the largest number of
   common neighbours:* $S = \arg\max_{u,v} \langle u, v \rangle : u, v \in V(G^t)$
2. *Choose the first vertex pair from the set S.*

Cos is the second novel heuristic introduced here. It is derived from DOTPROD
by normalisation of the dot product. As opposed to DOTPROD, COS takes in
consideration the number of non-common neighbours as well. In the following
definition we provide an algorithm of type SCA3 that uses the COS heuristics
to choose the next vector.

Definition 6 (Non-merged/uncoloured vertex choice of the Cos$_{bms}$)

1. *Find those vertices (hyper or normal) that have the largest number of com-
   mon neighbours, and that have the fewest constraints:* $S = \arg\max_{u,v} \frac{\langle u,v \rangle}{\|u\|\|v\|}$
2. *Choose the first vertex pair from the set S.*

Pál Erdős $O(n/\log n)$ [6, page 245] works as follows. Take the first colour and assign it to the vertex v that has the minimum degree. Vertex v and its neighbours are removed from the graph. Continue the algorithm in the remaining sub-graph in the same fashion until the sub-graph becomes empty, then take the next colour and use the algorithm for the non-coloured vertices and so on until each vertex is assigned a colour. This approach guarantees $\mathcal{O}(n/log_\chi(n))$ number of colours in the worst case. However, exact algorithms has proved limits for the number of colours used in a colouring which makes the exact analysis possible, other algorithms without such a limit can shows better performance in several cases. All representations are suitable as a basis for this heuristic. It uses SCA2, where the choice for the next target r-th (hyper-)vertex/colour-set for merging and colouring is greedy.

Definition 7 (Non-merged/uncoloured vertex choice of the Erdős$_{bmt}$)

1. *Choose an uncoloured vertex with minimum uncoloured degree. $S = \arg\min_u \tau_b^{unc}(u)$*
2. *Choose the first vertex from the set S.*

4   Guiding Graph Colouring by Graph Properties

The DOTPROD and COS heuristics are perfectly suitable to implement SCA3, as well as SCA1 and SCA2. Given a merge representation of a graph, we have to select the two rows of all rows with the maximum DOTPROD or COS value. These two rows are then merged and the procedure continues on the merged representation matrix in the same fashion. The number of rows decreases by one in each merge operation, hence the dot products of rows in the t-th step forms a $(n - t) \times (n - t)$ matrix; this is also known as the Gram matrix. In case of COS, the Gram matrix becomes the metric tensor of the rows. Since min or max places are not necessarily unique in the Gram matrices, we need to introduce a further strategy to select one of them. Here, we use the easiest approach, which is to take the first one found. This approach uses vertex properties. In other words local information is used to decide on merges, which then changes graph properties globally.

SCA3 can be defined in an indirect way, where certain graph properties are evaluated during the selection of two rows for a merge operation. Here, an analysis of a supposed merge effect is performed. First, gather these graph properties into a vector, denote these by ξ. Determine which values are known in advance for the final merged graph. It is important to know these, because they will be the goal of the reformulated problem. Then, compute $\xi(G^0)$ and $\xi(G^{n-k})$. Now, the "only" task left is to find a path from $\xi(G^0)$ to $\xi(G^{n-k})$ along the merge steps in the vector space induced by $\xi(G^t)$.

Figure 4(a) shows an example of how the three largest eigenvalues ($\lambda_3 <= \lambda_2 <= \lambda_1$) of A^t form different paths of 20- and 37-colourings in a three dimensional vector space. The start of the path is $(\lambda_3, \lambda_2, \lambda_1)[A^0]$. The path ends at $(\lambda_3, \lambda_2, \lambda_1)[A^{n-k}]$, which are known values. The first value is trivial, because

$A^0 = A$ is given and the last is $(-1, -1, k - 1)$, since the final merged graph G^{n-k} is a K_k complete graph on k nodes for BMS. Where k is the number of colours used in the colouring. Hence, the goal is to reach $(-1, -1, \chi - 1)$, which corresponds to a solution. Analysis of the paths results guidance of colouring process by following the optimal path.

Consider a simplified example in one dimensional space. Take the BMS representation and let $\xi(G^t) = \lambda_1(A^t)$, i.e.,the spectral radius or spectral norm of A^t. If we examine the initial BMS, we can verify that $\lambda_1(A^0)$ is greater or equal than $\lambda_1(A^{n-\chi}) = \chi - 1$ (see [7]). Due to this fact $\lambda_1(A^t)$ is decreasing, as is shown in Figure 4(b).

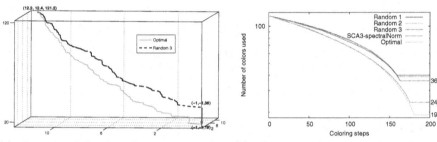

(a) 3D paths of the three largest eigenvalues (b) The spectral norm deepest descent implementation (SCA3-spectralNorm)

Fig. 4. Example of spectral properties of a graph during colouring; all while solving an equi-partite 20-colourable graph in the peak of the phase transition ($p = 0.64$)

This path is responsible for determining the colouring and the end of the path $k - 1$ defines the quality of the colouring. Unfortunately, the ideal path (between $\lambda_1(A^0)$ and $\chi - 1$) is of course unknown; the task of colouring is to find this path. A colouring path takes $n - k$ colouring steps, that is merges. An optimal colouring reaches χ, the smallest k-value, resulting the longest step-series, non-optimal colourings have shorter one as they get stuck when no further merging is possible, i.e., $k > \chi$. If we take into account the ideal path has the lowest end and takes the most steps, then it should be below all possible non-optimal colouring paths from a certain point. We can define a path in advance that follow this property. A trivial path between the initial point and the end is a linear path, where $\lambda_1(A^{t+1})$ is derived from $\lambda_1(A^t)$. The difference $\lambda_1(A^{t+1}) - \lambda_1(A^t)$ should approximate $\frac{\lambda_1(A^0) - \lambda_1(A^{n-\chi})}{n-\chi}$. Non-linear paths can be defined by analysis of more complicated graph properties and their behaviours in terms of spectrals.

To implement SCA3 in an intuitive way we need to re-compute the graph properties $\binom{n-t}{2}$ times in the t-th merging step. This number can be reduced considerably by updating only where possible. For the spectral norm this update, which consists of calculating the spectral norm from the original matrix, is not available in general. However, efficient approximations of the spectral norm for symmetric matrices exist [8], for which the updating is feasible. We shall use the method suggested in [8]: Equation 2.4 will be denoted by *Approx*$_1$. As the optimal path has the lowest end, $\chi - 1$, the aim is to reduce the approximated spectral norm in each step as much as possible.

5   An Evolutionary Algorithm Based on Merge Models

We embed the heuristics in an evolutionary algorithm (EA). The EA uses the BMT representation and the SCA1 contraction scheme. The genotype is a permutation of the rows of the BMT. The phenotype is a final BMT, where no rows can be merged further. Selection of which rows to merge is by an initial random ordering of all vertices. The strategy for the merging of selected vertices is one of the DOTPROD and COS local heuristics.

An intuitive way of measuring the quality of an individual (permutation) p in the population is by counting the number of rows remaining in the final BMT. This equals to the number of colours $k(p)$ used in the colouring of the graph, which needs to be minimised. When we know the optimal colouring is χ (in our experiments χ is defined in advance to verify quality), we can normalise the fitness function to $g(p) = k(p) - \chi$. This function gives a rather low diversity of fitnesses of the individuals in a population because it cannot distinguish between two individuals that use an equal number of colours. This problem is called the fitness granularity problem. We modify the fitness function to allow the use of first and second order structures introduced in Section 3.

The fitness relies on the heuristic that one generally wants to avoid highly constraint vertices and rows in order to have a higher chance of successful merges at a later stage, commonly called a succeed-first strategy. It works as follows. After the last merge the final BMT defines the groups of vertices with the same colour. There are $k(p) - \chi$ over-coloured vertices, i.e., merged rows. Generally, we use the indices of the over-coloured vertices to calculate the number of vertices that need to be minimized (see $g(p)$ above). But these vertices are not necessarily responsible for the over-coloured graph. Therefore, we choose to count the hyper-vertices that violate the fewest constraints in the final hyper-graph. To cope better with the fitness granularity problem we should modify $g(p)$ according to the constraints of the over-coloured vertices discussed previously. The fitness function used in the EA is then defined as follows. Let $\zeta^{unc}(p)$ denote the number of constraints, i.e., non-zero elements, in the rows of the final BMT that belong to the over-coloured vertices, i.e., the sum of the smallest $k(p) - \chi$ values of the right second order structure of the uncoloured vertices. This is the uncoloured portion of the (right-bottom) third order structure. The fitness function becomes $f(p) = g(p)\zeta^{unc}(p)$. Here, the cardinality of the problem is known, and used as a termination criterium ($f(p) = 0$) to determine the efficiency of the algorithm. For the case where we do not know the cardinality of the problem, this approach can be used by leaving out the normalisation step.

Below we show the outline of the evolutionary algorithm. It uses a generational model with 2-tournament selection and replacement, where it employs elitism of size one. This setting is used in all experiments. The initial population is created with 100 random individuals. Two variation operators are used to provide offspring. First, the 2-point order-based crossover (OX2) [9, in Section C3.3.3.1] is applied. Second, a simple swap mutation operator, which selects at random two different items in the permutation and then swaps. The probability of using OX2 is set to 0.4 and the probability for using the simple swap mutation is set to 0.6. These parameter settings are taken from previous experiments [2].

Definition 8 (Evolutionary Algorithm)

1. population = generate initial permutations randomly
2. repeat
   - evaluate each permutation p:
   -- merge $p_j - th$ unmerged vertex v into hyper-vertex r by DOTPROD or COS
   -- calculate $f(p) = (k(p) - \chi)\zeta^{unc}(p)$
   - population$_{xover}$ = xover(population, $prob_{xover}$)
   - population$_{mut}$ = mutate(population$_{xover}$, $prob_{mut}$))
   - population = select$_{2\text{-}tour}$(population $\cup$ population$_{xover}$ $\cup$ population$_{mut}$)
3. until termination condition.

The Erdős heuristic guarantees its performance. We omit this heuristics from the EA, as we would not be able to guarantee this property once embedded in the EA. A baseline version of the EA called EA-noheur serves a basis of the comparison and the DSATUR with backtracking as it is a commonly used reference method. Moreover, as this algorithm performs an exhaustive search it is useful to find the optimal solutions to problem instances, except for ones with a large k and with a graph density that positions them in the phase transition.

6   Experiments and Results

The test suites are generated using the well known graph k-colouring generator of Culberson [10]. It consists of k-colourable graphs with 200 vertices, where k is set to 3,5,10 and 20. For $k = 20$, ten vertices will form a colour set, therefore we do not use any larger chromatic number. The edge density of the graphs is varied in a region called the phase transition. This is where hard to solve problem instances are generally found[1], which is observed in the results as a typical easy-hard-easy pattern. The graphs are all generated to be equi-partite, which means that a solution should use each colour approximately as much as any other. The suite consists of groups where each group is a k-colouring with 10 unique instances. Using this test set we can ensure a fair comparison of the algorithms, since this set contains problems from easy to most difficult ones. Moreover, we would like to avoid comparison on some selected real life problems where the selection method can determine the outcome of the comparison of the performance, please see [11].

---

[1] Hardness can only be determined as a function of problem and algorithm; here we refer to the large amount of empirical evidence reported about these regions.

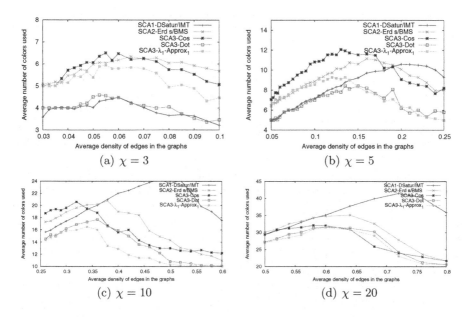

Fig. 5. Results of the sequential colouring heuristics; average number of colours through the phase transition

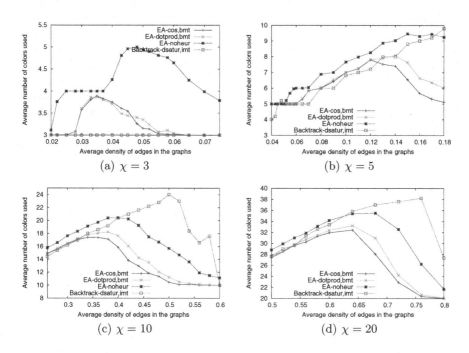

Fig. 6. Results of the EA; average number of colours through the phase transition

We compare three different heuristics embedded in SCA3 with each other with the two reference heuristics of SCA1 and SCA2, a DSATUR and the Erdős heuristic. The three combinations for SCA3 are max DOT, max COS and min λ_1-Approx$_1$.

Figure 5 shows the results of all combinations for different values of χ. The implementations of the linear approximations, which result in the deepest descent SCA3-λ_1 algorithms, perform the best except for very sparse graphs, where the DOT and DSATUR heuristic with local decisions performs better. The reason for the worse performance of the SCA3-λ_1 on sparse graphs is the low number of changes in the norm in the selection of candidate vertices pairs for merging. Because of the approximation used, several different values become equal, hence selecting too many candidates. The combination of SCA3 with COS does not always perform well, especially not for smaller chromatic numbers; however it can outperform baseline methods for dense graphs. As the chromatic number and the edge density increase COS gets better and can beat all other. DOTPROD's performance lies between that of COS and the SCA3-λ_1 algorithms; its strength lies with smaller chromatic numbers and sparse graphs.

While sequential algorithms make one run to get their result, EA experiments are performed several times due to the random nature of the algorithm. To cope with the increase in runs, a reduced number of instances is selected from the previous experiments. One set consists of five instances, except for 3 and 5 colourable instances where the set contains ten instances to allow for the low diversity in results for small chromatic numbers. On each instance we perform ten independent runs and calculate averages over the total number of runs.

Figure 6 shows the results for the EA with two different heuristics. Also shown are results for the reference EA without heuristics and the DSATUR algorithm. Similar to the previous experiments, the COS heuristics performs well, especially for larger k, and the DOTPROD is a close second. DSATUR is the strongest algorithm on 3-colourable graphs, where it always finds the optimum number of colours. However, the backtracking can help on very sparse graphs, DSatur quickly gets the last position as the chromatic number and hence the edge density increasing. During the experiments our methods remained below the Erdős heuristic which has a guaranteed performance in both experiments.

7   Conclusions

In this paper, we used compact representations for intermediate solutions to graph colouring problems that facilitate contractions of the graph. Together with the merge operations, they form a good basis for developing efficient graph colouring algorithms because of three beneficial properties: a significant reduction in constraint checks, access to heuristics that help guide the search, and a concise description of algorithms.

We created several combinations of sequential contraction algorithms that determine the order of merge operations with heuristics that rely on information derived from the compact representations. Furthermore, we incorporated these heuristics in an EA. We showed these heuristics are highly successful in achieving near optimal solutions to graph colouring instances around the phase transition, i.e., where difficult to solve problem instances are expected to occur. Compared to the method DSATUR, the evolutionary algorithm performs well on problems with larger k.

References

1. Craenen, B., Eiben, A., van Hemert, J.: Comparing evolutionary algorithms on binary constraint satisfaction problems. IEEE Transactions on Evolutionary Computation 7, 424–444 (2003)
2. Juhos, I., van Hemert, J.: Improving graph colouring algorithms and heuristics by a novel representation. In: Proceeding of the Evolutionary Computation in Combinatorial Optimization, 6th European Conference, pp. 123–134 (2006)
3. Culberson, J., Gent, I.: Frozen development in graph coloring. Theor. Comput. Sci. 265, 227–264 (2001)
4. Juhos, I., van Hemert, J.: Increasing the efficiency of graph colouring algorithms with a representation based on vector operations. Journal of Software 1, 24–33 (2006)
5. Dutton, R.D., Brigham, R.C.: A new graph coloring algorithm. Computer Journal 24, 85–86 (1981)
6. Graham, R.L., Grötschel, M., Lovász, L. (eds.): Handbook of combinatorics, vol. 1. MIT Press, Cambridge (1995)
7. Wilf, H.S.: The eigenvalues of a graph and its chromatic number, J. london. London Math. Soc., 330–332 (1967)
8. Merikoski, J.K., Kumar, R.: Lower bounds for the spectral norm. Journal of Inequalities in Pure and Applied Mathematics 6 (2005)
9. Bäck, T., Fogel, D., Michalewicz, Z. (eds.): Handbook of Evolutionary Computation. Institute of Physics Publishing Ltd, Bristol and Oxford University Press, Oxford (1997)
10. Culberson, J.: Iterated greedy graph coloring and the difficulty landscape. Technical Report TR 92-07, University of Alberta, Dept. of Computing Science (1992)
11. Coudert, O.: Exact coloring of real-life graphs is easy. In: DAC 1997: Proceedings of the 34th annual conference on Design automation, pp. 121–126. ACM, New York (1997)

A Hybrid Column Generation Approach for the Berth Allocation Problem

Geraldo R. Mauri[1,3], Alexandre C.M. Oliveira[2], and Luiz A.N. Lorena[3,*]

[1] Federal University of Espírito Santo - UFES, Brazil
[2] Federal University of Maranhão - UFMA, Brazil
[3] National Institute for Space Research - INPE, Brazil
mauri@lac.inpe.br,acmo@deinf.ufma.br,lorena@lac.inpe.br

Abstract. The Berth Allocation Problem (BAP) consists on programming and allocating ships to berthing areas along a quay. The BAP is modeled as a vehicle routing problem and a recently proposed evolutionary hybrid method denominated PTA/LP is used to solve it. The PTA/LP combines the Population Training Algorithm with Linear Programming to generate improving incoming columns in a column generation process. The computational results are obtained for a set of instances proposed in literature and new best known solutions are presented.

1 Introduction

The programming and allocation of ships to berths have a primary impact in the efficiency of the port operations [1]. A discussion about the decision problems that appear in a port is presented in [2].

The Berth Allocation Problem - BAP consists of optimally assigning ships to berthing areas along a quay in a port. The main decision to be made in that process accomplishes the choice of "where" and "when" the ships shall berth [3]. Managers want to minimize both port and user costs, which are related to the ships' service time. The BAP objective is usually to minimize the total service time of all ships.

The BAP can be modeled as a discrete problem considering the quay as a finite set of berths. In this case, the berths can be described as fixed length segments, or points if the spatial dimension is ignored [1,3]. Continuous models consider that ships can berth anywhere along the quay, where ships are of different lengths and the quay capacity varies dynamically.

In this paper, the problem is treated in discrete form considering the minimization of the time spent by ships arriving in a port, allocating and programming the ships mooring to berths, aiming to reduce the permanence time for ships inside the port. The remainder of the paper is organized as follows. Section 2 presents a brief literature review. The problem modeling is presented in Section 3. Section 4 describes the proposed model and the methods used to solve the BAP. Computational results are presented in Section 5, and the conclusions are summarized in Section 6.

---

* The authors acknowledge FAPESP and CNPq by partial research support.

J. van Hemert and C. Cotta (Eds.): EvoCOP 2008, LNCS 4972, pp. 110–122, 2008.

2   Literature Review

Cordeau et al. [3] presents a Tabu Search based heuristic to solve two different models for a discrete case of BAP. Only small instances could be solved optimally and the proposed Tabu Search always yields an optimal solution. The proposed heuristics could handle the various features of real-life problems, including time windows and favorite and acceptable berthing areas. The objective function could easily accommodate a weighted sum of the ship's service times.

Filho and Lorena [4] applied a heuristic column generation approach to graph coloring. They describe the principles of their Constructive Genetic Algorithm (CGA) and give a column generation formulation for the problem. The CGA is used to generate the initial columns and also to solve the sub-problems. The column generation is performed as long as the CGA finds columns with negative reduced costs. The master problem is solved by CPLEX [5].

Recently, Puchinger and Raidl [6] proposed new integer linear programming formulations for the three-stage two-dimensional bin packing problem. Based on these formulations, a branch-and-price algorithm was developed with a fast column generation performed by applying a hierarchy of four methods: a greedy heuristic, an evolutionary algorithm, a restricted pricing problem using CPLEX, and finally the complete pricing problem also using CPLEX.

3   BAP Modeling

This work considers the Berth Allocation Problem (BAP) modeled as a Multi-Depot Vehicle Routing Problem with Time Windows (MDVRPTW) [7,3] (discrete formulation). The ships are seen as customers, and the berths as depots at which one vehicle is located. There are m "vehicles", one for each depot, and each vehicle starts and finishes its tour at its depot. The ships are modeled as vertices in a multi-graph and every depot is divided into an origin and a destination vertices. The time windows can be imposed on every vertex, and its correspond to the availability period of the berth at the origin and destination vertices.

The model is given by a multi-graph $G^k = (V^k, A^k)$, $\forall k \in M$ where $V^k = N \cup \{o(k), d(k)\}$ and $A^k \subseteq V^k$ x V^k. The input data are given by:

- N: set of ships, n = $|N|$;
- M: set of berths, m = $|M|$;
- t_i^k: handling time of ship i at berth k;
- a_i: arrival time of ship i;
- s^k: start of availability time of berth k;
- e^k: end of availability time of berth k;
- b_i: upper bound for service time window for ship i;
- v_i: the value (cost) of service time for ship i.

The model variables are:

- $x_{ij}^k \in \{0,1\}$, $k \in M$, $(i,j) \in A^k$; $x_{ij}^k = 1$ if the ship j is scheduled after ship i at berth k;
- T_i^k, $k \in M$, $i \in N$: is the berthing time of ship i at berth k;
- $T_{o(k)}^k$, $k \in M$: is the starting operation time of berth k (the time when the first ship moors at the berth);
- $T_{d(k)}^k$, $k \in M$: is the ending operation time of berth k (the time when the last ship departs from the berth);
- $M_{ij}^k = \max\{b_i + t_i^k - a_j, 0\}$, $k \in M$, i and $j \in N$.

The BAP model is as follows:

Minimize:

$$Z = \sum_{i \in N} \sum_{k \in M} v_i \left[T_i^k - a_i + t_i^k \sum_{j \in N \cup \{d(k)\}} x_{ij}^k \right] \tag{1}$$

Subject to:

$$\sum_{k \in M} \sum_{j \in N \cup \{d(k)\}} x_{ij}^k = 1 \quad \forall i \in N \tag{2}$$

$$\sum_{j \in N \cup \{d(k)\}} x_{o(k)j}^k = 1 \quad \forall k \in M \tag{3}$$

$$\sum_{i \in N \cup \{o(k)\}} x_{i,d(k)}^k = 1 \quad \forall k \in M \tag{4}$$

$$\sum_{j \in N \cup \{d(k)\}} x_{i,j}^k - \sum_{j \in N \cup \{o(k)\}} x_{j,i}^k = 0 \quad \forall k \in M, \forall i \in N \tag{5}$$

$$T_i^k + t_i^k - T_j^k \leq (1 - x_{i,j}^k) M_{i,j}^k \quad \forall k \in M, \forall (i,j) \in A^k \tag{6}$$

$$T_i^k \geq a_i \quad \forall k \in M, \forall i \in N \tag{7}$$

$$T_i^k + t_i^k \sum_{j \in N \cup \{d(k)\}} x_{j,i}^k \leq b_i \quad \forall k \in M, \forall i \in N \tag{8}$$

$$T_{o(k)}^k \geq s^k \quad \forall k \in M \tag{9}$$

$$T_{d(k)}^k \leq e^k \quad \forall k \in M \tag{10}$$

$$x_{i,j}^k \in \{0,1\} \quad \forall k \in M, \forall (i,j) \in A^k \tag{11}$$

The objective function minimizes the elapsed time since the ships incoming, mooring and handling, considering a respective service cost. Constraints (2) state that each ship is only once assigned to one berth. Constraints (3) and (4) ensure that a ship will be the first handling by each berth and another will be the last. The flow conservation is given by constraint (5) and constraint (6) calculates the ships berthing time. Only the valid arches A^k ($\forall k \in M$) are considered in constraint (6), because some ships cannot be assisted by certain berth, because for instance, the type of available equipment in berth cannot be appropriate for handling some load types. The instance's data shown the berth capacity to attend the ships (handling time is different of zero). Constraints (7) and (8) state that the berthing time is posterior to the ship arrival time and completion time happens before the ship's time limit (time window). Constraint (9) and (10) ensure the non time windows violation in berths, and constraint (11) sets that decision variables $x_{i,j}^k$ will be binary.

4   The PTA/LP Method

Initially proposed by Mauri and Lorena [8], the PTA/LP is a heuristic method based on applying the Population Training Algorithm (PTA) and Linear Programming (LP) through the Column Generation technique. The PTA and LP are applied in an interactive way. The PTA uses the information of dual variables in a LP relaxation to generate improved incoming columns (low cost and good covering of the ships) in a column generation process. The LP relaxation is used for solve a Set Partitioning Problem (SPP) formed by these columns. The SPP is formulated as follows:

Minimize:

$$Z^* = \sum_{j=1}^{p} c_j x_j \tag{12}$$

Subject to:

$$\sum_{j=1}^{p} a_{ij} x_j = 1 \ i = 1, ..., n \tag{13}$$

$$x_j \in \{0, 1\}; \ j = 1, ..., p \tag{14}$$

The BAP is modeled as a matrix constructing with columns representing berths and lines the ships. Each element $a_{ij} \in \{0, 1\}$, $i \in N = 1..n$ and $j \in P = \{1..p\}$. n is the number of ships (lines) and p the number of generated columns. $a_{ij} = 1$ if the column j attends the ship i, and 0 otherwise. This is a classic formulation constantly used in several works found in the literature. The c_j represents the cost of column j (defined in eq. 16) and x_j is equal to 1 if column j belongs to the problem solution and 0 otherwise.

In BAP specific case, each berth has its own features, and sometimes a ship type could not be attended by a berth. Just a berth (or none) of each available "type" in quay must be used (each column belonging to the final problem solution should represent a different berth, without repetitions). Then, a new constraint must be inserted in SPP (eq. 15) to forming a set partitioning problem with an additional constraint (SPP^+).

$$\sum_{j=1}^{p} b_{ij} x_j \leq 1 \ i = 1, ..., m \tag{15}$$

Each element $b_{ij} \in \{0, 1\}$, $i \in M = \{1..m\}$ and $j \in P = \{1..p\}$. m is the number of available berths, and $b_{ij} = 1$ if the column j represents the berth i.

Now seeing in an evolutionary computation context, each column is represented through an "individual" formed by integers, where the first position indicates the berth referring to a column, and the other positions represent the ships attended by this berth (column). For the columns' cost calculation, the time windows constraints in the BAP model (7-10) are relaxed and moved to objective function considering weight factors (vector $w = [w_0, w_1, w_2]$). This approach is used to facilitate and accelerate the generation of new columns, because the computational cost (time) to generate the columns with the restrictions (7-10) is higher. The columns are evaluated in a relaxed way and its cost will receive a high weight for violations in time windows. The cost of each column (individual) is given by

$$c_k = w_0 \sum_{i \in B^k} v_i \left(T_i^k - a_i + t_i^k \right) + \tag{16}$$

$$w_1 \sum_{i \in B^k} \left(\max \left(0, a_i - T_i^k \right) + \max \left(0, T_i^k + t_i^k - b_i \right) \right) +$$

$$w_2 \left(\max \left(0, s^k - T_{o(k)}^k \right) + \max \left(0, T_{d(k)}^k + e^k \right) \right)$$

The generation of all necessary columns to build and solve SPP^+ (eq. 12-15) can be a challenge. So, the LP relaxation is used with PTA to generate a suitable set of columns for commercial solvers (a number of columns to be solved by CPLEX - see [5]). More details on PTA/LP can be seen in [9] and [10].

4.1   The Population Training Algorithm

The Population Training Algorithm - PTA is a kind of evolutionary technique first employed in [11] and derived from the Constructive Genetic Algorithm (CGA) proposed by Lorena and Furtado [12]. The CGA has a number of innovative features compared to traditional genetic algorithms. These include a "ranked" population of dynamic size composed of "schemata" and "structures". The schemata and structures are directly evaluated in a common basis, using a double fitness process, called *fg-fitness*.

The schemata are not used in PTA and the fg-fitness will be performed by heuristics. An individual is considered well adapted if it cannot be better regarding the employed training heuristic. The adaptation in the population training is, therefore, used to guide the search to promising areas.

The two functions used in evolutionary training are defined by $g(k) =$ "quality" of column (individual) k (eq. 19), and $f(k) = Best\ g(k')|k' \in Neighborhood(k)$. The $f(k)$ value is obtained through training heuristic (Fig. 3) and the evolutionary process is developed privileging the individuals presenting small differences $[g(k) - f(k)]$ and small $g(k)$, assigning to them the following *ranks*:

$$\delta(k) = d \times [g_{\max} - g(k)] - [g(k) - f(k)] \tag{17}$$

g_{max} is the cost of worst individual (column) created in initial population and d is a constant percentage of g_{max}. The population is dynamically controlled by an evolution parameter denominated α, and updated as:

$$\alpha = \alpha + Step \times PS \times \frac{\delta_{bst} - \delta_{wst}}{RG} \tag{18}$$

Step is a constant that controls the evolutionary process speed and *PS* is the current population size. $(\delta_{bst} - \delta_{wst})$ is the variation among the ranks of the best and worst individuals, respectively, and RG is an estimated number of remaining generations to finish the process.

The parameter α is compared to the ranks (eq. 17), and if $\alpha \geq \delta(k)$ the individual k is eliminated from the population. The population at the evolution time α is dynamic in size and can be emptied during the process.

```
1.  CREATE (m empty berths);
2.  CREATE (a list L with all the ships);
3.  ORDER (the list L by ships incoming time);
4.  FOR (each ship j in L, j = 1,2,...,n) DO
5.      SELECT (a berth i, i = 1,2,...,m);
6.      IF (the berth i was unable to handling the ship j)
7.          RETURN (to step 5);
8.      ELSE
9.          ASSIGN (the ship j to berth i);
10.     END-IF;
11. END-FOR;
```

Fig. 1. Distribution heuristic

The initial population is generated through two heuristics: *distribution heuristic* and *programming heuristic*. The distribution heuristic attributes the ships to the berths. This heuristic is based on the distribution heuristic presented by Mauri and Lorena [13] and the FCFS-G heuristic presented by Cordeau et al. [3]. The programming heuristic makes the ships schedule in the berths. The distribution heuristic runs for "initial population size" times.

The distribution heuristic creates initially m empty berths. The n ships are organized by incoming order on port and distributed to the berths in a random way. In this process the selected berth must always be able to assist the selected ship. This heuristic ensures that each ship will be assigned to a berth that must be able to attend it. The berthing times may present overlapping and/or time windows violations, for ships or berths. The Figure 1 describes the distribution heuristic.

```
1.  GIVEN (any column k)
2.  FOR (each ship i assigned to k) DO
3.      T_i^k = { max(a_i, s^k), i = 1
                { max(a_i, T_{i-1}^k + t_{i-1}^k), i > 1
4.  END-FOR;
5.  CALCULATE (c_k, g(k), f(k) and δ(k))
```

Fig. 2. Programming heuristic

In programming heuristic the berthing time for each ship and the solution objective function of column k (c_k) are computed. The functions $g(k)$ and $f(k)$ and rank $\delta(k)$ are also computed. The Figure 2 presents the programming heuristic.

A simple local search heuristic is used as training function $f(k)$, and several alternative individuals (columns) in a neighborhood are evaluated. This heuristic is described in Figure 3.

```
1.  GIVEN (any column k);
2.  k' ← k;
3.  f* = g(k')
4.  FOR (neighborhood size times)
5.      i ← any ship attended by column k';
6.      j ← another ship attended by column k';
7.      CHANGE (the attendance sequence for ships i and j);
8.      EXECUTE (the programming heuristic for column k');
9.      IF (g(k') < f*);
10.         f* ← g(k');
11.     END-IF;
12. END-FOR;
13. f(k) ← f*;
```

Fig. 3. Training heuristic

The used mutation is also based in a local search implemented through a simple change of the handling positions of two ships (randomly selected) assisted by a column (individual). This process is described in Figure 4.

```
1.  GIVEN (any column k);
2.  i ← any ship attended by column k;
3.  j ← another ship attended by column k;
4.  CHANGE (the attendance sequence for ships i and j);
5.  EXECUTE (the programming heuristic for column k);
```

Fig. 4. Mutation

The crossover generates new individuals as follows: two individuals are selected (base and guide) and a new individual is created similar to the base. Each ship assisted by the guide individual is inserted in the new individual if the corresponding berth can attend it. The handling sequence of the new individual

```
1.  GIVEN (a base column k);
2.  GIVEN (a guide column k');
3.  k'' ← clone(k);
4.  FOR (each ship i attended by column k') DO
5.      IF (the berth referring to column k'' was able to attend the ship i)
6.          INSERT (the ship i in column k'');
7.      END-IF;
8.  END-FOR;
9.  ORDER (the attendance sequence for column k'');
10. EXECUTE (the programming heuristic for column k'');
11. INSERT (the column k'' in population);
```

Fig. 5. Crossover

```
1.  CREATE (an initial population);
2.  WHILE (the generation maximum number not be reached)
3.      SELECT (a base individual);
4.      SELECT (a guide individual);
5.      k ← CROSSOVER (base,guide);
6.      IF (rand() < mutation probability) MUTATION (k); END-IF;
7.      CALCULATE (δ(k));
8.      IF (δ(k) > α) INSERT (k in population); END-IF;
9.      ORDER (the population by the individuals rank);
10.     UPDATE (α);
11.     FOR (every k ∈ population) DO
12.         IF (δ(k) ≤ α) ELIMINATE (k); END-IF;
13.     END-FOR;
24. END-WHILE;
```

Fig. 6. PTA algorithm

is ordered by the ships' arrival time on the port. The crossover is presented in Figure 5.

These operators are enclosed to PTA and its pseudo-code is shown in Figure 6. It is interesting to notice that using these processes the PTA will form populations of several sizes, guided by the objective of selecting low cost columns with an enough covering of the ships. The best columns should include a varied number of ships. This fact is featured by using the training heuristic that will guide the evolutionary process.

4.2   PTA and LP Interaction

The interaction of PTA with LP is made through the fitness function (function g) of the individuals in PTA. This function is defined using the dual variables of LP. The function g is defined as follows:

$$
g(k) = \begin{cases} \dfrac{c_k}{\sum\limits_{i=1}^{n} \lambda_i a_{ik} + \sum\limits_{i=1}^{m} \lambda_i b_{ik}} \ for \ \left(\sum\limits_{i=1}^{n} \lambda_i a_{ik} + \sum\limits_{i=1}^{m} \lambda_i b_{ik} > 0 \right) \\[2em] c_k \ for \ \left(\sum\limits_{i=1}^{n} \lambda_i a_{ik} + \sum\limits_{i=1}^{m} \lambda_i b_{ik} \leq 0 \right) \end{cases} \tag{19}
$$

c_k is the cost of column k (eq. 16) and λ_i is the dual variable corresponding to constraint i. Using the concepts of the column generation technique, the reduced cost of column k (θ_k) inserted in SPP^+ can be calculated through the following equation:

$$
\theta_k = c_k - \left(\sum_{i=1}^{n} \lambda_i a_{ik} + \sum_{i=1}^{m} \lambda_i b_{ik} \right) \tag{20}
$$

We can observe through equations (19) and (20) that for negative costs ($\theta_k < 0$) the value of function g will be situated inside of the interval $[0, 1]$. Therefore, the training heuristic that defines the corresponding function f values (best g in a neighborhood) will assign small differences (g - f) for columns that have negative reduced costs. For positive costs ($\theta_k \geq 0$) the value of the g function will be the respective cost (a "high" value). So, the population is indirectly trained for individuals with negative reduced costs, improving the ship's covering for SPP^+, avoiding the generation of an excessive number of columns and consequently speeding up the process of column generation. The Figure 7 presents the pseudo-code of PTA/LP.

```
1.  CREATE (an initial set of columns);
2.  SOLVE (LP);
3.  FOR (iterations number or maximum number of columns are not reached)
4.      EXECUTE (PTA);
5.      REMOVE (invalid columns);
6.      CALCULATE (reduced cost for new columns);
7.      ADD (columns with negative reduced cost);
8.      SOLVE (LP);
9.  END-FOR;
10. CONVERT (LP for ILP);
11. SOLVE (ILP);
```

Fig. 7. PTA/LP algorithm

In PTA/LP, an initial set of columns addressed to the problem is randomly created. This set must contain columns that form a feasible solution for SPP^+. These columns are generated running the distribution heuristic (Fig. 1) followed by the programming heuristic (Fig. 2) for each column. The solution formed by these columns will be probably invalid, because the columns can present time windows violations. However, the columns with high costs (due to the weights) will be removed from new SPP^+ solutions when improved columns were generated by PTA.

A SPP^+ is formed by the initial set of columns and the LP relaxation is solved by CPLEX. New columns are generated through PTA considering the values of the dual variables to build the fitness functions. The valid columns (that do not present violations in time windows) that present negative reduced costs are added to current SPP^+ and it is solved again through the LP relaxation. These processes are repeated by a certain number of iterations or while a maximum number of generated columns is not reached.

The final SPP^+ is converted to an integer linear problem and solved by CPLEX (through the CPLEX callable library - see [5]). A feasible solution for SPP^+ is obtained, and this solution should be valid and closed of optimal for the BAP model (eq. 1-11).

5   Computational Experience

Several experiments were performed over 30 different instances (60 ships and 13 berths). These instances were randomly generated by Cordeau et al. [3]. All the computational tests were accomplished in a PC with *AMD Athlon* 64 3500 of 2.2 GHz processor with 1GB of RAM and the code was implemented in C++.

Table 1. PTA/LP details

Instance name	Number of generated columns	SPP^+ solved by LP	SPP^+ solved by ILP	Processing time (s) PTA/LP	ILP	Total
i01	26664	1409.00	1409	72.14	2.47	74.61
i02	12752	1261.00	1261	58.92	1.83	60.75
i03	70000	1129.00	1129	94.62	40.83	135.45
i04	54612	1302.00	1302	103.16	7.02	110.17
i05	70019	1207.00	1207	72.20	52.50	124.70
i06	25990	1261.00	1261	74.22	4.12	78.34
i07	70023	1279.00	1279	86.73	27.47	114.20
i08	70005	1299.00	1299	48.77	8.30	57.06
i09	37846	1444.00	1444	91.86	4.61	96.47
i10	70005	1213.00	1213	61.81	37.59	99.41
i11	43507	1369.00	1369	95.34	4.00	99.34
i12	18508	1325.00	1325	77.39	3.30	80.69
i13	70017	1360.00	1360	62.55	27.39	89.94
i14	26221	1233.00	1233	69.05	4.91	73.95
i15	70002	1295.00	1295	71.28	2.91	74.19
i16	30063	1365.00	1365	169.81	0.55	170.36
i17	70033	1283.00	1283	32.89	13.67	46.58
i18	36108	1345.00	1345	81.78	2.23	84.02
i19	16135	1367.00	1367	122.00	1.19	123.19
i20	20528	1328.00	1328	74.25	8.05	82.30
i21	48386	1341.00	1341	103.52	4.56	108.08
i22	54140	1326.00	1326	104.17	1.20	105.38
i23	70010	1266.00	1266	41.59	2.12	43.72
i24	70008	1260.00	1260	75.81	3.09	78.91
i25	41210	1376.00	1376	95.09	1.48	96.58
i26	70011	1318.00	1318	70.00	31.11	101.11
i27	37022	1261.00	1261	77.38	5.48	82.86
i28	70004	1360.00	1360	51.52	1.39	52.91
i29	70001	1280.00	1280	196.36	7.00	203.36
i30	7837	1344.00	1344	69.62	1.39	71.02
Average	48256	1306.87	1306.87	83.53	10.46	93.99

Table 2. Comparison against other methods

Instance name	TS Z	CPLEX Z	CPLEX Gap	PTA/LP Z*	Improvements (%) A	Improvements (%) B
i01	1415	-	-	1409	0.43	-
i02	1263	2606	3.82	1261	0.16	106.66
i03	1139	2565	4.00	1129	0.89	127.19
i04	1303	4353	8.62	1302	0.08	234.33
i05	1208	2672	4.89	1207	0.08	121.38
i06	1262	-	-	1261	0.08	-
i07	1279	2887	4.73	1279	0.00	125.72
i08	1299	5177	11.69	1299	0.00	298.54
i09	1444	-	-	1444	0.00	-
i10	1212	-	-	1213	-0.08	-
i11	1378	-	-	1369	0.66	-
i12	1325	3206	5.48	1325	0.00	141.96
i13	1360	-	-	1360	0.00	-
i14	1233	-	-	1233	0.00	-
i15	1295	4672	9.77	1295	0.00	260.77
i16	1375	4320	8.97	1365	0.73	216.48
i17	1283	-	-	1283	0.00	-
i18	1346	3681	6.94	1345	0.07	173.68
i19	1370	2400	3.04	1367	0.22	75.57
i20	1328	-	-	1328	0.00	-
i21	1346	-	-	1341	0.37	-
i22	1332	3489	7.31	1326	0.45	163.12
i23	1266	-	-	1266	0.00	-
i24	1261	4867	10.13	1260	0.08	286.27
i25	1379	1993	2.67	1376	0.22	44.84
i26	1330	2520	3.62	1318	0.91	91.20
i27	1261	3209	5.70	1261	0.00	154.48
i28	1365	-	-	1360	0.37	-
i29	1282	4809	9.43	1280	0.16	275.70
i30	1351	-	-	1344	0.52	-
Average	1309.67	3495.65	6.52	1306.87	0.21	170.46

The control parameters used by PTA/LP are presented as follows. The initial population size was set to 10; the *Step* parameter was set to 0.001; the maximum number of generations was 70; the base percentage and the mutation probability was set to 10 and 60 respectively; the neighborhood size was set to 6, and the parameter d was set to 0.01; the maximum number of columns was limited to 70000, and the iterations number was set to 10000. In all of the experiments the values of g_{max} were obtained from the largest g evaluation on individuals generated in the initial population. The initial value of α was set to 0 and the weights were set to $w = [1,10,10]$.

The Table 1 presents some details of the PTA/LP performance. The solution value of the last SPP^+ (formed by all the generated columns) was the same when solved by LP and ILP. This fact indicates that optimal solutions are found for the SPP^+ formed by the generated columns subset (these solutions should be close of the original problems optimal). The interaction time for PTA and LP and the time for final SPP^+ resolution through ILP were relatively low resulting in a competitive total time of processing for PTA/LP.

In Table 2 the column "A" presents the improvement obtained by PTA/LP over Tabu Search (TS). The column "B" presents the improvement of PTA/LP over CPLEX. The solutions obtained by PTA/LP were compared against the best known solutions for the used instances. These best solutions were obtained through a Tabu Search heuristic presented in [3]. Besides, the CPLEX 10.0.1 [5] was also used in an isolated way to solve the model described in Section 3. The CPLEX was unable to find solutions for several instances (see Table 2). The CPLEX and Tabu Search, respectively, spent 1 hour (3600 seconds) and approximately 120 seconds of processing time for solving each instance [3], while PTA/LP spent an average of 93.99 seconds for each instance. This fact shows the PTA/LP competitiveness over Tabu Search and CPLEX.

6   Conclusions

This work presented a new hybrid column generation technique to solve the BAP. The PTA integrated with a traditional column generation technique solves column generation sub-problems in an implicit way. The definition of the PTA *fg-fitness* using dual variables information is the essential feature for PTA/LP performance. The computational results were very good and obtained in reasonable processing times compared against the Tabu Search and CPLEX.

The proposed approach doesn't guarantee to find of optimal solutions for BAP, because the column generation sub-problem was solved through a heuristic method. However, the results show good quality solutions, which are probably close to the optimal, suggesting the application to real problems of Brazilian ports and other similar problems.

References

1. Imai, A., Nishimura, E., Papadimitriou, S.: Berthing ships at a multi-user container terminal with a limited quay capacity. Transportation Research - Part E (2006)
2. Vis, I.F.A., Koster, R.D.: Transshipment of containers at a container terminal: An overview. European Journal of Operational Research 147, 1–16 (2003)
3. Cordeau, J.F., Laporte, G., Legato, P., Moccia, L.: Models and tabu search heuristics for the berth allocation problem. Transportation Science 39, 526–538 (2005)
4. Filho, G.R., Lorena, L.A.N.: Constructive genetic algorithm and column generation: an application to graph coloring. In: Proceedings of APORS 2000 - The Fifth Conference of the Association of Asian-Pacific Operations Research Societies within IFORS (2000)
5. ILOG France: ILOG CPLEX 10.0 - User's Manual (2006)
6. Puchinger, J., Raidl, G.R.: Models and algorithms for three-stage two-dimensional bin packing. European Journal of Operational Research. Feature Issue on Cutting and Packing (2006)
7. Cordeau, J.F., Laporte, G., Mercier, A.: A unified tabu search heuristic for vehicle routing problems with time windows. Journal of the Operational Research Society 52, 928–936 (2001)
8. Mauri, G.R., Lorena, L.A.N.: Método interativo para resolução do problema de escalonamento de tripulações. In: XXXVI Brazilian Symposium of Operational Research (2004)
9. Mauri, G.R.: Novas heurísticas para o problema de escalonamento de tripulações. Master Thesis in Applied Computing. Brasilian Institute for Space Research (2005)

10. Mauri, G.R., Lorena, L.A.N.: A new hybrid heuristic for driver scheduling. International Journal of Hybrid Intelligent Systems 1(4), 39–47 (2007)
11. Oliveira, A.C.M., Lorena, L.A.N.: 2-opt population training for minimization of open stack problem. In: Bittencourt, G., Ramalho, G.L. (eds.) SBIA 2002. LNCS (LNAI), vol. 2507, pp. 313–323. Springer, Heidelberg (2002)
12. Lorena, L.A.N., Furtado, J.C.: Constructive genetic algorithm for clustering problems. Evolutionary Computation 3(9), 309–327 (2001)
13. Mauri, G.R., Lorena, L.A.N.: Simulated annealing aplicado a um modelo geral do problema de roteirização e programação de veículos. In: XXXVIII Brazilian Symposium of Operational Research (2006)

Hybrid Metaheuristic for the Prize Collecting Travelling Salesman Problem

Antonio Augusto Chaves and Luiz Antonio Nogueira Lorena*

National Institute for Space Research - INPE
Laboratory of Computing and Applied Mathematics
São José dos Campos, Brazil
`chaves;lorena@lac.inpe.br`

Abstract. The Prize Collecting Traveling Salesman Problem (PCTSP) can be associated to a salesman that collects a prize in each city visited and pays a penalty for each city not visited, with travel costs among the cities. The objective is to minimize the sum of travel costs and penalties, while including in the tour enough cities to collect a minimum prize. This paper presents one solution procedure for the PCTSP, using a hybrid metaheuristic known as Clustering Search (CS), whose main idea is to identify promising areas of the search space by generating solutions and clustering them into groups that are then explored further. The validation of the obtained solutions was through the comparison with the results found by CPLEX.

1 Introduction

This paper presents a new hybrid metaheuristic to solve the Prize Collecting Traveling Salesman Problem (PCTSP). The PCTSP is a generalization of the Traveling Salesman Problem (TSP), where a salesman collects a prize p_i in each city visited and pays a penalty γ_i for each city not visited, considering travel costs c_{ij} between the cities. The problem is to minimize the sum of travel costs and penalties paid, while including in the tour enough cities to collect a minimum prize (p_{min}), defined a priori. In this tour, each city can be visited at most one time.

The solution of PCTSP is difficult due to a large number of possible solutions. Since the PCTSP generalizes the TSP it is also a NP-hard problem, as the TSP is a particular case of PCTSP where the minimum prize is the same as the sum of prizes of all nodes.

In this paper, the PCTSP is solved using a hybrid metaheuristic, known as Clustering Search (CS), which was proposed by Oliveira and Lorena [1]. The CS consists of detecting promising areas of the search space using an algorithm that generates solutions to be clustered. These promising areas may then be explored through local search methods as soon as they are discovered. The algorithm used to generate the solutions was a method combining Greedy Randomized Adaptive Search Procedure (GRASP) [2] and Variable Neighborhood Search (VNS) [3].

---

* The authors acknowledge CNPq by partial research support.

J. van Hemert and C. Cotta (Eds.): EvoCOP 2008, LNCS 4972, pp. 123–134, 2008.

The commercial solver CPLEX [4] has been used to solve the formulation of the PCTSP, in order to validate the computational results of CS algorithm.

The remainder of the paper is organized as follows. Section 2 reviews previous works about PCTSP. Section 3 describes the metaheuristic that was used in this paper, and section 4 present the CS algorithm applied to PCTSP. Section 5 presents the computational results and section 6 concludes the paper.

2   Literature Review

The PCTSP was introduced by Egon Balas [5,6] as a model for scheduling the daily operations of a steel rolling mill. The author presented some structural properties of the problem and two mathematical formulations.

Fischetti and Toth [7] developed several bounding procedures, based on different relaxations. A branch and bound algorithm was also developed that was applied to small size problems.

Goemans and Williamson [8] provided a 2-aproximation procedure to a version of the PCTSP, without the minimum prize constraint. Dell'Amico et al. [9] developed a Lagrangean heuristic, which uses a Lagrangean relaxation for generating starting solutions and an Extension and Collapse procedure to seek improving these solutions.

Chaves and Lorena [10] proposed new heuristics based in the CS to solve the PCTSP, first using a genetic algorithm as generator of solutions for the clustering process, and later changing the genetic algorithm for other metaheuristics.

Feillet et al. [11] presents a survey on TSP with profits that include the PCTSP, identifying and comparing the complexity of different classes of applications, modelling approaches and exact or heuristic solution techniques. Furthermore, this paper shows that the literature is full of solution algorithms that prove to be very efficient and effective.

3   Clustering Search

The Clustering Search (CS) algorithm generalizes the Evolutionary Clustering Search (ECS), proposed by Oliveira and Lorena [1,12], that employs clustering for detecting promising areas of the search space. It is particularly interesting to find out such areas as soon as possible to change the search strategy for them. An area can be seen as a search subspace defined by a neighborhood relationship in metaheuristic coding space. In the ECS, a clustering process is executed simultaneously to an evolutionary algorithm, identifying groups of individuals that deserve special attention. In the CS, the evolutionary algorithm was substituted by distinct metaheuristics, such as Simulated Annealing, GRASP, Tabu Search and others.

The CS attempts to locate promising search areas by framing them by clusters. A cluster can be defined as a tuple $\mathcal{G} = \{C; r; \beta\}$ where C, r and β are, respectively, the center and the radius of the area, and a search strategy associated to the cluster.

The center of the cluster C is a solution that represents the cluster, identifying its location inside the search space. Initially, the centers are obtained randomly, but progressively, they tend to fall along really promising points in the close subspace. The radius r establishes the maximum distance, starting from the center, for which a solution can be associated to the cluster. The search strategy β is a systematic search intensification, in which solutions of a cluster interact among themselves along the clustering process, generating new solutions.

The CS consists of four conceptually independent components with different attributions: a search metaheuristic (SM); an iterative clustering (IC); an analyzer module (AM); and, a local searcher (LS).

The search metaheuristic (SM) component works as a full-time solution generator. The algorithm is executed independently of the remaining components and must be able to provide a continuous generation of solutions to the clustering process. Clusters are maintained, simultaneously, to represent these solutions. This entire process works like an infinite loop, in which solutions are generated along the iterations.

The iterative clustering (IC) component aims to gather similar solutions into groups, identifying a representative cluster center for them. To avoid extra computational effort, IC is designed as an online process, in which the clustering is progressively fed by solutions generated in each iteration of SM. A maximum number of clusters $\mathcal{NC}$ is an upper bound value that prevents an unlimited cluster creation. A distance metric must be defined, a priori, allowing a similarity measure for the clustering process.

The analyzer module (AM) component provides an analysis of each cluster, at regular intervals, indicating a probable promising cluster. A cluster density, δ, is a measure that indicates the activity level inside the cluster. For simplicity, δ_i counts the number of solutions generated by SM and allocated to the cluster i. Whenever δ_i reaches a certain threshold, indicating that some information template has become predominantly generated by SM, that information cluster must be better investigated to accelerate the convergence process on it.

Finally, the local search (LS) component is a local search module that provides the exploitation of a supposed promising search area framed by the cluster. This process is executed each time AM finds a promising cluster and the local search is applied on the center of the cluster. LS can be considered as the search strategy β associated with the cluster, i.e., a problem-specific local search to be applied into the cluster.

4   CS Algorithm for PCTSP

A version of CS for the PCTSP is presented in this section. A solution was represented through a computational structure that contains the visited nodes, the not visited nodes and the objective value.

4.1   The GRASP/VNS Metaheuristic

The component SM, responsible for generating solutions to the clustering process, was a metaheuristic that combines GRASP [2] and VNS [3].

The GRASP is basically composed of two phases: a *construction phase*, in which a feasible solution is generated, and a *local search phase*, in which the constructed solution is improved.

At each iteration of the construction phase, let the set of candidate elements be formed by all nodes that do not visited yet. The selection of the next node for incorporation is determined by the evaluation of all candidate elements according to a greedy evaluation function (1). The evaluation of the nodes by this function leads to the creation of a restricted candidate list (RCL) formed by the best elements. The element to be incorporated into the partial solution is randomly selected from those in the RCL. Once the selected node is incorporated to the partial solution, the candidate list is updated. The construction phase stops when the minimum prize to be collected. The greedy evaluation function for adding a node k between the nodes i and j is

$$g(k) = c_{ij} + \gamma_k - c_{ik} - c_{kj} \qquad (1)$$

that it is composed by the cost of the arc (i, j), the penalty of the node k and the costs of the arcs (i, k) and (k, j), respectively.

The local search phase of GRASP uses the VNS, which is a metaheuristic going on a systematic change of the neighborhood within a local search algorithm.

Initially a set of neighborhood structures is defined through random movements. The VNS proposed implement nine neighborhood structures, through the following movements:

- m_1: add one node to the tour;
- m_2: drop one node from the tour;
- m_3: swap position from two nodes of the tour;
- m_4: add two nodes to the tour;
- m_5: drop two nodes from the tour;
- m_6: swap position from four nodes of the tour;
- m_7: add three nodes to the tour;
- m_8: drop three nodes from the tour;
- m_9: swap position from six nodes of the tour;

Starting from the current solution, at each iteration, randomly a neighbor is selected in the kth neighborhood of the incumbent solution. That neighbor is then submitted to some local search method. If the solution obtained is better than the incumbent, update the incumbent and continue the search of the first neighborhood structure. Otherwise, the search continues to the next neighborhood. The VNS stopped when the maximum number of iterations since the last improvement is satisfied.

In this paper we use the method Variable Neighborhood Descent (VND) [3] as a local search of the VNS, and it is composed by three different improvement methods, that combine two movements: Add-step and Drop-step [13]. The Add-step movement consists of a node addition that provides the best value of the addition function. The Drop-step movement consists of a node removal (from the tour) that provides the best value of the removal function. In both movements, if the function value was positive then the objective function will be better after the movement. The main aspect to be observed is that all moves are executed preserving feasibility.

The improvement methods of the VND are:

- **SeqDrop:** to apply a sequence of Drop-step movements while the objective function value is being decreased;
- **AddDrop:** to apply one Add-step movement and one Drop-step movement;
- **SeqAdd:** to apply a sequence of Add-step movements while the objective function value is being decreased.

Whenever some improvement method obtains a better solution, the VND returns to the first improvement method. The stopping condition of the VND is that there are no more improvements for the incumbent solution.

4.2   The Clustering Process

The IC is the CS's core, working as a classifier, keeping in the system only relevant information, and guiding search intensification in the promising search areas. A maximum number of clusters ($\mathcal{NC}$) is defined a priori. The ith cluster has its own center C_i and a radius r, like the other clusters.

Solutions generated by GRASP/VNS are passed to IC, that attempts to group these solutions as known information in a cluster, chosen according to a distance metric. The solution activates the closest center C_i (cluster center that minimizes the distance metric), causing some kind of disturbance on it. In this paper, the metric distance is the number of different edges between the GRASP/VNS and the center of the cluster solutions. When there are a larger number of different edges between them it increases the dissimilarity.

The disturbance is an assimilation process, in which the center of the cluster is updated by the newly generated solution. In this paper, this process is the path-relinking method [14], that generates several points (solutions) along the path that connects the solution generated by GRASP/VNS and the center of the cluster. Since each point is evaluated by the objective function, the assimilation process itself is an intensification mechanism inside the clusters. The new center C_i is the best-evaluated solution obtained in the path.

The AM is executed whenever a solution is assigned to a cluster, verifying if the cluster can be considered promising. A cluster becomes promising when the density δ_i reaches a certain threshold, given for:

$$\delta_i \geq \mathcal{PD}.\frac{\mathcal{NS}}{|Clus|} \tag{2}$$

where, $\mathcal{NS}$ is the number of solutions generated in the interval of analysis of the clusters, $|Clus|$ is the number of clusters, and $\mathcal{PD}$ is the desirable cluster density beyond the normal density, obtained if $\mathcal{NS}$ was equally divided to all clusters. The center of a promising cluster is improved through the LS.

The LS was implemented by a 2-Opt procedure [15], which seeks to improve the center of the promising cluster. The 2-Opt is based on resequencing of the route always leads to a better solution, since it may possibly decrease travel costs, while leaving prizes and penalties unchanged. It amounts to simply considering the set of nodes currently visited by the route and trying to shorten the length of the route through these nodes. In this paper, the 2-Opt consists in 2-changes over a route, deleting two arcs and replacing them by two other arcs to form a new route. This method continues while there is improvement in the route through this movement.

The whole CS pseudo-code is presented in Figure 1.

```
procedure CS
    { SM component }
    for (number of iterations is not satisfied) do
        {construction phase of GRASP}
        s = ∅
        while (solution not built) do
            compute candidate list (C)
            RCL = C * α
            e = select at random a value of RCL
            s = s ∪ {e}
        end while
        {local search phase of GRASP - VNS}
        kmax = number of neighborhoods
        while (stop condition is not satisfied) do
            k = 1
            while (k ≤ kmax) do
                generate at random s' ∈ N^k(s)
                s" = apply VND with s' as starting point
                if ( f(s") < f(s) ) then
                    s = s"
                    k = 1
                else
                    k = k + 1
            end while
        end while

        { IC component }
        calculate the distance of the solution GRASP/VNS (s) and the clusters
        insert the solution in the most similar cluster (Ci)
        apply the assimilation process – path-relinking(s, Ci)

        { AM component }
        verify if the cluster can be considered promising. If so, the LS component is
        applied to it

        { LS component }
        apply the 2-Opt heuristic to the promising cluster
    end for
end procedure
```

Fig. 1. CS pseudo-code

5 Computational Results

The CS algorithm was coded in C++ and it was run on a 3 GHz Pentium 4. The experiments were accomplished with objective of evidencing the flexibility of the method in relation to the algorithm used to feed the clustering process, and also to validate the proposed approach, showing that the clustering search algorithm can be competitive to solve the PCTSP.

There are no available instances for the PCTSP in the literature. In this paper, test instances were randomly generated as in Dell'Amico et al. [9]. We generated problems with $n = (20, 40, 60, 80, 100, 200, 300, 400, 500)$ vertices, travel costs $c_{ij} \in [1, M]$ with $M \in \{1000, 10,000\}$, prizes $p_i \in [1, 100]$, and penalties $\gamma_i \in [1, N]$ with $N \in \{100, 1000, 10,000\}$. The value of minimum prize (p_{min})

Table 1. PCTSP: Symmetric random instances, $c_{ij} \in [1, 1000]$; $\gamma_i \in [1, 100]$. Times in seconds.

		CPLEX			CS				GRASP/VNS			
σ	n	BI	Gap	RT	BS	AS	AT	DE	BS	AS	AT	DE
	20	903	0.00	0.80	**903**	**903.0**	**0.03**	**0.00**	903	903.0	0.07	0.00
	40	996	0.00	20.46	**996**	**996.0**	**2.28**	**0.00**	996	996.0	4.41	0.00
	60	1314	0.00	474.94	**1314**	**1314.0**	**13.73**	**0.00**	1314	1321.3	25.36	0.56
	80	**1384**	**0.00**	**26692.93**	1386	1392.8	83.03	0.64	1531	1548.0	215.46	11.85
0.2	100	1514	1.85	100,000.00	**1508**	**1526.4**	**196.10**	**1.22**	1552	1562.3	122.38	3.60
	200	-	-	-	**1816**	**1834.4**	**502.94**	**1.01**	1898	1922.3	464.14	5.86
	300	-	-	-	**2281**	**2313.0**	**1069.11**	**1.40**	2439	2506.7	845.10	9.89
	400	-	-	-	**2504**	**2554.2**	**1212.85**	**2.00**	2671	2691.3	1138.07	7.48
	500	-	-	-	**3233**	**3281.1**	**1355.62**	**1.48**	3382	3401.3	1936.49	5.21
	20	1123	0.00	14.45	**1123**	**1123.0**	**0.30**	**0.00**	1123	1140.0	0.35	1.51
	40	996	0.00	21.84	**996**	**996.0**	**2.34**	**0.00**	996	1008.0	4.26	1.20
	60	1314	0.00	468.51	**1314**	**1314.0**	**11.42**	**0.00**	1314	1339.0	18.00	1.90
	80	**1384**	**0.00**	**32121.21**	1388	1396.6	72.74	0.91	1497	1519.7	296.15	9.80
0.5	100	1514	1.75	100,000.00	**1513**	**1534.6**	**181.00**	**1.43**	1562	1584.3	84.83	4.71
	200	-	-	-	**1816**	**1844.2**	**572.46**	**1.55**	1902	1943.7	485.48	7.03
	300	-	-	-	**2171**	**2250.5**	**1213.63**	**3.66**	2428	2452.1	1127.12	12.94
	400	-	-	-	**2489**	**2579.7**	**1490.49**	**3.64**	2694	2744.3	1048.74	10.26
	500	-	-	-	**3159**	**3200.7**	**1784.58**	**1.32**	3246	3298.4	2216.91	4.41
	20	1354	0.00	8.79	**1354**	**1354.0**	**0.15**	**0.00**	1354	1354	0.40	0.00
	40	1129	0.00	44.69	**1129**	**1137.0**	**2.95**	**0.71**	1156	1186.2	4.67	5.07
	60	1319	0.00	474.80	**1319**	**1344.2**	**17.46**	**1.91**	1379	1387.7	9.28	5.21
	80	**1384**	**0.00**	**27498.29**	1396	1400.2	63.86	1.17	1468	1485.3	218.82	7.32
0.8	100	1575	6.08	100,000.00	**1519**	**1537.6**	**186.54**	**1.22**	1563	1589.0	105.63	4.61
	200	-	-	-	**1768**	**1797.2**	**805.01**	**1.65**	1908	1926.7	630.53	8.97
	300	-	-	-	**2148**	**2213.0**	**1528.29**	**3.03**	2314	2444.0	1211.34	13.78
	400	-	-	-	**2455**	**2494.3**	**1668.90**	**1.60**	2599	2669.3	1536.90	8.73
	500	-	-	-	**3214**	**3324.2**	**1708.36**	**3.42**	3438	3471.6	1842.49	8.02

has been generated as $\left\lceil \sigma \sum_{i=1}^{n} p_i \right\rceil$ with $\sigma \in \{0.2, 0.5, 0.8\}$. These test instances are available in http://www.lac.inpe.br/~lorena/instancias.html.

The following parameters' values for approach CS were adjusted through several executions. The following parameters obtained the best results:

- number of solutions generated at each analysis of the clusters $\mathcal{NS} = 200$;
- maximum number of clusters $\mathcal{NC} = 20$;
- density pressure $\mathcal{PD} = 2.5$;
- percentage of the best elements in the RCL, $\alpha = 0.2$.

The formulation presented in [10] was solved using the solver CPLEX 10.0.1, and the results are presented in following Tables. The CPLEX solved the PCTSP up to 100 nodes, founding the optimal solution in a reasonable execution time for small instances. However, for the instances with 80 nodes, the CPLEX took several hours execution to find the optimal solution, and, for any instances with 100 nodes the CPLEX did not get to close the gap between lower and upper

Table 2. PCTSP: Symmetric random instances, $c_{ij} \in [1, 10,000]$; $\gamma_i \in [1, 1000]$. Times in seconds.

		CPLEX			CS				GRASP/VNS			
σ	n	BI	Gap	RT	BS	AS	AT	DE	BS	AS	AT	DE
	20	11677	0.00	2.65	**11677**	**11677.0**	**0.04**	**0.00**	11677	11667.0	0.09	0.00
	40	10776	0.00	21.97	**10776**	**10776.0**	3.19	**0.00**	10776	10896.0	7.23	1.11
	60	**14236**	**0.00**	**1151.37**	14243	14314.1	7.11	0.55	14684	15033.2	41.46	5.60
	80	**14484**	**0.00**	**68464.35**	14609	14760.9	114.86	1.91	15022	15327.1	134.18	5.82
0.2	100	14841	10.79	100,000.00	**13620**	**14015.0**	104.17	2.90	14328	14510.9	79.97	6.54
	200	-	-	-	**15303**	**15628.2**	528.53	2.13	16250	16560.3	412.80	8.22
	300	-	-	-	**21869**	**22158.0**	662.13	1.32	22760	23898.2	653.38	9.28
	400	-	-	-	**24390**	**25099.6**	1354.40	2.91	25685	26639.7	1215.72	9.22
	500	-	-	-	**31090**	**31558.7**	1643.15	1.51	32965	33666.3	1323.48	8.29
	20	12900	0.00	9.84	**12900**	**12900.0**	**0.17**	**0.00**	12900	12996.9	0.23	0.75
	40	10776	0.00	21.95	**10776**	**10776.0**	5.58	**0.00**	10776	10861.1	8.03	0.79
	60	**14236**	**0.00**	**1152.99**	14349	14421.9	13.73	1.31	15005	15065.3	51.12	5.82
	80	**14484**	**0.00**	**68464.35**	14512	14830.0	111.92	2.39	15458	15744.9	208.18	8.71
0.5	100	14841	10.79	100,000.00	**13900**	**14089.1**	118.35	1.36	14447	14696.0	61.04	5.73
	200	-	-	-	**15190**	**15440.4**	664.17	1.64	16132	16676.3	502.56	9.78
	300	-	-	-	**22731**	**23211.7**	696.75	2.11	23855	24504.3	840.55	7.80
	400	-	-	-	**23898**	**24525.3**	1755.43	2.63	25233	25494.1	1395.65	6.68
	500	-	-	-	**30275**	**30842.0**	2173.11	1.87	32932	33669.3	1532.14	11.21
	20	16559	0.00	6.28	**16559**	**16559.0**	**0.07**	**0.00**	12559	12559.0	0.12	0.00
	40	10776	0.00	31.32	**10776**	**10776.0**	4.66	**0.00**	10776	10844.0	5.83	0.63
	60	14864	0.00	18508.17	**14864**	15017.1	32.74	1.03	15215	15888.2	65.40	6.89
	80	**14484**	**0.00**	**70205.22**	14740	14793.9	92.77	2.14	15195	15329.9	142.87	5.84
0.8	100	17316	23.54	100,000.00	**13704**	**13971.9**	95.31	1.96	14393	14483.1	104.33	5.68
	200	-	-	-	**15200**	**15376.2**	716.94	1.16	15891	16249.3	515.52	6.90
	300	-	-	-	**22168**	**22467.7**	1496.77	1.35	23616	24187.3	1135.30	9.11
	400	-	-	-	**22790**	**23688.3**	1953.06	3.94	26059	26327.4	1493.63	15.52
	500	-	-	-	**30385**	**30707.0**	2336.85	1.06	32608	33273.1	1527.11	9.50

Table 3. PCTSP: Symmetric random instances, $c_{ij} \in [1, 1000]$; $\gamma_i \in [1, 10,000]$. Times in seconds.

		CPLEX			CS				GRASP/VNS			
σ	n	BI	Gap	RT	BS	AS	AT	DE	BS	AS	AT	DE
	20	1192	0.00	1.21	**1192**	**1192.0**	**0.23**	**0.00**	1192	1212.7	0.17	1.73
	40	1449	0.00	180.33	**1449**	**1449.0**	**6.52**	**0.00**	1449	1506.2	5.93	3.95
	60	1666	0.00	173.17	**1666**	1670.2	**34.94**	0.25	1687	1702.7	30.85	2.20
	80	**1794**	**0.00**	**41884.95**	1807	1814.6	144.96	1.15	1842	1846.3	136.71	2.92
0.2	100	**1601**	**0.00**	**22767.73**	1628	1647.0	281.07	2.87	1680	1696.3	189.99	5.95
	200	-	-	-	1898	1946.8	**911.18**	2.57	2060	2091.3	481.61	10.19
	300	-	-	-	2246	2334.0	**920.96**	3.92	2462	2501.0	969.01	9.37
	400	-	-	-	2880	2933.7	1370.23	1.86	3049	3105.3	1455.80	7.82
	500	-	-	-	3385	3428.5	2455.34	1.29	3491	3688.1	2039.97	8.95
	20	1192	0.00	1.22	**1192**	**1192.0**	**0.24**	**0.00**	1192	1218.7	0.20	2.24
	40	1449	0.00	180.12	**1449**	**1449.0**	**6.05**	**0.00**	1449	1544.8	8.43	6.61
	60	1666	0.00	172.99	**1666**	1670.8	**36.33**	0.29	1684	1690.0	33.03	1.44
	80	**1794**	**0.00**	**32442.24**	1813	1821.6	131.57	1.54	1855	1876.1	112.92	4.57
0.5	100	**1601**	**0.00**	**22730.51**	1655	1665.3	220.50	4.02	1695	1711.3	141.33	6.89
	200	-	-	-	1968	2012.0	**529.79**	2.24	2052	2087.7	516.25	6.08
	300	-	-	-	2300	2382.3	**1152.44**	3.58	2447	2484.3	1118.08	8.01
	400	-	-	-	2842	2935.3	1348.65	3.28	3063	3123.3	1496.11	9.90
	500	-	-	-	3274	3332.6	2054.47	1.79	3500	3587.4	1878.52	9.57
	20	1192	0.00	1.21	**1192**	**1192.0**	**0.25**	**0.00**	1192	1230.7	0.23	2.40
	40	1449	0.00	180.55	**1449**	**1449.0**	**14.07**	**0.00**	1449	1534.8	9.80	5.92
	60	1666	0.00	172.85	**1666**	1669.6	**46.41**	0.22	1687	1691.3	30.85	1.52
	80	**1794**	**0.00**	**27699.33**	1801	1815.2	110.43	1.18	1863	1868.0	83.50	4.12
0.8	100	**1601**	**0.00**	**22722.64**	1626	1658.8	235.88	3.61	1700	1712.3	193.64	6.95
	200	-	-	-	1978	1999.8	**609.86**	1.10	2050	2062.3	665.13	4.26
	300	-	-	-	2319	2371.7	**1100.99**	2.27	2473	2536.2	969.01	9.37
	400	-	-	-	2837	2849.0	1554.54	0.42	2876	2907.3	1864.01	2.48
	500	-	-	-	3305	3333.2	2134.51	0.85	3491	3494.5	1985.14	5.73

bounds in 100,000 seconds. Beside that, the CPLEX did not get to find a feasible solution for instances with $n \geq 200$ in 100,000 seconds.

Tables 1-4 give the results for the PCTSP. The entries in the tables are:

- value of parameter σ;
- number of vertices (n) in the original graph;
- the best integer solutions (BI) found by the CPLEX, the Gap and running time (RT) of CPLEX. The values of Gap equal zero define that the optimal has been achieved;
- best solution (BS), average solution (AS), average running time (AT) to find the best solution during the CS execution and the Deviation (DE), that reflects the relative error of the average solution for the CS algorithm, and

Table 4. PCTSP: Symmetric random instances, $c_{ij} \in [1, 10, 000]$; $\gamma_i \in [1, 100]$. Times in seconds.

		CPLEX			CS				GRASP/VNS			
σ	n	BI	Gap	RT	BS	AS	AT	DE	BS	AS	AT	DE
	20	3011	0.00	9.77	**3011**	**3011.0**	**0.01**	**0.00**	3011	3046.2	0.05	1.17
	40	3506	0.00	26.06	**3506**	**3531.6**	**0.67**	0.73	3506	3544,4	2.21	1.10
	60	**4251**	**0.00**	**130.31**	4277	4287.2	14.86	0.85	4617	4693.2	9.91	10.40
	80	**4903**	**0.00**	**2226.51**	4903	5044.3	22.61	2.88	5347	5543.3	22.36	13.05
0.2	100	**5635**	**0.00**	**10824.22**	5702	5802.0	59.71	2.96	6095	6125.7	40.44	8.71
	200	-	-	-	9035	**9129.0**	303.41	1.04	9669	9865.3	246.28	9.19
	300	-	-	-	14592	14875.2	319.14	1.94	15380	15540.0	444.85	6.50
	400	-	-	-	16651	16850.3	640.17	1.20	17297	17564.7	567.84	5.49
	500	-	-	-	20612	21305.7	836.68	3.37	21986	22231.0	730.29	7.85
	20	4313	0.00	7.97	**4313**	**4313.0**	**0.17**	**0.00**	4313	4652.1	0.28	7.86
	40	4694	0.00	59.86	**4694**	**4694.0**	4.29	**0.00**	4694	4736.6	8.82	0.91
	60	**6120**	**0.00**	**700.35**	6232	6361.7	14.05	3.95	6739	6937.1	10.57	13.35
	80	**6319**	**0.00**	**72518.87**	6528	6628.1	63.69	4.89	7180	7303.7	88.78	15.58
0.5	100	**6869**	**0.00**	**69562.82**	7710	7833.7	119.52	14.04	8206	8655.0	120.07	26.00
	200	-	-	-	10293	10578.0	438.80	2.76	11165	11337.0	231.34	0.51
	300	-	-	-	15312	15698.3	549.65	2.52	16872	17646.1	664.34	15.24
	400	-	-	-	17263	17535.7	841.70	1.58	18256	18552.3	744.61	7.47
	500	-	-	-	20896	21623.7	1056.08	3.48	22136	22300.9	801.02	6.72
	20	7797	0.00	14.47	**7797**	**7797.0**	**0.11**	**0.00**	7797	7958.6	0.22	2.07
	40	9070	0.00	43.26	**9070**	**9171.6**	**7.55**	1.12	9224	9302.6	7.27	2.56
	60	**9459**	**0.00**	**10854.96**	9664	9810.5	32.77	3.72	9934	10165.1	22.20	7.46
	80	**9699**	**0.00**	**98073.17**	9991	10048.0	97.02	3.60	11118	11315.2	108.70	16.96
0.8	100	**10002**	0.31	**100,000.00**	10641	10724.1	157.96	7.22	11267	11470.0	131.20	14.68
	200	-	-	-	12650	13024.0	616.84	2.96	13448	13853.1	406.51	9.51
	300	-	-	-	18253	18740.8	1100.30	2.67	19532	20128.3	908.09	10.27
	400	-	-	-	18501	18955.7	1574.56	2.46	19996	21119.3	1027.42	14.15
	500	-	-	-	**23234**	**23590.3**	1920.92	**1.53**	24239	25364.9	1451.63	9.17

the best solution found by CPLEX or CS, and it is calculated by (AS - BS or BI)/(BS or BI) ×100; and

- best solution (BS), average solution (AS), average running time (AT) to find the best solution during the GRASP/VNS execution and Deviation (DE).

The best solutions found (BS), the averages of solutions (AS) and the averages running times to find the best solution (AT) were considered to compare the approaches. The values in boldface show the best objective function values and execution times for each instance.

Table 1 gives the results for all values of n, σ, $c_{ij} \in [1, 1000]$ and $\gamma_i \in [1, 100]$. One can see that the approach CS has better results in 89% of the tests, has found the optimal solutions for instances up to 60 nodes, and solutions better than the CPLEX for instances with 100 nodes. The running times of CS algorithm were very competitive related to the CPLEX. The GRASP/VNS, without the clustering process, has worse results than CS in quality of solutions and deviation. The CS algorithm was very robust producing small deviations. The same conclusions can be drawn for instances with $c_{ij} \in [1, 10, 000]$ and $\gamma_i \in [1, 1000]$ presented in Table 2.

Table 3 reports the results of computational experiments with $c_{ij} \in [1, 1000]$ and $\gamma_i \in [1, 10, 000]$. In this case, the CS algorithm has better results in 77% of the tests, having found the optimal solutions for instances up to 60 nodes. The running times of the CS algorithm were very competitive related to the CPLEX and again the results of the CS were better than the GRASP/VNS without the clustering process.

The results of Table 4 refer to instances with $c_{ij} \in [1, 10, 000]$ and $\gamma_i \in [1, 100]$. The results for these instances were worse than the other three classes of instances, and, the CS algorithm did not get to find the optimal solutions for instances with 60, 80 and 100 nodes, but the solutions were closer to optimal. The CS, in this case, has better results only in 66,6% of the tests. On the other hand, the results of the CS were better than the GRASP/VNS.

Finally, note that the results of Tables 1 and 3 with $c_{ij} \in [1, 1000]$ did not have large changes when σ increases. And, for Tables 2 and 4 with $c_{ij} \in [1, 10, 000]$ the results were very different when σ goes from 0.2 to 0.8. In spite of that, the CS algorithm found results with small deviations for all classes of instances, independently of the value of σ.

6   Conclusions

This paper has presented a solution for the Prize Collecting Traveling Salesman Problem using Clustering Search (CS). The CS uses the concept of hybrid algorithms, combining metaheuristics with a clustering process.

The idea of the CS is to avoid applying a local search heuristic to all solutions generated by a metaheuristic, which can make the search process impracticable because of time consumption, mainly when the heuristic has a high computational cost. The CS detects the promising regions in the search space during the solution generation process and applies the local search heuristics only in these regions, i.e., to detect promising regions becomes an interesting alternative preventing the indiscriminate application of such heuristics.

This paper reports results of different classes of instances to the PCTSP found by CPLEX, CS and GRASP/VNS methods. CS algorithm got better results than GRASP/VNS and it founds good values comparing to CPLEX. CS has two advantages over CPLEX: execution time, and the cost of a commercial solver.

The results show that the CS approach is competitive for the resolution of the PCTSP in reasonable computational times. For instances up to 60 nodes, the optimal solutions were found. Besides, the CS obtained better results than CPLEX for some instances with 100 nodes. Therefore, these results validate the CS application to the PCTSP.

Further studies can be done which analyze others metaheuristics to generate solutions for the clustering process of the CS, such as the Ant Colony System,

Tabu Search and Simulated Annealing, and apply the CS in other generalizations of the TSP, such as Profitable Tour Problem and Quota TSP.

References

1. Oliveira, A.C.M., Lorena, L.A.N.: Detecting promising areas by evolutionary clustering search. In: Bazzan, A.L.C., Labidi, S. (eds.) SBIA 2004. LNCS (LNAI), vol. 3171, pp. 385–394. Springer, Heidelberg (2004)
2. Feo, T., Resende, M.: Greedy randomized adaptive search procedures. Journal of Global Optimization 6, 109–133 (1995)
3. Mladenovic, N., Hansen, P.: Variable neighborhood search. Computers and Operations Research 24, 1097–1100 (1997)
4. ILOG France: ILOG CPLEX 10.0 - User's Manual (2006)
5. Balas, E.: The prize collecting travelling salesman problem. In: Anais..., ORSA/TIMS Meeting (1986)
6. Balas, E.: The prize collecting travelling salesman problem. Networks 19, 621–636 (1989)
7. Fischetti, M., Toth, P.: An additive approach for the optimal solution of the prize collecting traveling salesman problem. In: Vehicle Routing: Methods and Studies, 319–343 (1988)
8. Goemans, M.X., Williamson, D.P.: A general aproximation technique for constrained forest problems. SIAM Journal on Computing 24(2), 296–317 (1995)
9. Dell'Amico, M., Maffioli, F., Sciomanchen, A.: A lagrangian heuristic for the prize collecting travelling salesman problem. Annals of Operations Research 81, 289–305 (1998)
10. Chaves, A.A., Lorena, L.A.N.: Hybrid algorithms with detection of promising areas for the prize collecting traveling salesman problem. In: Proceedings of International Conference on Hybrid Intelligent Systems, pp. 49–54. IEEE Computer Society, Los Alamitos, California (2005)
11. Feillet, D., Dejax, P., Gendreau, M.: Traveling salesman problems with profits. Transportation Science 2(39), 188–205 (2005)
12. Oliveira, A.C.M., Lorena, L.A.N.: Hybrid evolutionary algorithms and clustering search. In: Grosan, C., Abraham, A., Ishibuchi, H. (eds.) Hybrid Evolutionary Systems - Studies in Computational Intelligence. SCI Series, pp. 81–102. Springer, Heidelberg (2007)
13. Gomes, L.M., Diniz, V.B., Martinhon, C.A.: An hybrid grasp+vnd metaheuristic fo the prize collecting traveling salesman problem. In: XXXII Brazilian Symposium of Operational Research (SBPO), pp. 1657–1665 (2000)
14. Glover, F.: Tabu search and adaptive memory programing: Advances, applications and challenges. In: Interfaces in Computer Science and Operations Research, pp. 1–75 (1996)
15. Croes, G.: A method for solving travelling salesman problems. Operations Research 6, 791–812 (1958)

An ILS Based Heuristic for the Vehicle Routing Problem with Simultaneous Pickup and Delivery and Time Limit

Anand Subramanian[1] and Lucídio dos Anjos Formiga Cabral[2]

[1] Universidade Federal da Paraíba, Departamento de Engenharia de Produção, Centro de Tecnologia - Campus I - Bloco G - Cidade Universitária, Castelo Branco, 58051-970 - João Pessoa - PB, Brazil
[2] Universidade Federal da Paraíba Departamento de Estatística, Centro de Ciências Exatas e da Natureza - Campus I - Cidade Universitária, Castelo Branco, 58051-900 - João Pessoa - PB, Brazil

Abstract. This paper deals with the Vehicle Routing Problem with Simultaneous Pick-up and Delivery and Time Limit (VRPSPDTL). Due to the combinatorial nature of the problem, heuristics methods are commonly used to generate good quality solutions within an acceptable computational time. Accordingly, an Iterated Local Search (ILS) procedure that uses a Variable Neighborhood Descent (VND) method to perform the local search is proposed. The algorithm was applied to test problems and, in most cases, was found to produce better results than those reported in the literature.

1 Introduction

The Vehicle Routing Problem with Pick-up and Delivery (VRPPD), i.e., where people or objects should be collected and distributed, constitutes an important category of the well-known Vehicle Routing Problem (VRP). In the late 80's, Min [1] proposed a new variation for the VRPPD where, in certain situations, the pick-up and delivery services must be carried out simultaneously for each customer, characterizing one of the most important variants of this class: the Vehicle Routing Problem with Simultaneous Pick-up and Delivery (VRPSPD).

The Vehicle Routing Problem with Simultaneous Pick-up and Delivery and Time Limit (VRPSPTL) was first dealt with by Salhi and Nagy [2], where the authors consider a maximum time constraint, that is, a vehicle must not exceed an established time limit. This can be treated similarly as the maximum distance constraint. The total amount of time consumed by a vehicle is given by the time traveled plus the time spent attending the customers (drop time). The total distance traveled is given by the vehicle speed multiplied by the time traveled. Thus, the values of the time traveled and the total distance are equal when a unit speed is assumed. Therefore the total time can be calculated as the total traveled distance plus the drop times. When the drop time is assumed to be the same for all clients, the total time spent attending to all customers can be obtained

J. van Hemert and C. Cotta (Eds.): EvoCOP 2008, LNCS 4972, pp. 135–146, 2008.
© Springer-Verlag Berlin Heidelberg 2008

directly by multiplying the number of clients visited by the pre-determined drop time.

Some applications of the VRPSPDTL can be observed in the beverage industry, where filled bottles are delivered while the empty ones are collected; in grocery stores, where pallets or containers are collected for re-use in merchandise transportation, etc. It is important to mention that some clients can demand that the pick-up and delivery services should not be carried out separately, since, in certain cases, this may result in additional costs and operational efforts for these customers. In addition, some companies are obliged by law to accomplish the daily working time regimen of a vehicle driver and, due to this requirement, some routes must be carried out within a time limit.

Thus, one should consider not only the Distribution Logistics, but also the management of the reverse flow. It is in this context, that the concept of Reverse Logistics arises, which can be defined as the process of planning, implementing and controlling the return of raw materials, inventories under process, finished products and information related to the point of consumption until the point of origin. Therefore, the Distribution Logistic and Reverse Logistic should act together with an aim to guarantee the synchronization between the pick-up and delivery operations, as well as their impact on the company's supply chain, resulting in the customer's satisfaction and minimization of the operational efforts.

However, this is not a simple task, since the VRPSPDTL is NP-hard, and the determination of the optimum solution, by means of an exact method, in an acceptable time, is almost impossible. Due to combinatorial nature of the problem, heuristic techniques have been often applied in order to obtain good quality solutions in an acceptable time. Hence, this paper proposes a multistart Iterated Local Search (ILS) heuristic which uses a greedy approach for generating the initial solution and the Variable Neighborhood Descent (VND), with most of the VRP familiar neighborhood structures, in the local search phase. To the best of our knowledge, this is the first time that the ILS has been applied to solve a variant of the VRPPD.

This paper is organized as follows. Section 2 provides a brief literature review of some related works. Section 3 brings the main aspects of the ILS metaheuristic. Section 4 illustrates solution procedure. Section 5 contains the results obtained and a comparison with the ones found in the literature. Finally, Section 6 presents the concluding remarks.

2   Literature Review

The VRPSPD was first proposed by Min [1], where the author shows a real-life application through a case study carried over in public library's distribution system. Since then, little work has been done related to this variant. Only a decade later, Salhi and Nagy [2] suggested some insertion heuristics, also capable of solving the problem with multi-depots and time limit constraints. Recently the same authors have developed another procedure in [3] which involves solutions with certain degree of feasibility.

Dethloff [4] treats VRPSPD under various aspects of reverse logistics and proposes a constructive procedure (insertion heuristic) based on the cheapest feasible insertion criterion, radial surcharge, and the residual capacity, where the last one is an adaptation of the load-base approach. The VRPSPDTL is also considered in his work.

Vural [5] makes use of two methods based on Genetics Algorithms and it appears that this was the first work where a metaheuristic was applied to solve the VRPSPD. Gokçe [6] developed a four phase heuristics based on the Ant Colony metaheuristic.

Crispim and Brandão [7] present a hybrid procedure where Tabu Search (TS) and VND are combined. Montané and Galvão [8] proposed a TS algorithm involving multiple neighborhood structures: reallocation, interchange (swap), crossover and 2-opt. The same metaheuristic was implemented by Gribkovskaia et al. [9], for the case where only one vehicle is considered.

Ropke and Pisinger [10] developed a large neighborhood heuristic associated with a procedure similar to the VNS metaheuristic. The authors also solved several variants of the VRPPD including the VRPSPDTL.

Chen and Wu [11] proposed an insertion heuristic to generate initial solutions and a local search procedure based on the record-to-record travel approximation and tabu lists. Chen [12] developed a hybrid heuristic that combines the principles of the Simulated Annealing and TS metaheuristics.

Bianchessi and Righini [13] suggest some constructive algorithms and local search heuristics as well as a TS procedure that uses a variable neighborhood structure, in which the node-exchange-based and arc-exchange-based movements were combined. Wassan et al. [14] made use of a constructive procedure based on the sweep algorithm and propose a reactive tabu search with the following neighborhood structures: reallocation of a client (shift), exchanging two clients between two different routes (exchange) and reversing the route direction (reverse). They also dealt with the VRPSPDTL.

Dell'Amico et al. [15] make use of an exact approach based on the branch-and-price technique. The problem is treated in two different ways, both under the dynamic programming scope. The same technique is applied by Angelelli and Mansini [16], where the authors consider the VRPSPD with time-windows constraints.

3   Iterated Local Search

Let us assume that a local optimum solution has been found by a local search algorithm. Instead of restarting the same procedure from a completely new solution, the ILS metaheuristic applies a local search repeatedly to the initial solutions achieved by perturbing the local optimum solutions previously visited. According to Lourenço et al. [17], the essential idea of ILS resides in the fact that it focuses on a smaller subset, instead of considering the total space of solutions, defined by the local optimums of a given optimization procedure. To implement the ILS, four procedures should be specified: (i) `GenerateInitialSolution`, where an initial solution is constructed; (ii) LocalSearch, which improves the solution initially obtained; (iii) Perturb, where a new starter point is generated through a perturbation of a solution found in the LocalSearch; (iv) AcceptanceCriterion, that determines from which solution the search should continue.

As stated by Stützle [18], the modification realized in the perturbation phase is used in order to escape from a current local optimum solution. Frequently, the movement is randomly chosen within a larger neighborhood in comparison to the one utilized in the local search procedure, or a movement that the local search algorithm cannot undo in just one step. In principle, any local search method can be used, however, its performance, in terms of the solution quality and computational effort, strongly depends on the chosen algorithm. The acceptance criterion is used to decide the next solution that should be perturbed. The selection of this criterion is important because it controls the balance between intensification and diversification. The search history is employed for deciding if some local optimum solution, found previously, should be chosen. The ILS procedure can lead to good samples of the search space as far as the perturbations are not too large or too small. If it is small, not many new solutions will be explored, while if it is too large, it will adopt almost randomly starting points.

```
Procedure ILS-VND(MaxIter, MaxIterILS, seed, γ, v)
  LoadData( );
  f(s*) := ∞;
  for k := 1,..., MaxIter do
      s := GenerateInitialSolution(γ, v, seed);
      s' := s;
      iterILS := 0;
      while iterILS < MaxIterILS do
          s := VND(N(.), f(.), r, s); {r = n⁰ of neighborhoods}
          if f(s) < f(s')
              s' := s;
              f(s') := f(s);
              iterILS := 0;
          end if;
          if s := Perturb(s');
          iterILS := iterILS + 1;
      end while;
      if f(s') < f(s*)
          s* := s';
          f(s*) := f(s');
      end if;
  end;
  return;
  end ILS-VND;
```

4   Solution Procedure

The proposed algorithm (ILS-VND) works as follows. The procedure is executed *MaxIter* times, where a initial solution is generated by a greedy heuristic and then it is improved by a procedure based on the ILS metaheuristic which realizes a local search by means of a VND heuristic. The pseudocode is described in the previous page, where s^* corresponds do the best solution, v is the number of vehicles imputed and γ is a parameter treated in details on Subsection 4.1.

4.1   Constructive Procedure

The method employed for building a feasible initial solution involves a greedy approach and is an adaptation of Dethlof's [4] insertion heuristic, but without considering the residual capacity. The pseudocode is given below.

```
Procedure GenerateInitialSolution(seed, γ, v)
  s := ∅;
  Inicialize the Candidate List(CL);
  Let s = {s¹,...,sᵛ} be the set composed by v empty routes;
  t := 1;
  while t ≤ v;
      sᵗ := e ∈ CL selected at random;
      Update CL;
      t := t + 1;
  end while;
  while CL ≠ ∅ do;
      Evaluate the value of each cost g(e) for e ∈ CL;
      gᵐⁱⁿ := min{g(e)|e ∈ CL};
      n := client e associated to gᵐⁱⁿ;
      s := s ∪ {n};
      Update CL;
  end while;
  return s;
end GenerateInitialSolution;
```

To begin with, the number of vehicles v to be considered for constructing the initial solution is pre-determined. Then, all routes are filled with a client e, chosen at random from the Candidate List (CL). Later, the clients belonging to the CL are evaluated according to the insertion criterion expressed by the Equation (1).

$$g\left(e^v\right) = (C_{ik} + C_{kj} - C_{ij}) - \gamma\left(C_{0k} + C_{k0}\right) \tag{1}$$

The first part of (1) is related to the well-known cheapest feasible insertion criterion, which consists of a greedy approach that takes into account the least additional cost regarding the insertion of the node k between the nodes i and j of the route v. Naturally, only the feasible insertions are admitted. The second part corresponds to a surcharge used to avoid late insertions of customers remotely located. The distance from the depot and back is weighted by a factor $\gamma \in [0, 1]$. The client e associated to $g^{\min}$ is then added to the solution s. The constructive procedure ends when all the clients have been added to the solution s.

4.2 Local Search

The local search phase, responsible to improve the initial solution is performed by a heuristic based on the VND algorithm. Mladenović and Hansen [19] proposed the variable neighborhood descent method which systematically modifies the neighborhood structures that belong to a set $N = \{N^{(1)}, N^{(2)}, N^{(3)}, \ldots, N^{(r)}\}$, in a deterministic way.

In the proposed algorithm, a set of six neighborhood structures are used to perform movements between clients of different routes. Just the feasible movements are admitted, i.e., the ones that do not violate the maximum load or maximum time constraints. Therefore, every time an improvement occurs, one should check whether this new solution is feasible or not. The N set of neighborhoods is described next.

Shift(1,0) – $N^{(1)}$ – A client c is transferred from a route $r1$ to a route $r2$. The vehicle load is checked as follows. All nodes located before the insertion's position have their loads added by q_c (delivery demand of the client c), while the ones located after have their loads added by p_c (pick-up demand of the client c). It is worth mentioning that certain devices to avoid unnecessary infeasible movements can be employed. For instance, before checking the insertion of c in some certain route, a preliminary verification is performed in $r2$ to evaluate the vehicle load before leaving, $\sum_{i \in r2} q_i + q_c$, and when arriving, $\sum_{i \in r2} p_i + p_c$, the depot. If the load exceeds the vehicle capacity Q, then all the remaining possibilities of inserting c in this route will be always violated.

Crossover – $N^{(2)}$ – The arc between adjacent nodes $c1$ and $c2$, belonging to a route $r1$, and the one between $c3$ and $c4$, from a route $r2$, are both removed. Later, an arc is inserted connecting $c1$ and $c4$ and another is inserted linking $c3$ and $c2$. The procedure for testing the vehicle load is more complex in comparison to Shift(1,0). At first, the initial (l_0) and final (l_f) vehicle loads of both routes are calculated. If the values of l_0 and l_f do not exceed the vehicle capacity Q then the remaining loads are verified through the following expression: $l_i = l_{i-1} + p_i - q_i$. Hence, if l_i surpass Q, the movement is infeasible.

Swap(1,1) – $N^{(3)}$ – Permutation between a node $c1$ from a route $r1$ and a node $c2$, from a route $r2$. The loads of the vehicles of both routes are examined in the same manner. For example, in case of $r2$, all clients situated before the position that $c2$ was found (now replaced by $c1$), have their values added by q_{c1} and subtracted by q_{c2}, while the load of the clients positioned after $c1$ increases by p_{c1} and decreases by p_{c2}.

Shift(2,0) – $N^{(4)}$ – Two consecutive nodes, $c1$ and $c2$, are transferred from a route $r1$ to a route $r2$. The vehicle load is tested likewise Shift(1,0).

Swap(2,1) – $N^{(5)}$ – Permutation of two consecutive nodes, $c1$ and $c2$, from a route $r1$ by a node $c3$ from a route $r2$. The load is verified by means of an extension of the approach used in the neighborhoods Shift(1,0) and Swap(1,1).

Swap(2,2) – $N^{(6)}$ – Permutation between two consecutive nodes, $c1$ and $c2$, from a route $r1$ by another two consecutive $c3$ and $c4$, belonging to a route $r2$. The load is checked just as Swap(1,1).

```
Procedure VND(N(.), f(.), r, s)
   {Let r be the number of neighborhood structures}
   k := 1; {current neighborhood}
   while k ≤ r do;
        Find the neighbor s' of s ∈ N^k;
        if f(s') < f(s)
            then
                  s := s';
                  f(s) := f(s');
                  k := 1;
                  {intensification in the modified routs}
                  s' := Or-opt(s);
                  s'' := 2-opt(s');
                  s''' := Exchange(s'');
                  s'''' := Reverse(s''');
                  if f(s'''') ≤ f(s)
                        s := s'''';
                        f(s) := f(s'''');
                  end if;
            else
                  k := k + 1;
        end if;
   end while;
   return s;
   end VND;
```

In case of improvement of the current solution, one should aim to further refine the quality of the routes that contributed to reduce the objective function, that is, those which participated in the last betterment move. Hence, four different neighborhoods are explored.

Or-opt – Introduced by Or [20], where one, two or three consecutive clients are removed and inserted in another position of the route.

2-opt – A pair of arcs is removed and another one is inserted.

Exchange – Permutation between two nodes

Reverse – This movement reverses the route direction if the value of the maximum load of the corresponding route is reduced.

4.3   Perturbation Mechanism

Just one mechanism was used for perturbing a local optimum solution, namely, the double-bridge movement. Introduced by Martin *et al.* [21] this perturbation was originally developed for the Traveling Salesman Problem (TSP) and consists in cutting four edges of a given route and inserting another four as shown in Fig. 1. In our case, whenever the **Perturb**() function is called, the double-bridge movement is randomly applied in all routes. Lourenço *et al.* [17] states that several applications of the ILS for the TSP have employed this type of perturbation, and it has been noted to be effective for different instance sizes.

It is important to point out that other perturbation mechanisms, involving more than one route, were tried. For example, exchanging paths of two different routes or successive Swap(1,1) moves. However, they were unsuccessful in most of the cases. The main reason is due to difficulty of finding a feasible move, since these perturbations often led to infeasible solutions, mostly because of the maximum time constraint.

Fig. 1. Double-Bridge Perturbation. Four edges are removed and four new ones are inserted.

5   Computational Results

The algorithm was implemented in C++ programming language, using the Borland C++ Builder 6.0 compiler and executed in a PC Intel Core 2 Duo 2.13 GHz with 1024 MB of RAM memory and operation system Windows XP - Professional Edition. The heuristic was applied to 14 test problems proposed by Salhi and Nagy [2]. In all instances the drop time was admitted to be the same for all customers. Some authors (Salhi and Nagy [2], Dethloff [4], Ropke and Pisinger [10]) have tested these instances including the drop time, while Montané and Galvão [8] did not include it in their work. Recently, Wassan *et al.* [14] admitted both situations. On the other hand, Nagy and Salhi [3] treated the problem including drop times but allowing two visits to the same client. Thus, a straightforward comparison cannot be made with their results.

For all the scenarios considered, the number of iterations ($MaxIter$) and perturbations allowed ($MaxIterILS$), was 15 and 30 respectively. Thirty executions were performed for each one of the different parameterizations of γ. Preliminary tests showed that $\gamma \in [0, 0.5]$ yielded superior results. In order to show the influence of the variation of this parameter, a graph was constructed (Fig. 2) in which γ was considered to be 0.05, 0.15, 0.25, 0.35 and 0.45. The gap shown

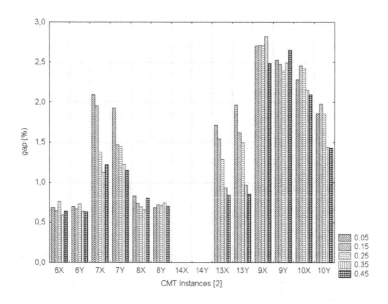

Fig. 2. Influence of the parameter γ on the average solutions of the ILS-VND

in the graph corresponds to the deviation between the average solution and the best solution found by the algorithm in each case.

It can be seen from the graph that the gap tends to be smaller for $\gamma = 0.25$ and $\gamma = 0.45$ as compared to $\gamma = 0.05$, $\gamma = 0.15$ and $\gamma = 0.35$. It should be stressed that the value of γ that yielded the best average solution frequently, but not always, led to the best solution determined by the ILS-VND.

The results found by the ILS-VND when the drop time was considered are shown in Table 1, where v_i represents the number of vehicles initially imputed and v_f the number of vehicles associated with the final solution. The average gap between the best solutions and the average solutions was 1.23%, with the highest value in the instance CMT9X (2.53%). Table 2 shows a comparison of our results with the best ones reported in the literature. We observe that, out of 14 problems, the ILS-VND improved the results in 12 instances and equaled 2, with an average gap of -1.85%.

Table 3 illustrates the results obtained by the ILS-VND without considering the drop time. In this case, the average gap between the best solutions and the average solutions was 1.83%, with the highest gap in the instance CMT13Y (3.86%). From Table 4, it can be verified that among the 14 test problems, the ILS-VND produced better results in 8 cases and the same result in a single instance. The average gap was -0.54%.

Analyzing the results, it can be verified that the developed heuristic considerably depends on the choice of the parameter γ. On the other hand, the double-bridge perturbation seemed to work quite well along with the different set of neighborhoods structures adopted.

Table 1. Results obtained in Salhi and Nagy's instances (with drop time)

Problem	No. of clients	v_i	v_f	Best Sol.	Time (s)	Avg. Sol.	Gap (%)	γ	Avg. Time (s)
CMT6X	50	7	6	555.43	2.47	558.71	0.59	0.35	2.61
CMT6Y	50	7	6	555.43	2.44	558.93	0.63	0.45	2.48
CMT7X	75	14	11	901.22	3.72	911.35	1.12	0.35	3.83
CMT7Y	75	14	11	901.22	3.88	910.88	1.07	0.40	3.79
CMT8X	100	10	9	865.50	19.53	870.72	0.60	0.50	20.13
CMT8Y	100	10	9	865.50	23.14	870.42	0.57	0.20	25.28
CMT14X	100	11	10	821.75	29.62	821.75	0.00	0.00	32.85
CMT14Y	100	11	10	821.75	28.33	821.75	0.00	0.00	31.87
CMT13X	120	12	11	1545.96	95.00	1566.97	1.36	0.20	102.57
CMT13Y	120	12	11	1542.86	105.66	1557.83	0.97	0.35	99.29
CMT9X	150	16	14	1167.23	69.20	1196.81	2.53	0.20	67.87
CMT9Y	150	17	14	1167.69	65.36	1194.08	2.26	0.10	66.89
CMT10X	199	21	18	1407.66	151.41	1437.31	2.11	0.20	157.55
CMT10Y	199	20	18	1413.88	148.11	1434.05	1.43	0.45	158.21

Table 2. Comparison between ILS-VND and literature results (with drop time)

Problem	No. of clients	Literature Best	v	ILS-VND	v	Gap(%)
CMT6X	50	556.06^W	6	**555.43**	6	-0.11
CMT6Y	50	558.17^W	6	**555.43**	6	-0.49
CMT7X	75	**901**^{RP}	-	**901.22**	11	0.00
CMT7Y	75	903.36^W	11	**901.22**	11	-0.24
CMT8X	100	**866**^{RP}	-	**865.50**	9	-0.06
CMT8Y	100	873^{RP}	-	**865.50**	9	-0.86
CMT14X	100	823.95^W	10	**821.75**	10	-0.26
CMT14Y	100	823.34^W	10	**821.75**	10	-0.19
CMT13X	120	1576^D	11	**1545.96**	11	-1.91
CMT13Y	120	1576^D	11	**1542.86**	11	-2.10
CMT9X	150	1197^{RP}	-	**1167.23**	14	-2.49
CMT9Y	150	1213.11^W	15	**1167.69**	14	-3.74
CMT10X	199	1462^{RP}	-	**1407.66**	18	-3.72
CMT10Y	199	1419.79^W	18	**1413.88**	18	-0.42

*(W) Wassan *et al.*; (RP) Ropke and Pisinger; (D) Dethloff.

Table 3. Results obtained in Salhi and Nagy's instances (without drop time)

Problem	No. of clients	v_i	v_f	Best Sol.	Time (s)	Avg. Sol.	Gap (%)	γ	Avg. Time (s)
CMT6X	50	3	3	466.77	3.56	467.62	0.18	0.30	4.13
CMT6Y	50	3	3	466.77	3.20	466.84	0.01	0.25	4.19
CMT7X	75	7	6	686.52	12.59	698.03	1.68	0.45	12.38
CMT7Y	75	7	6	688.46	11.62	698.97	1.53	0.10	12.93
CMT8X	100	5	6	721.40	27.33	727.60	0.86	0.35	31.04
CMT8Y	100	5	6	721.40	34.64	728.09	0.93	0.50	31.82
CMT14X	100	6	6	663.50	34.08	672.82	1.40	0.20	34.06
CMT14Y	100	6	5	662.22	37.09	672.56	1.56	0.20	32.48
CMT13X	120	5	4	846.85	63.95	878.27	3.71	0.50	61.50
CMT13Y	120	5	4	848.45	71.19	881.23	3.86	0.40	57.11
CMT9X	150	8	7	855.74	100.64	870.77	1.76	0.35	102.88
CMT9Y	150	8	7	856.74	111.47	871.21	1.69	0.30	103.71
CMT10X	199	11	10	1037.37	271.45	1058.31	2.02	0.05	284.52
CMT10Y	199	11	10	1036.59	225.45	1059.56	2.22	0.15	242.11

Table 4. Comparison between ILS-VND and literature results (without drop time)

Problem	No. of clients	Literature Best	v	ILS-VND	v	Gap(%)
CMT6X	50	471.89^{W}	3	**466.77**	3	-1.08
CMT6Y	50	467.70^{W}	3	**466.77**	3	-0.20
CMT7X	75	$\mathbf{663.95}^{\text{W}}$	6	686.52	6	3.40
CMT7Y	75	$\mathbf{662.50}^{\text{W}}$	6	688.46	6	3.92
CMT8X	100	$\mathbf{720}^{\text{MG}}$	5	721.40	5	0.19
CMT8Y	100	$\mathbf{721}^{\text{MG}}$	5	**721.40**	5	0.06
CMT14X	100	$\mathbf{644.70}^{\text{W}}$	5	663.50	6	2.92
CMT14Y	100	$\mathbf{659.52}^{\text{W}}$	6	662.22	5	0.41
CMT13X	120	858.48^{W}	5	**846.85**	4	-1.35
CMT13Y	120	880.56^{W}	4	**848.45**	4	-3.65
CMT9X	150	880.61^{W}	7	**855.74**	7	-2.82
CMT9Y	150	886.84^{W}	7	**856.74**	7	-3.39
CMT10X	199	1079.99^{W}	10	**1037.37**	10	-3.95
CMT10Y	199	1058.09^{W}	10	**1036.59**	10	-2.03

*(W)Wassan *et al.*; (MG) Montané and Galvão.

6   Conclusion

This paper dealt with the Vehicle Routing Problem with Simultaneous Pick-up and Delivery and Time Limit Constraint. An algorithm based on the ILS metaheuristic embedded with a VND procedure for performing the local search, was proposed. To the best of our knowledge, this was the first time the ILS has been applied to solve this problem.

The algorithm was tested in the instances proposed by Salhi and Nagy [2]. In most cases, it yielded better results than those reported in the literature. When the drop time was considered, the average gap between the ILS-VND solutions and the best ones found in the literature was -1.85% and when it was not included, the deviation was -0.54%.

References

1. Min, H.: The multiple vehicle routing problem with simultaneous delivery and pick-up points. Transportation Research 23(5), 377–386 (1989)
2. Salhi, S., Nagy, G.: A cluster insertion heuristic for single and multiple depot vehicle routing problems with backhauling. Journal of the Operational Research Society 50, 1034–1042 (1999)
3. Nagy, G., Salhi, S.: Heuristic algorithms for single and multiple depot vehicle routing problems with pickups and deliveries. European Journal of Operational Research 162(1) (2005)
4. Dethloff, J.: Vehicle routing and reverse logistics: the vehicle routing problem with simultaneous delivery and pick-up. OR Spektrum 23, 79–96 (2001)
5. Vural, A.V.: A ga based meta-heuristic for capacited vehicle routing problem with simultaneous pick-up and deliveries. Master's thesis, Graduate School of Engineering and Natural Sciences, Sabanci University (2003)
6. Gökçe, E.I.: A revised ant colony system approach to vehicle routing problems. Master's thesis, Graduate School of Engineering and Natural Sciences, Sabanci University (2004)

7. Crispim, J., Brandao, J.: Metaheuristics applied to mixed and simultaneous extensions of vehicle routing problems with backhauls. Journal of the Operational Research Society 56(7), 1296–1302 (2005)
8. Montané, F.A.T., Galvão, R.D.: A tabu search algorithm for the vehicle routing problem with simultaneous pick-up and delivery service. European Journal of Operational Research 33(3), 595–619 (2006)
9. Gribkovskaia, I., Halskau, Ø., Laporte, G., Vlcek, M.: General solutions to the single vehicle routing problem with pickups and deliveries. European Journal of Operational Research 180, 568–584 (2006)
10. Ropke, S., Pisinger, D.: A unified heuristic for a large class of vehicle routing problems with backhauls. Technical Report 2004/14, University of Copenhagen (2004)
11. Chen, J.F., Wu, T.H.: Vehicle routing problem with simultaneous deliveries and pickups. Journal of the Operational Research Society 57(5), 579–587 (2005)
12. Chen, J.F.: Approaches for the vehicle routing problem with simultaneous deliveries and pickups. Journal of the Chinese Institute of Industrial Engineers 23(2), 141–150 (2006)
13. Bianchessi, N., Righini, G.: Heuristic algorithms for the vehicle routing problem with simultaneous pick-up and delivery. Computers & Operations Research 34(2), 578–594 (2007)
14. Wasasn, N.A., Wassan, A.H., Nagy, G.: A reactive tabu search algorithm for the vehicle routing problem with simultaneous pickups and deliveries. Journal of Combinatorial Optimization (2007)
15. Dell'Amico, M., Righini, G., Salanim, M.: A branch-and-price approach to the vehicle routing problem with simultaneous distribution and collection. Transportation Science 40(2), 235–247 (2006)
16. Angelelli, E., Mansini, R.: A branch-and-price algorithm for a simultaneous pick-up and delivery problem. In: Quantitative Approaches to Distribution Logistics and Supply Chain Management, pp. 249–267. Springer, Heidelberg (2002)
17. Lourenço, H.R., Martin, O.C., Stützle, T.: Iterated Local Search. In: Handbook of Metaheuristics, pp. 321–353. Kluwer Academic Publishers, Norwell, MA (2002)
18. Stützle, T.: Local Search Algorithms for Combinatorial Problems. PhD thesis, Darmstadt University of Technology, Germany (1998)
19. Mladenović, N., Hansen, P.: Variable neighborhood search. Computers & Operations Research 24(11), 1097–1100 (1997)
20. Or, I.: Traveling salesman-type combinational problems and their relation to the logistics of blood banking. PhD thesis, Northwestern University, USA (1976)
21. Martin, O., Otto, S.W., Felten, E.W.: Large-step markov chains for the traveling salesman problem. Complex Systems 5, 299–326 (1991)

An Immune Genetic Algorithm Based on Bottleneck Jobs for the Job Shop Scheduling Problem

Rui Zhang and Cheng Wu

Department of Automation, Tsinghua University, Beijing 100084, P.R. China
zhang-r@mails.thu.edu.cn

Abstract. An immune genetic algorithm based on bottleneck jobs is presented for the job shop scheduling problem in which the total weighted tardiness must be minimized. Bottleneck jobs have significant impact on final scheduling performance and therefore need to be considered with higher priority. In order to describe the characteristic information concerning bottleneck jobs, a fuzzy inference system is employed to transform human knowledge into the bottleneck characteristic values which reflect the features of both the objective function and the current optimization stage. Then, an immune operator is designed based on these characteristic values and a genetic algorithm combined with the immune mechanism is devised to solve the job shop scheduling problem. Numerical computations for problems of different scales show that the proposed algorithm achieves effective results by accelerating the convergence of the optimization process.

1 Introduction

The job shop scheduling problem has long been an important research field since the 1950s and most shop scheduling problems have been shown to be $\mathcal{NP}$-hard [1]. Therefore, it's considerably difficult to obtain the optimal solution even for instances of 10 machines and 10 jobs. In recent years, global search strategies such as genetic algorithms (GA) played a significant role in solving small-scale job shop scheduling problems [2,3,4]. However, when the problem size grows, the $\mathcal{NP}$-hard property of job shop problems means that the search space will increase exponentially and the performance of heuristic search methods alone will hardly be satisfactory. To address this difficulty, two major approaches may be roughly identified in the existing literature:

(1) Traditional crossover and mutation operations have been modified in order to produce better offspring, and some new operators, such as the Estimation of Distribution Algorithms (EDA) [5], are devised to replace the traditional genetic operators so as to enhance the searching ability of GA;

J. van Hemert and C. Cotta (Eds.): EvoCOP 2008, LNCS 4972, pp. 147–157, 2008.

(2) Problem-specific or instance-specific information is extracted to accelerate the converging speed of GA.

In this latter approach, immune genetic algorithm (IGA) is designed to utilize the characteristic information based on the specific problem to improve the individuals in each generation. IGA has been proved efficient for the traveling salesman problem (TSP) [6]. However, for other optimization problems, how to effectively extract and describe the characteristic information remains a challenging but promising research topic.

In job shop scheduling, the concept of 'bottleneck' has attracted wide attention from many researchers [7,8,9]. In most problems, the final scheduling performance could be notably improved if these bottleneck machines or bottleneck jobs are well scheduled. However, how to describe the bottleneck indices in a quantitative manner and how to design bottleneck identification procedures for different scheduling objectives and different optimization stages are still difficult and crucial problems in the scheduling community. In this paper, we devise a fuzzy inference system based on human knowledge to evaluate the bottleneck characteristic value for each job and then use this information in an immune mechanism to promote the optimization efficiency of GA. Computational results for problems of different sizes show that the proposed algorithm is effective.

The paper is organized as follows. The discussed job shop scheduling problem is formulated in section 2. Section 3 gives the detailed algorithm. The computational results are provided in section 4. Finally, conclusions are given in section 5.

2   Problem Formulation

Job shop is one of the most frequently adopted models when dealing with scheduling problems. In a Job Shop Scheduling Problem (JSSP), a set of n jobs $\{J_i\}_{i=1}^{n}$ are to be processed on a set of m machines $\{M_k\}_{k=1}^{m}$ under the following basic assumptions:

- There is no machine breakdown and no preemption of operations is allowed;
- All jobs are released at time zero, and the transportation time between different machines and the setup time for different jobs are all neglected;
- Each machine can process only one job at a time and each job can be processed by only one machine at a time.

Each job has a fixed processing route which traverses all the machines in a predetermined order. Besides, a preset due-date is given for each job.

JSSP can also be described by a disjunctive graph $G(O, A, E)$ [10], in which $O \equiv \{0, 1, \cdots, *\}$ represents the set of nodes (including two dummy nodes, 0 and $*$); A is the set of conjunctive arcs and $E = \bigcup_{k \in M} E_k$ is the set of disjunctive arcs if we denote by E_k the disjunctive arcs that correspond to machine M_k. Then the discussed JSSP can be formulated as follows:

$$
\begin{cases}
\min \ WT = \sum_{j \in C(J)} w_j \left(t_j + p_j - d_j \right)^+ \\
\text{s.t.} \\
\qquad t_i + p_i \leq t_j, \quad (i,j) \in A, \\
\qquad t_i + p_i \leq t_j \vee t_j + p_j \leq t_i, \quad (i,j) \in E_k, k \in M, \\
\qquad t_i \geq 0, \quad i \in O.
\end{cases}
$$

In this formulation, $C(J)$ is the set of the final operations of all the jobs; w_j and d_j are respectively the tardiness weight and the due-date of the job which operation j belongs to; p_j and t_j are the processing time and the starting time of operation j, respectively; $(x)^+ = \max\{x, 0\}$.

The scheduling objective considered in this paper is to determine the processing sequence of operations on each machine such that the total weighted tardiness is minimized. The problem is noted as $J//\sum T_i$ in accordance with the three-field notation [11].

3   The Algorithm

3.1   The Bottleneck Characteristic Values

In the proposed IGA algorithm, the immune operator is designed based on the bottleneck characteristic information. The bottleneck characteristic value (JBN) of each job is evaluated by a fuzzy inference system described below.

In each generation of GA, when an individual is decoded and evaluated, we obtain the completion time of each job under the corresponding schedule. With this information, here we define:

- The relative distance between job i's completion time and its due-date: $g_i = \left(\tilde{F}_i - d_i \right) \big/ d_i$, where d_i and $\tilde{F}_i$ respectively denote job i's due-date and completion time under the current schedule;
- The relative slack time of job i: $h_i = \left(d_i - C_i^{(0)} - \sum_{j \in J'_i} p_j \right) \big/ d_i$, where $C_i^{(0)}$ refers to the completion time of the currently considered operation of job i and also the release time of J'_i which is the set of its succeeding operations in job i. Note that h_i corresponds to specific operations of job i and thus reflects the different processing stages of the job;
- Normalized tardiness weight of each job: $v_i = \bar{w}_i = (w_i - w_{\min})/(w_{\max} - w_{\min})$, where $w_{\max} = \max_{1 \leq i \leq n} w_i$, $w_{\min} = \min_{1 \leq i \leq n} w_i$.

Based on these variables, a fuzzy controller is designed to calculate the JBN value.

1. Input / output variables
   The fuzzy controller takes g_i, h_i and v_i as input variables, and outputs the JBN value for each job. In the fuzzy inference system, the four input/output linguistic variables are respectively denoted by G, H, V and B, and are divided into three fuzzy sets as follows.

- $G, H = \{NL, Z, PL\}$, i.e. {Negative, Around Zero, Positive};
- $V = \{S, M, L\}$, i.e. {Small, Medium, Large};
- $B = \{NB, MB, B\}$, i.e. {Not a bottleneck job, Maybe a bottleneck job, A bottleneck job}.

2. Definition of membership functions

In this fuzzy controller, all membership functions are chosen to have symmetrical triangular distributions, so we only have to determine the two end points. For example, if one of the linguistic variables discussed above is described as $\{(-\infty, a, b); (b, c); (c, d, +\infty)\}$, then the three corresponding membership functions are displayed in Fig. 1.

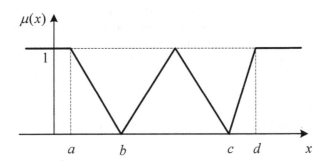

Fig. 1. The membership functions adopted by the fuzzy controller

3. The fuzzy rules

Bottleneck jobs are those jobs that need to be considered with higher priority in the optimization process because such jobs can have a significant role to play in improving the overall performance measures.

In practical shop scheduling practices, there exists certain human experience that indicates which jobs should have higher priority under different circumstances. After further abstraction, this kind of knowledge can be expressed in terms of fuzzy rules which have the form of "If..., then...". For example, "If $G = PL$, $H = NL$ and $V = L$, then $B = B$" means that if under the current schedule, a certain job has relatively large tardiness and relatively small slack time, and its weight is large, then this job should be regarded as a bottleneck job.

According to such priori knowledge, we obtain the fuzzy rule table shown in Table 1 by enumerating all possible and feasible combinations of the input variables. These rules try to reflect the basic properties of bottleneck jobs from different perspectives.

4. The inference system and defuzzification

The fuzzy inference system based on the 18 rules in Table 1 adopts the Mamdani model, in which the T-norm is implemented by the "min" operator. Since the output bottleneck characteristic value should be a quantifiable number, here we use the smallest-of-maximum Z_{SOM} [12] as the defuzzification method.

Table 1. The fuzzy rule table

$\begin{smallmatrix}&&V\\G,H&&\end{smallmatrix}$	S	M	L
NL, PL	NB	NB	NB
Z, Z	MB	B	B
Z, PL	NB	NB	MB
PL, NL	B	B	B
PL, Z	MB	B	B
PL, PL	NB	MB	MB

By applying the above fuzzy inference process to a selected individual in the current population, we obtain the bottleneck characteristic values $\{JBN_i\}_{i=1}^{n}$ for each job which are then used to design an immune operator that accelerates the optimization process of GA.

Particularly, the bottleneck characteristic values defined in this paper have the following important features:

(1) The definition of bottleneck reflects the characteristics of the scheduling objective function. Substantially different from the existing bottleneck identification methods based on machine workload or idle time, our proposed approach aims at the final scheduling performance and depicts the bottleneck in attempts to improve the performance measures. Therefore, bottleneck jobs should vary when the optimization objective changes.
(2) The bottleneck characteristic values reflect the characteristics of different optimization stages. In the evolutionary process of GA, the individuals keep changing under genetic operators. The proposed bottleneck characteristic values are valid only for the current state of an individual, and depict the bottleneck jobs in attempts to improve this individual in the subsequent optimization process. Therefore, bottleneck jobs should vary as the population evolves.

3.2   The Immune Genetic Algorithm Based on Bottleneck Jobs

To solve the job shop scheduling problem with the objective of minimizing total weighted tardiness, we design an immune genetic algorithm that utilizes the bottleneck characteristic information acquired by the aforementioned fuzzy inference system. The key steps in the algorithm are the construction and the application of the immune operator.

1. Encoding
   The encoding scheme is based on operation priority lists. An individual relates each machine with a priority list of n operations to be processed on this machine. Note that after the decoding procedure, the actual processing order of operations in the final feasible schedule may differ from these priority lists.

2. Initialization

Here we design a heuristic method to produce the initial population. For each operation, the SLACK index is defined as $\theta_i = d_i - \sum_{j \in J_i} p_j$, where $\sum_{j \in J_i} p_j$ is the total processing time of job i.

Then, we define a priority index α for each operation as

$$\alpha_i = \left(2 - \bar{\theta}_i\right)\left(1 + \bar{w}_i\right), \ \bar{\theta}_i, \bar{w}_i \in [0, 1],$$

where $\bar{\theta}$ and $\bar{w}$ denote normalized values, i.e., $\bar{x}_i = (x_i - x_{\min})/(x_{\max} - x_{\min})$, $x_{\min} = \min_{i \in N} x_i$, $x_{\max} = \max_{i \in N} x_i$, $x \in \{\theta, w\}$.

Finally, we sort all the operations in a non-increasing order of α, and by assigning the operations to their processing machines according to their order in this sequence, we obtain the priority lists as one initial solution for GA. The other individuals in the population are formed by random permutation of operations based on this solution.

3. Mutation

For a selected individual under the mutation probability, we first calculate the potential weighted tardiness of each machine in the corresponding schedule as

$$PWT_i = \sum_{j \in N^i} w_j \cdot \left[\left(C_j - LDD_j\right)^+ - \left(C_{i\_pre(j)} - LDD_{i\_pre(j)}\right)\right],$$

where N^i is the set of operations to be processed on machine i; C_j is the completion time of operation j and w_j is the weight of its corresponding job; i_pre(j) refers to the immediate job predecessor of operation j; LDD_j is defined as $LDD_j = d_j - \sum_{l \in JS(j)} p_l$ ($JS(j)$ is the set of job successors of operation j) and refers to the latest finishing time of operation j in the event that no tardiness is allowed.

PWT_i reflects the performance of the current schedule on machine i, and therefore if we focus the optimization operations more on those machines with higher PWT values, it is more likely to obtain performance improvements. So proportional method (Roulette Wheel Selection) is adopted here for the selection of machines, i.e., machine j is selected with probability $P_j^M = PWT_j / \sum_{k=1}^m PWT_k$. Finally, the SWAP operator is applied to exchange two operations in the selected machine's priority list, which implements the mutation operator.

4. Crossover

Under the crossover probability, each crossover is performed using the LOX operator to a randomly selected individual and the best individual in the current population. In particular, we first select a machine according to P_j^M defined above and then apply LOX to the two operation lists on the selected machine.

5. Decoding and evaluation

The decoding process concerns iteratively scheduling the ready operations (whose preceding operations have been scheduled) according to their priority order and as early as possible.

6. The immune operator
   (a) Selection of the vaccine
   In practical scheduling experiences, the jobs (those with larger JBN values) which are likely to cause deterioration to the scheduling performance measures are usually scheduled with a higher preference while the relatively safer jobs (those with smaller JBN values) are sometimes postponed to certain extent. Although this priori knowledge cannot guarantee the global optimality of the obtained schedule, it is a sensible choice when only a small fraction of the jobs are considered at a local level.
   Hence, in a satisfactory schedule for the original problem, there should exist a large number of operations whose processing orders are consistent with this priori knowledge, i.e., urgent operations are processed before moderate operations. Such properties may be used as characteristic information in the course of problem solving and act as an approach to abstracting a vaccine.
   (b) The vaccination process
   Step 1: Under the immune probability, choose an individual from the current population, apply the above-mentioned PWT-based roulette wheel selection to select a machine, and randomly choose an operation O_k from its priority list. Calculate $\{JBN_i\}_{i=1}^n$ for all the operations on this machine.
   Step 2: Evaluate $i_1 = \arg\max_{i \in N^p(k)} \left\{ (JBN_k - JBN_i)^+ \right\}$, where $N^p(k)$ denotes the set of operations that are before O_k in the priority list.
   Step 3: Evaluate $i_2 = \arg\max_{i \in N^s(k)} \left\{ (JBN_i - JBN_k)^+ \right\}$, where $N^s(k)$ denotes the set of operations that are after O_k in the priority list.
   Step 4: Let $i^* = \arg\max_{i \in \{i_1, i_2\}} \left\{ |JBN_k - JBN_i| \right\}$, and then swap O_k and O_{i^*} in the priority list.
   (c) The function of the immune operator
   The vaccination is performed after the crossover and mutation procedure to a selected portion of individuals in the current population. When the vaccination for an individual is finished, we evaluate the fitness of the new individual, and if its fitness is improved, the new individual will be kept down to the next generation, otherwise this vaccination is discarded and the individual will be restored.

4 Computational Results

4.1 Testing Problem Generation and the Algorithm Parameters

To test the effectiveness of our proposed algorithm (referred to as **IGA-BJ**), we randomly generate different-scale job shop scheduling problem instances[1]

---

[1] Because the discussed optimization objective is total weighted tardiness in this paper, the standard JSSP benchmark instances (from OR-Lib) which don't include due-date and tardiness weight information are not used here.

in which the route of each job is a random permutation of m machines and the (integral) processing time of each operation follows a uniform distribution $\mathcal{U}(1, 99)$. The due-date of each job is obtained by a series of simulation runs which apply different priority rules (such as SPT, SRPT, etc.) to the machines and finally we take the average completion time of each job among these simulations as its due-date. The (integral) tardiness weight of each job is generated from a uniform distribution $\mathcal{U}(1, 10)$.

The membership functions of the fuzzy inference system for calculating JBN values are:

- $\{(0.1, 0.3); (0.3, 0.7); (0.7, 1)\}$ for linguistic variable V;
- $\{(-1, -0.2); (-0.2, 0.2); (0.2, 1, +\infty)\}$ for linguistic variable G;
- $\{(-\infty, -1, -0.2); (-0.2, 0.2); (0.2, 1)\}$ for linguistic variable H and
- $\{(0, 0.1); (0.1, 0.6); (0.6, 1)\}$ for output B.

The parameters for IGA-BJ are chosen as:

(1) the mutation probability $p_m = 0.5$,
(2) the crossover probability $p_c = 0.8$,
(3) the immune probability $p_i = 0.8$,
(4) the population size $PS = 30$, and
(5) the number of generations $GN = 20$.

4.2   Numerical Computations and Comparison

We compare the proposed IGA-BJ with a standard genetic algorithm using operation-based encoding scheme (SGA) and the hybrid optimization strategy for job shop problems (GASA) presented in [13]. The GASA approach incorporates simulated annealing mechanism into the crossover operator and therefore accepts offspring individuals with probability $\min\{1, \exp(-\Delta f/t)\}$. The population size for SGA and GASA is chosen as $PS' = 40$, and other parameters are identical with those of IGA-BJ.

The results for 10 different-scale problems are shown in Table 2. In this table, we see that for small-scale instances (whose number of operations doesn't exceed 200), the difference between the three algorithms is trivial. But as the problem size grows, the advantage of the two hybrid approaches becomes apparent. For most problem instances, GASA and IGA-BJ achieve better results than SGA. Moreover, IGA-BJ obtains smaller $\overline{WT}$ than GASA for 8 out of the 10 instances, with the largest improvement rate as high as 14% (for instance 8). It is also observed that for problems where the ratio of job number to machine number (i.e., n/m) is relatively large, IGA-BJ results in more improvement, which shows the effectiveness of the immune operator based on bottleneck jobs in the course of GA search.

If we define the relative performance improvement ratio for IGA-BJ as

$$PIR_{\text{IGA-BJ}} = \frac{f_{\text{GASA}} - f_{\text{IGA-BJ}}}{f_{\text{SGA}} - f_{\text{IGA-BJ}}},$$

Table 2. Performance of IGA-BJ on different-scale scheduling instances

No.	Size ($n \times m$)	SGA			IGA-BJ			GASA		
		$\overline{WT}$[a]	WT_{min}[b]	Time[c]	$\overline{WT}$	WT_{min}	Time	$\overline{WT}$	WT_{min}	Time
1	10×10	21.8	19	3.0	21.4	18	6.3	21.5	18	4.2
2	20×5	180.7	172	2.6	184.3	170	6.1	185.5	166	3.5
3	10×20	92.1	85	9.4	90.4	82	19.5	89.5	83	14.7
4	20×10	284.2	271	9.0	280.8	256	17.9	283.6	263	13.1
5	20×15	278.4	264	16.3	273.0	260	29.4	272.8	252	26.4
6	50×6	191.4	183	15.8	178.4	152	28.1	189.3	159	24.5
7	20×20	563.5	525	43.9	521.7	517	74.8	551.9	531	64.0
8	40×10	2240.8	1976	41.6	1860.3	1841	67.6	2149.4	1828	63.4
9	50×10	1786.3	1583	59.3	1579.2	1428	80.3	1594.0	1479	78.4
10	100×5	9254.8	8747	58.0	8699.7	8530	78.2	8907.3	8604	77.9

[a] $\overline{WT}$ refers to the average total weighted tardiness of the obtained schedules by the algorithm in 10 consecutive runs.

[b] WT_{min} refers to the best (minimum) total weighted tardiness of the obtained schedules by the algorithm in 10 consecutive runs.

[c] Time (in seconds) is the average running time of the program on a P4-2.0 GHz / Windows XP platform.

where f_X refers to the average optimization result obtained by algorithm X (as is shown in Table 2), then the above computational results can be transformed into a histogram in Fig. 2.

From Fig. 2, we see that IGA-BJ achieves better improvement for medium-scale and larger-scale problems. When the total number of operations reaches 500, however, the exponential expansion of search space means that it's increasingly difficult for pure heuristic search algorithms to obtain high-quality

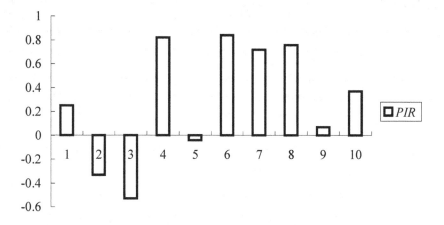

Fig. 2. The relative performance improvement of IGA-BJ

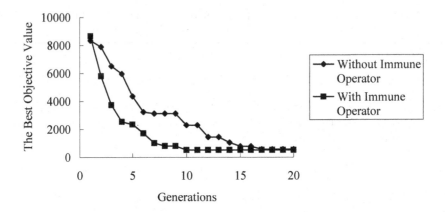

Fig. 3. Typical converging curves

schedules. In this case, decomposition-based optimization algorithms (e.g., [14] and [10]) might be used in future research and the proposed IGA-BJ can serve as an optimization module to deal with each subproblem in the decomposition-optimization framework.

Finally, to provide a closer observation for the impact of the immune operator on the optimization process, we provide two typical converging curves that occur in the course of solving problem 7 in Fig. 3.

In the future research, we will further statistically compare the performance difference between our algorithm and some of the most recent meta-heuristic approaches, such as [15], [16] and [4].

5   Conclusion

In this paper, a genetic algorithm based on bottleneck characteristic information for solving job shop scheduling problems is proposed. The defined bottleneck characteristic values reflect both the properties of the objective function and the most crucial jobs at different stages of the optimization process. The bottleneck information is extracted and used as an immune operator to increase the converging speed of GA. Numerical computational results show that the proposed algorithm is effective. Moreover, the results of this paper are also promising for due-date-related scheduling problems in practical manufacturing environment.

References

1. Lenstra, J.K., Kan, A.H.G.R., Brucker, P.: Complexity of machine scheduling problems. Annals of Discrete Mathematics 7, 343–362 (1977)
2. Cheng, R., Gen, M.: A tutorial survey of job-shop scheduling problems using genetic algorithms—Part I: representation. Computers & Industrial Engineering 34(4), 983–997 (1996)

3. Cheng, R., Gen, M.: A tutorial survey of job-shop scheduling problems using genetic algorithms—Part II: hybrid genetic search strategies. Computers & Industrial Engineering 36(2), 343–364 (1999)
4. Gao, J., Sun, L., Gen, M.: A hybrid genetic and variable neighborhood descent algorithm for flexible job shop scheduling problems. Computers & Operations Research (in press)
5. Larrañaga, P., Lozano, J.A.: Estimation of Distribution Algorithms: A New Tool for Evolutionary Optimization. Kluwer Academic Publishers, Boston (2002)
6. Jiao, L., Wang, L.: A novel genetic algorithm based on immunity. IEEE Transactions on Systems, Man and Cybernetics, Part A 30(5), 552–561 (2000)
7. Adams, J., Balas, E., Zawack, D.: The shifting bottleneck procedure for job shop scheduling. Management Science 34(3), 391–401 (1988)
8. Roser, C., Nakano, M., Tanaka, M.: Shifting bottleneck detection. In: Proceedings of the Winter Simulation Conference, pp. 1079–1086 (2002)
9. Varela, R., Vela, C.R., Puente, J., Gomez, A.: A knowledge-based evolutionary strategy for scheduling problems with bottlenecks. European Journal of Operational Research 145(1), 57–71 (2003)
10. Wu, S.D., Byeon, E.S., Storer, R.H.: A graph-theoretic decomposition of the job shop scheduling problem to achieve scheduling robustness. Operations Research 47(1), 113–124 (1999)
11. Jain, A.S., Meeran, S.: Deterministic job-shop scheduling: Past, present and future. European Journal of Operational Research 113(2), 390–434 (1999)
12. Jang, J.S.R., Sun, C.T., Mizutani, E.: Neuro-Fuzzy and Soft Computing. Prentice-Hall, Englewood Cliffs (1997)
13. Zhang, C.Y., Li, P.G., Rao, Y.Q., Li, S.X.: A new hybrid GA/SA algorithm for the job shop scheduling problem. In: Raidl, G.R., Gottlieb, J. (eds.) EvoCOP 2005. LNCS, vol. 3448, pp. 246–259. Springer, Heidelberg (2005)
14. Singer, M.: Decomposition methods for large job shops. Computers and Operations Research 28(3), 193–207 (2001)
15. Zhang, C.Y., Li, P.G., Rao, Y.Q., Guan, Z.L.: A very fast TS/SA algorithm for the job shop scheduling problem. Computers & Operations Research 35, 282–294 (2008)
16. Huang, K.L., Liao, C.J.: Ant colony optimization combined with taboo search for the job shop scheduling problem. Computers & Operations Research 35, 1030–1046 (2008)

Improved Construction Heuristics and Iterated Local Search for the Routing and Wavelength Assignment Problem

Kerstin Bauer[1], Thomas Fischer[1], Sven O. Krumke[2], Katharina Gerhardt[2], Stephan Westphal[2], and Peter Merz[1]

[1] Department of Computer Science
University of Kaiserslautern, Germany
{k_bauer,fischer,pmerz}@informatik.uni-kl.de
[2] Department of Mathematics
University of Kaiserslautern, Germany
{krumke,gerhardt,westphal}@mathematik.uni-kl.de

Abstract. This paper deals with the design of improved construction heuristics and iterated local search for the Routing and Wavelength Assignment problem (RWA). Given a physical network and a set of communication requests, the static RWA deals with the problem of assigning suitable paths and wavelengths to the requests. We introduce benchmark instances from the SND library to the RWA and argue that these instances are more challenging than previously used random instances. We analyze the properties of several instances in detail and propose an improved construction heuristic to handle 'problematic' instances. Our iterated local search finds the optimum for most instances.

1 Introduction

The *Routing and Wavelength Assignment* problem (RWA) deals with *Wavelength Division Multiplexed* (WDM) optical networks, where communication requests between nodes in a network have to be fulfilled by routing them on optical fiber links with a given capacity. Chlamtac *et al.* [1] showed the static RWA in general networks to be NP-complete since it contains the graph-coloring problem.

The problem is defined as follows: Given is a graph $G(V, E, W)$ with nodes V, arcs E and wavelengths W. An arc $e \in E$ is an optical fiber link in the physical network, where each wavelength $\lambda \in W$ is eligible. A request $r_i = (v_i^s, v_i^t, d_i)$ connects nodes v_i^s and v_i^t having a demand of $d_i \in \mathbb{N}^+$. For each unit of demand a lightpath between the request's endpoints has to be established. A *lightpath* is an optical path between two nodes created by the allocation of the same wavelength throughout the path of optical fiber links providing a 'circuit-switched' interconnection. Lightpaths have to fulfill two constraints: The *wavelength conflict constraint* defines that each wavelength on a physical link is used by at most one lightpath at the same time. The *wavelength continuity constraint* requires a lightpath to use the same wavelength on each link. There are problem variants

J. van Hemert and C. Cotta (Eds.): EvoCOP 2008, LNCS 4972, pp. 158–169, 2008.
© Springer-Verlag Berlin Heidelberg 2008

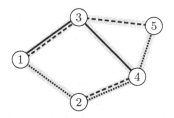

Fig. 1. Example for the RWA, using three wavelengths (different line styles) to route the requests $\{(1,5,2),(1,4,1),(2,4,1)\}$. The first request uses the dashed lightpath $\langle 1,3,5\rangle$ and dotted path $\langle 1,2,4,5\rangle$, the second request uses the solid path $\langle 1,3,4\rangle$ and the last request uses the dashed path $\langle 2,4\rangle$. Physical links are shaded in light gray.

relaxing these constraints: The first constraint can be relaxed by *multiple-fiber links* [2], the latter constraint can be relaxed by introducing *wavelength converters* [3] which change the wavelength of a lightpath at selected nodes.

The RWA can be seen as a *static* (*static lightpath establishment*, SLE) or a *dynamic* (*dynamic lightpath establishment*, DLE) problem. Furthermore, there are different types of *cost functions* available for the RWA. In the static case, a set of requests to be routed in parallel is known *a priori* and the objective is to find lightpaths for all requests minimizing the number of used wavelengths. In the dynamic case, time-bounded requests turn up over time, the routing has to be decided on-line, and the objective is to maximize the number of routed requests. In this paper, we focus on the static minimization problem.

First, we present related work on the static RWA and iterated local search. In Sec. 2 we give a formulation for the static RWA's lower bounds. Section 3 describes the used benchmark instances and our improved construction heuristics. In Sec. 4 we discuss a local search for the RWA, embedded into our iterated local search in Sec. 5. Finally, we draw conclusions and present ideas for future work.

1.1   Related Work

Ozdaglar and Bertsekas [4] present a Linear Programming (LP) approach to the RWA. Here, the RWA is represented as a *multicommodity network flow* problem with additional constraints. Additional constraints vary for setups with no, sparse, or full wavelength conversion. The LP formulations are similar to our formulation (Sec. 2), except that the authors use a piecewise linear cost function and limit the maximum number of wavelengths, whereas we assume constant link costs of 1 and impose no limit on the wavelengths to guarantee feasibility.

A memetic algorithm is presented by Sinclair [5] including a mutation operator, recombination (exchange a subset of paths between parents, reassign wavelengths) and two local search operators. The fitness function is rather complicated considering link length, link usage, node degree, and more. The first local search (called 'path reroute') reroutes a request in a wavelength with smaller index using one of the k-shortest paths, whereas the second local search ('path shift-out') assigns a path to another wavelength, but shifts out conflicting paths first. The network model uses the concept of several fibers on one physical link. Although this approach is said to perform well, the large population size (500) and the vast number of generations (100 000) question its efficiency.

Another approach for the RWA on general graphs is splitting the problem into subproblems for routing and wavelength assignment, respectively. An extensive overview on this approach is provided in [6] by Zang *et al.* and in [7] by Choi *et al.*

The static RWA is an NP-complete problem, for which we present an iterated local search algorithm. *Iterated Local Search* (ILS) [8] describes a class of algorithms that build a sequence of solutions using an embedded *local search* (LS) algorithm. To leave local optima, results of the embedded algorithm may be perturbated e. g. by a random mutation which moves the current solution to a different part of the search space allowing the LS to find other (hopefully better) local optima. LS [9] is an algorithm class defining for each solution a *neighborhood*, which is a subset of the solution space and contains all solutions which differ from the current solution in some selected aspect. In each iteration, the algorithm evaluates the current solution's neighborhood to choose a new solution and thus performs a random walk in the search space. Both neighborhood and moving strategy influence the LS's performance. Furthermore, LS algorithms are incomplete (finding optimal solutions is not guaranteed) and get stuck in local optima requiring diversification.

2   Lower Bounds

To evaluate our solutions' quality, we determined lower bounds by solving a relaxation of the RWA. Disregarding the wavelength continuity constraint, the RWA reduces to a multicommodity flow problem as follows. For each request $r \in R$ we have commodity which needs to be routed by means of flow through the network G. The mass balance d_v^r required for request r at node v is $-d_i$ if $v = s_i$, $+d_i$, if $v = t_i$ and zero otherwise. Let variable $x_{(i,j)}^r$ indicate the amount of flow from request r sent over the edge $(i,j) \in E$. Then, the problem of minimizing the maximum flow sent over an edge can be stated as:

$$\min \max_{(u,v) \in E} \sum_{r \in R} x_{(u,v)}^r + \sum_{r \in R} x_{(v,u)}^r$$

$$\sum_{u \in V:(u,v) \in E} x_{(u,v)}^r - \sum_{u \in V:(v,u) \in E} x_{(v,u)}^r = d_v^r \qquad \forall v \in V, r \in R \qquad (1)$$

$$x_{(u,v)}^r + x_{(v,u)}^r \leq d_v^r \qquad \forall r \in R, (i,j) \in E \qquad (2)$$

$$x_{(u,v)}^r \in \mathbb{N} \qquad \forall r \in R, (u,v) \in E$$

The objective is to minimize the maximum load of any edge in the network. Constraints (1) are flow-conservation constraints ensuring each request in the demand is fulfilled. Constraints (2) ensure that the flow for a request traverses an edge (u,v) only in one direction.

3   Benchmark Instances and Construction Algorithms

Solutions can be represented by a mapping from requests to sets of wavelength-path combinations. As our objective has the prerequisite of fulfilling all requests completely, an equivalent representation can be achieved by setting the requests' demands to 1, but allowing multiple requests for the same node pair. Therefore w. l. o. g., for the course of this paper every request has a demand of exactly one.

A proposal for a construction algorithm is given in [10]. Requests are processed iteratively by assigning each request a wavelength and a path depending on one of the strategies below. Initially, one wavelength is available, but the number of wavelengths is increased when no path can be found for a request.

First Fit (FF_RWA). Requests are ordered randomly and will be routed in the first wavelength with a feasible (shortest) path.

Best Fit (BF_RWA). Requests are ordered randomly and will be routed in the wavelength with the shortest feasible path.

First Fit Decreasing (FFD_RWA). Requests are sorted non-increasingly by the length of each request's shortest path in G and will be routed in the first wavelength with a feasible path.

Best Fit Decreasing (BFD_RWA). Requests are sorted non-increasingly by the length of each request's shortest path in G and will be routed in the wavelength with the shortest feasible path.

Experimental results in [10] indicate that BFD_RWA finds best results. Here, all test instances were constructed by the Erdős-Rényi random graph model $G(n, p)$ [11], where each possible edge (i, j) is chosen with probability $p = \frac{\delta}{n-1}$ independently from all other edges (where $n = |V|$ and δ is the expected node degree) and only connected graphs are accepted. In a second step the graph is made bidirectional by replacing each undirected edge by a pair of antiparallel directed edges. The demand between each pair of nodes (i, j) ((i, j) and (j, i) are considered as different node pairs) was set to 1 by a given probability between 0.2 and 1.0. Similar instances have been used in other publications, e. g. [12,13].

Our own preliminary experiments with this graph model and demand matrix (in our setting, however, edges can be used in both directions and thus are not replaced by a pair of unidirectional edges) indicate that this type of instance is completely uninteresting due to two facts: First, the demand is so small that construction algorithms as above already give (near-)optimal solutions. In a preliminary experiment, we generated a $G(n, p)$ instance with 50 nodes, $\delta = 3$, and probability for a request between a node pair of $p = 0.4$, where the BFD_RWA construction heuristic found a solution whose quality equals the lower bound (Sec. 2). Second, the underlying graph is 'pathological' in most cases. We call a graph pathological if it is only 1-edge connected, i. e. it contains at least one bridge. To illustrate the high probability of getting a pathological graph by the Erdős-Rényi random graph model, consider a graph $G(100, \delta=3)$. Here it is rather improbable that a graph with a connectivity of degree > 1 will be generated. The probability of one special node being a leaf is already $\left(1 - \frac{3}{99}\right)^{98} \cdot \left(\frac{3}{99}\right)^1 \cdot 99 = 0.147$ indicating that the probability of an at least 2-edge connected graph is very small [14]. Every request depending on such a bridge automatically requires an additional wavelength increasing the

Table 1. Properties of instances from SNDlib [15]. 'Pairs' describes the number of node pairs communicating with each other, 'Requests' summarizes the demand volume.

Instance	Nodes	Edges	Pairs	Requests	Lower Bound
atlanta[†]	15	22	210	6840	1256
france[†]	25	45	300	10008	1060
germany50	50	88	662	2365	147
janos-us-ca[†]	39	122	1482	10173	1288
newyork	16	49	240	1774	85
nobel-eu	28	41	378	1898	304
nobel-germany	17	26	121	660	85
nobel-us	14	21	91	5420	670
norway	27	51	702	5348	543
pdh	11	34	24	4621	214
polska	12	18	66	9943	1682
zib54	54	81	1501	12230	705

[†] Instance has been modified, see text for details.

total number of wavelengths, although previously allocated wavelengths still have a lot of free links.

Instead of generating random instances as described above, we used standard benchmark instances from the SNDlib collection [15]. This collection consists of 22 networks and a set of associated models which can be used as problem instances for the Survivable Network Design problem. The data was derived from industrial and research background. A network is described by nodes,

(physical) links, demands, and other, for our problem irrelevant planning data. From this collection we chose 12 networks as listed in Tab. 1. We restricted our selection as the other instances were either uninteresting for sophisticated algorithms or exceeded our available system capacities. To include some large instances, we scaled down the demand matrices of instances atlanta, france, and janos-us-ca by a factor of 20, 10, and 200, respectively. For the benchmark instances, capacity constraints and predefined admissible paths were not used and edge costs were fixed to 1. Edges were interpreted to be bidirectional and all demands between node pairs (i, j) and (j, i) were summed up to a unidirec-

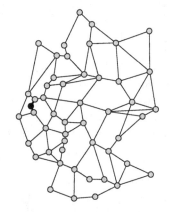

Fig. 2. Instance germany50 based on 50 cities in Germany

tional demand. The graph of instance nobel-us corresponds to the well-known NSF network with 14 nodes.

Our preliminary experiments indicated that the structure of the graph is not the only limiting factor for the solution quality. Additional problems arise if the requests' load is unbalanced distributed among the node pairs. Given a node with degree δ, in each wavelength this node can handle at most δ requests, especially requests starting or ending in this node. In instance **germany50** (Fig. 2), the highlighted node has a node degree of 2 and 43 requests start here with a summarized demand of 293. This leads to the lower bound $\lceil \frac{293}{2} \rceil = 147$, which equals the solution found by our iterated local search (Sec. 5) and the lower bound given by the multicommodity flow solution (Tab. 1).

We classify requests depending on overloaded edges as 'evil' requests. The influence of evil requests can best be observed for instances where an unfavorable combination of graph and request set leads to some heavily overloaded edges. These observations motivate a new approach of the construction heuristic. In this new approach, evil requests are preferably routed during the construction phase. For our experiments we consider three different sorting strategies. Other combinations of the components Len, Anti and Evil are also possible but will not be considered further.

Len. Requests are ordered non-increasingly by the length of their shortest path, equals BFD_RWA.

AntiEvil. Evil requests are routed first, otherwise using a random sorting.

LenAntiEvil. Requests are first sorted by Len and within each set of requests with equal shortest path length by AntiEvil.

AntiEvilLen. Requests are first sorted by AntiEvil and within the evil and non-evil requests, respectively, by Len.

Shuffle Requests are ordered in a random fashion.

To locate evil requests for a given graph and set of requests, we propose the following method. In the first step, an initial solution is constructed with the standard BFD_RWA algorithm and all edges $e \in E'$ heavily used in marginally used wavelengths are taken as candidates for overloaded edges. We define marginally used wavelengths as wavelengths whose usage lies below a multiple k of the average path length. In a second step requests are marked as evil requests, if they cannot be routed in $G \setminus E'$ (Fig. 3). Here, $u(s, \lambda)$ calculates the actual load of wavelength λ (number of used physical links), use(s, W', e) describes how often edge e is used in solution s restricted to wavelengths W', and routeable(G, r) checks if request r is physically routeable in G.

1: **function** FINDEVILREQUESTS(Graph G, Requests R, $k \in \mathbb{R}$)
2:     $s \leftarrow$ BFD_RWA(G, R) ▷ Construct a BFD_RWA solution
3:     $n \leftarrow k \cdot \lceil \text{avg}_{r \in R} \, l(p_{G,s}(r)) \rceil$ ▷ n is the k times the average path length in s
4:     $E \leftarrow \emptyset$ ▷ Set of possibly overloaded edges
5:     $W' \leftarrow \{\lambda \in W : u(s, \lambda) < n\}$ ▷ Set of marginally used wavelengths
6:     **for all** $e \in E$ **do**
7:         **if** use$(s, W', e) > |W'| - 2$ **then**   ▷ is e used in $> |W'| - 2$ wavel. of W'?
8:             $E' \leftarrow E' \cup \{e\}$ ▷ store overloaded edge
9:     **return** $\{r \in R : \neg \text{routeable}(G \setminus E', r)\}$

Fig. 3. Determination of Evil Requests

Table 2. Construction heuristic's results (minimum and average over 50 runs)

Instance	Len min	Len avg	LenAntiEvil min	LenAntiEvil avg	AntiEvilLen min	AntiEvilLen avg	AntiEvil min	AntiEvil avg	Shuffle min	Shuffle avg
atlanta	<u>1415</u>	<u>1473.7</u>	1417	1474.0	1424	1518.0	1460	1556.5	1457	1524.8
france	<u>1091</u>	<u>1115.1</u>	1095	1120.2	1094	1131.7	1110	1175.3	1105	1164.6
germany50	185	193.2	185	190.8	165	166.4	<u>161</u>	<u>164.3</u>	163	172.9
janos-us-ca	1781	1806.7	1764	1791.3	<u>1495</u>	<u>1507.2</u>	1523	1567.3	1522	1611.7
newyork	92	96.9	<u>91</u>	97.4	<u>91</u>	<u>96.4</u>	95	100.5	93	101.3
nobel-eu	<u>304</u>	304.2	<u>304</u>	<u>304.0</u>	<u>304</u>	<u>304.0</u>	<u>304</u>	305.5	<u>304</u>	310.3
nobel-germany	<u>91</u>	<u>94.0</u>	92	94.1	<u>91</u>	94.5	93	98.8	92	99.7
nobel-us	827	892.1	812	885.7	836	890.1	<u>805</u>	<u>877.7</u>	835	898.6
norway	<u>555</u>	564.8	557	<u>564.7</u>	559	568.5	570	586.8	570	585.0
pdh	267	298.0	275	296.4	271	298.7	<u>257</u>	297.0	268	<u>291.6</u>
polska	1750	<u>1836.1</u>	<u>1747</u>	1861.6	1788	1943.1	1842	2010.7	1855	1971.9
zib54	824	903.7	<u>816</u>	<u>884.2</u>	825	892.2	827	947.6	854	983.3

3.1 Experimental Results

All problem instances in Tab. 1 were solved by the construction heuristics Len, LenAntiEvil, AntiEvilLen, AntiEvil and Shuffle. Each experiment was repeated with 50 seeds, the results (minimum and average) are summarized in Tab. 2.

As can be seen in Tab. 2, there exists no clear preference between the sorting strategies Len and LenAntiEvil (except for germany50 and janos-us-ca, the confidence intervals overlap). Instances germany50, janos-us-ca, nobel-us, and pdh perform better with the sorting strategy AntiEvil than with Len or LenAntiEvil regarding the best solution, however, only for janos-us-ca and germany50 there is a significant difference (99 %). E. g. the best solution for germany50 using AntiEvil is 13.0 % better than using Len or LenAntiEvil. We observed that for these instances the least used wavelengths in solutions built using Len contain only a few, similar paths indicating that the underlying request matrix is unbalanced and thus yields some heavily overloaded edges. E. g. in one selected solution for germany50, a request having a demand of 76 uses 32 wavelengths used by no other request to route 63 lightpaths. As long as there exist few evil requests, AntiEvilLen nearly matches with the sorting strategy Len and thus has quite a similar but slightly worse performance. For the instances where AntiEvil performs well, no clear preference can be made between AntiEvil and AntiEvilLen. Thus, for unknown instances our experiments indicate first to try both strategies Len and AntiEvil and then depending on which strategy performs better to construct the final solutions either with Len/LenAntiEvil or with AntiEvil/AntiEvilLen, respectively. However, there may be a large variance between different solutions from the same construction heuristic. E. g. for nobel-us, the variance on the number of wavelengths ranges between 24.2 and 40.5 (not shown in Tab. 2). This suggests to construct several solutions to confirm the decision for the best construction heuristic.

1: **procedure** SHIFTPATHS(Solution s, Requests R)
2:    **for all** $r \in R$ **do**   ▷ for each request ...
3:       $W' \leftarrow \{\lambda \in W | \Psi_{G,s}(r, \lambda) \neq \emptyset\}$ ▷ find wavelengths in which r can be routed
4:       $\lambda \leftarrow \arg\max_{\lambda \in W'} u(s, \lambda)$ ▷ find wavelength λ with maximum load
5:       **if** $u(s, \lambda) > u(s, \lambda_s(r))$ **then**   ▷ compare usage of λ and request's wavel.
6:          $\lambda_s(r) \leftarrow \lambda$ ▷ set request's new wavelength
7:          $p_{G,s}(r) \leftarrow \Psi_{G,s}(r, \lambda)$ ▷ request's path set to shortest path in new wavel.

Fig. 4. Local Search moving path to wavelengths with higher load

4   Local Search

Although there is a large variety in solution quality among different request sorting strategies and random seeds, in many cases the results are considerably worse than the lower bounds motivating our local search (LS).

The general idea of our LS is to shift requests from less used wavelengths to more often used wavelengths by looking for alternative (possibly longer) paths. The algorithm (Fig. 4), which operates on a solution s and the set of requests R, works as follows: For each request $r \in R$, the set of all wavelengths $W' \subseteq W$ in which r is routeable is determined. Among all wavelengths in W', the wavelength λ with highest load is chosen. If the load for λ is larger than the load for the request's current wavelength, then the request's wavelength is set to λ and the request's path is updated with the shortest path in λ. Function $\Psi_{G,s}(r, \lambda)$ calculates the shortest path in G for request r routed in λ and function $u(s, \lambda)$ is introduced in Sec. 3. We define $\Psi_{G,s}(r, \lambda) = \emptyset$, iff there exists no such path.

4.1   Experimental Results

To evaluate the effectiveness of the LS, we performed multistart experiments using the same setup as described in Sec. 3.1. We applied our LS SHIFTPATHS to the 50 initial solutions of each construction heuristic setup until a local optimum was reached. The results are summarized in Tab. 3 (best of 50).

When comparing our multistart LS algorithm to the construction heuristics, the former performs only slightly better than the underlying construction heuristic. E. g. for instance germany50, the construction heuristic's best solution using Len is 185 (average 193.2), but the multistart LS's best solution is only 184. A significance analysis (99 % confidence interval) shows that the multistart LS improves the construction heuristics only in 5 cases (france+Shuffle, germany50+AntiEvil/AntiEvilLen, norway+AntiEvil/Shuffle).

This observation matched our expectations, as both the construction heuristics and the LS follow similar strategies. The only difference is that the construction heuristics prefer shortest paths, whereas the LS allows to reroute requests using longer paths if the load on wavelengths gets changed in respect of the optimization criterion. Differences in the quality of solutions created by the various construction heuristics cannot be compensated by the multistart LS.

Table 3. Multistart local search's results (best of 50 runs)

Instance	Len	LenAntiEvil	AntiEvilLen	AntiEvil	Shuffle
atlanta	1414	1417	1424	1418	1430
france	1086	1092	1085	1102	1096
germany50	184	185	164	159	160
janos-us-ca	1781	1764	1494	1497	1495
newyork	92	90	91	92	92
nobel-eu	304	304	304	304	304
nobel-germany	91	91	91	91	89
nobel-us	827	804	808	787	802
norway	554	554	555	556	560
pdh	263	259	266	254	256
polska	1748	1747	1787	1827	1854
zib54	800	787	785	785	810

5   Iterated Local Search

As shown above, local search alone is not sufficient to considerably improve the quality of the initial solutions. Thus, in order to escape from local optima, we introduce an iterated local search (ILS) combining the local search with a mutation (perturbation) operator which randomly changes paths within an already existing solution (Fig. 5). In each mutation step, two wavelengths λ_1 and λ_2 are randomly chosen and an arbitrary request whose path p is routed in the wavelength with lower usage (here, λ_2) is taken. Paths preventing p from being routed in wavelength λ_1 are removed from the solution and p is routed in λ_1. The function $\text{rem}(s, \lambda_1, p)$ determines the paths to be removed and stores the paths' requests in R'. To restore a valid solution, all requests in R' are rerouted in the first possible wavelength.

We applied different mutation strategies to the ILS, where the strength is defined by a percentage of the number of paths. Mutation strategies were either constant (ranging between 1 % and 25 %) or variable, where the mutation strength started with an initial high mutation rate (either 10 % or 25 %) and decreased linearly (step width 1 % or 2 %) until reaching a strength of 1 %.

Within our ILS, we accept new solutions after mutation and local search iff it is better (by length-lex ordering on wavelength usage vectors) than the previous best solution, otherwise the previous best solution is restored. Length-lex ordering sorts vectors first by length and then by lexicographic ordering.

5.1   Experimental Results

Due to similarities of Len and LenAntiEvil, we restrict the following discussion to Len, AntiEvil, AntiEvilLen, and Shuffle. The number of generations was limited to 50 as for most instances optimal solutions were found within this bound. Experiments were repeated 50 times.

```
 1: procedure MUTATE(Solution s, Requests R, strength)
 2:     for i = 1 ... strength do
 3:         (λ₁, λ₂) ← RandomWavelengths
 4:         if u(s, λ₁) < u(s, λ₂) then
 5:             swap(λ₁, λ₂)
 6:         r ← rInWL(s, λ₂, R)          ▷ get a request r with path p in wavelength λ₂
 7:         p ← p_{G,s}(r)
 8:         R' ← rem(s, λ₁, p)           ▷ remove paths stopping p from being routed in λ₁
 9:         p_{G,s}(r) ← p
10:         λ_s(r) ← λ₁
11:         for all r ∈ R' do                        ▷ for all currently unrouted paths
12:             λ_s(r) ← arg min_{λ∈W} Ψ_{G,s}(r, λ) ≠ ∅
13:             p_{G,s}(r) ← Ψ_{G,s}(r, λ)    ▷ route request on the first possible wavelength
```

Fig. 5. Mutation shifting path between wavelengths

Whereas LS without mutation is not powerful enough to improve initial solutions, the ILS always improves them to the optimum for many instances (reaching the LB from Sec. 2 and Tab. 1). Best results were achieved with a variable mutation strategy, where the initial strength was set to 25 % decreasing by 2 % each generation ($25\%{\downarrow}_{2\%}$, Tab. 4). In each setup, the ILS resulted in significantly better results than the multistart LS (99 % confidence intervals).

Regarding the convergence towards the optimal solution, we observed two different patterns. In the first case, the optimal solution was reached in a few generations by setups with strong mutation, whereas setups with weak mutation converged much slower. Instance atlanta (Fig. 6a) is an example for this behavior, where the optimal solution is reached after about 15 generations for strong mutation (25 %), after 40 generations for mutation strength 5 % and not

Table 4. Iterated Local Search's results with minimum and average over 50 runs for the mutation strategy starting at 25 % and decreasing by step width 2 % ($25\%{\downarrow}_{2\%}$)

Instance	LB	Len min	avg	AntiEvilLen min	avg	AntiEvil min	avg	Shuffle min	avg
atlanta	1256	1256	1256.5	1256	1256.4	1256	1256.3	1256	1256.2
france	1060	1061	1062.5	1060	1062.4	1060	1062.7	1061	1062.9
germany50	147	147	147.8	147	147.6	147	148.0	147	147.5
janos-us-ca	1288	1343	1351.4	1337	1347.5	1340	1357.8	1342	1358.4
newyork	85	85	85.1	85	85.2	85	85.0	85	85.2
nobel-eu	304	304	304.0	304	304.0	304	304.0	304	304.0
nobel-germany	85	85	86.2	86	86.5	85	86.4	86	86.4
nobel-us	670	684	689.3	685	689.3	684	689.3	684	688.6
norway	543	543	543.0	543	543.1	543	543.1	543	543.0
pdh	214	215	217.3	215	217.2	216	217.3	215	217.0
polska	1682	1682	1682.2	1682	1682.1	1682	1682.1	1682	1682.1
zib54	705	709	711.3	709	711.5	708	710.0	708	710.2

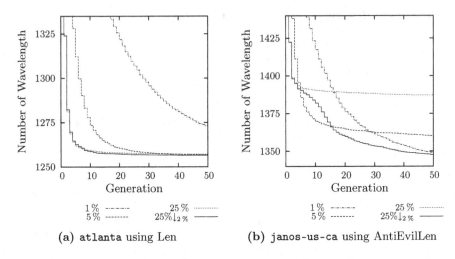

(a) `atlanta` using Len   **(b)** `janos-us-ca` using AntiEvilLen

Fig. 6. Performance plots for two ILS setups using different mutation strategies

reached for weak mutation (1 %) within 50 generations. In the second case, the optimal solution was not reached by our ILS. Again, our mutation strategies performed differently. Stronger mutations let the ILS converge faster, but, how-ever, get stuck in weak local optima. Weaker mutation lead to slow convergence, but end eventually in better solutions. Using dynamic mutation, our ILS con-verges fast during the initial phase and later is able to continuously improve its current solution. An example for this behavior is shown in Fig. 6b for prob-lem instance `janos-us-ca`. The bump for line '25%↓$_{2}$%' is due to the dynamic mutation strategy and varies for different parameters.

For 8 out of 12 benchmark instances our algorithm is able to find optimal solutions, as it reaches the lower bound. Here, we can argue that the use of wavelength converters will not result in solutions with less wavelengths, as the lower bound is based on a relaxation assuming wavelength converters at ev-ery node. For `pdh` and `zib54`, near-optimal solutions are found, and with more generations our algorithm can find optimal solutions. Only for `janos-us-ca` and `nobel-us` our ILS is not able to reach solutions close to the lower bound. Reasons may be that these two problem instances are harder than the other instances or that the lower bounds are considerably below the optimal solution.

6   Conclusion

In this paper we improved an existing construction heuristic and developed an iterated local search for the static RWA. We adopted problem instances from the SND library for the RWA and argued that these benchmark instances are more interesting than previously used random instances.

As for some benchmark instances the BFD_RWA construction heuristic per-formed badly, we subsequently suggested alternative request sorting strategies resulting in considerably better initial solutions. To further improve these solutions, we introduced an ILS which finds provable optimal solutions for eight instances and near-optimal solutions for two more instances.

Future work will focus on run-time optimizations for large instances. We are evaluating a multi-level approach based on the scaling mechanism we used for large problems in this paper. Furthermore, using don't-look-bits on wavelengths or requests to restrict the search space is another promising concept.

References

1. Chlamtac, I., Ganz, A., Karmi, G.: Lightnet: Lightpath based solutions for wide bandwidth wans. In: INFOCOM, pp. 1014–1021 (1990)
2. Xu, S., Li, L., Wang, S.: Dynamic routing and assignment of wavelength algorithms in multifiber wavelength division multiplexing networks. IEEE J. Sel. Areas Comm. 18(10), 2130–2137 (2000)
3. Chlamtac, I., Faragó, A., Zhang, T.: Lightpath (Wavelength) Routing in Large WDM Networks. IEEE J. Sel. Areas Comm. 14(5) (1996)
4. Ozdaglar, A.E., Bertsekas, D.P.: Routing and wavelength assignment in optical networks. IEEE/ACM Trans. Netw. 11(2) (2003)
5. Sinclair, M.C.: Minimum cost routing and wavelength allocation using a genetic-algorithm/heuristic hybrid approach. In: Proc. 6th IEE Conf. Telecom. (1998)
6. Zang, H., Jue, J.P., Mukherjee, B.: A Review of Routing and Wavelength Assignment Approaches for Wavelength-Routed Optical WDM Networks. Optical Networks Magazine 1(1), 47–60 (2000)
7. Choi, J.S., Golmie, N., Lapeyrere, F., Mouveaux, F., Su, D.: A Functional Classification of Routing and Wavelength Assignment Schemes in DWDM networks: Static Case. In: Proc. of the 7th International Conference on Optical Communications and Networks, OPNET 2000, pp. 1109–1115 (2000)
8. Lourenço, H.R., Martin, O.C., Stützle, T.: Iterated Local Search. In: Glover, F., Kochenberger, G. (eds.) Handbook of Metaheuristics. International Series in Operations Research & Management Science, vol. 57, pp. 321–353 (2002)
9. Hoos, H.H., Stützle, T.: Stochastic Local Search: Foundations and Applications. The Morgan Kaufmann Series in Artificial Intelligence. Morgan Kaufmann, San Francisco (2004)
10. Skorin-Kapov, N.: Routing and Wavelength Assignment in Optical Networks using Bin Packing Based Algorithms. EJOR 177(2), 1167–1179 (2007)
11. Erdős, P., Rényi, A.: On Random Graphs I. Publ. Math. Debrecen 6, 290–297 (1959)
12. Manohar, P., Manjunath, D., Shevgaonkar, R.K.: Routing and Wavelength Assignment in Optical Networks From Edge Disjoint Path Algorithms. IEEE Communications Letters 6(5), 211–213 (2002)
13. de Noronha, T.F., Resende, M.G.C., Ribeiro, C.C.: A Random-Keys Genetic Algorithm for Routing and Wavelength Assignment. In: Proc. of the Seventh Metaheuristics International Conference (MIC 2007) (2007)
14. Bollobás, B.: Random Graphs. Cambridge University Press, Cambridge (2001)
15. Orlowski, S., Pióro, M., Tomaszewski, A., Wessäly, R.: SNDlib 1.0–Survivable Network Design Library. In: Proc. of the 3rd International Network Optimization Conference (INOC 2007), Spa, Belgium (2007), http://sndlib.zib.de

Improving Metaheuristic Performance by Evolving a Variable Fitness Function*

Keshav Dahal, Stephen Remde, Peter Cowling, and Nic Colledge

MOSAIC Research Group, University of Bradford, Bradford, BD7 1DP, United Kingdom
{s.m.remde, p.i.cowling, k.p.dahal, n.j.colledge}@bradford.ac.uk
http://mosaic.ac/

Abstract. In this paper we study a complex real world workforce scheduling problem. We apply constructive search and variable neighbourhood search (VNS) metaheuristics and enhance these methods by using a variable fitness function. The variable fitness function (VFF) uses an evolutionary approach to evolve weights for each of the (multiple) objectives. The variable fitness function can potentially enhance any search based optimisation heuristic where multiple objectives can be defined through evolutionary changes in the search direction. We show that the VFF significantly improves performance of constructive and VNS approaches on training problems, and "learn" problem features which enhance the performance on unseen test problem instances.

Keywords: Variable Fitness Function, Evolution, Heuristic, Meta-heuristic.

1 Introduction

Search gets stuck when local optima are reached, and there are no better neighboring solutions, but the solution is not globally optimal. While the global fitness function is ideally suited to an approach guaranteed to find an optimal solution, it is not adequate in assessing the fitness of a local move. Many metaheuristics allow escape from these local optima however they may ultimately fail at a higher level because of the nature of the global fitness function.

The variable fitness function seeks to tackle this problem by redefining the fitness function so it may change over the course of the search. The result is that the local fitness function is different from the global fitness function and can be more effective than the global fitness function to assess local moves. [1] shows the variable fitness function's effectiveness at enhancing local search heuristics and in this paper we attempt to show its ability to enhance a metaheuristics and to learn reusable information to guide the search of a difficult optimization problem. The problem we study is a complex real world workforce scheduling problem which contains many scheduling problems from the literature as subproblems. Like many other real world problems it has many features that are hard to understand and model, and objectives that are

---

* This work was funded by EPSRC and @Road Ltd, a Trimble Company under an EPSRC CASE studentship, which was made available through and facilitated by the Smith Institute for Industrial Mathematics and System Engineering.

J. van Hemert and C. Cotta (Eds.): EvoCOP 2008, LNCS 4972, pp. 170–181, 2008.

non-linear in nature. This can make it hard for a person to define a global fitness function, let alone one to describe how to assess the quality of local moves. We aim to show the variable fitness function's ability to enhance the search when we solve this workforce scheduling problem using constructive search and Variable Neighborhood Search (VNS).

In the next section, related work is discussed. In section 3 the variable fitness function is defined and section 4 describes the problem. In section 5 the computational experiments are presented and the results analysed. Finally, section 6 will draw conclusions.

2 Related Work

The variable fitness function (VFF) is a new search enhancement technique that can be used to enhance any search based optimization heuristic provided that a) the problem is multi-objective (or multiple objectives can be defined in some way) and b) we have CPU time available to use this process offline (although the resulting VFF can be used very quickly online). First presented in [1], the variable fitness function provides a simple scheme for encoding a piecewise linear function into a genetic algorithm and a method for evolving these functions. The variable fitness functions are then used to determine the local fitness function at each step in the local search.

Guided Local Search [2] also modifies the fitness function to change the direction when a local optimum has been found. Features of a solution are identified and penalties for solutions exhibiting these features are increased when the solution is stuck in a local optimum. A feature which occurs in a local optimum has its penalty score increased slightly, and these penalties are used to modify the fitness function, attempting to force the search to move in another direction. The primary differences between the VFF and Guided Local Search approaches are in their approaches to modifying the fitness function (evolutionary versus reinforcement learning), the fitness function objects that are being tuned (objectives versus features) and, most importantly, the ability of the Variable Fitness Function to be applied with no CPU time overhead for unseen test instances.

The problem we study is based on a mobile workforce scheduling problem presented in [3,4]. It is a complex real workforce scheduling problem identified by @Road Ltd. and shares many complexities found in various other scheduling problems such as the Resource Constrained Project Scheduling Problem (RCPSP) and its variants [5], Job Shop Scheduling Problem (JSSP) [6] and its variants, along with other problems such as Vehicle Routing [7] and the Traveling Salesman Problem [8]. In [3], the problem's multi objective nature was used to show the trade off between diversity and solution quality when using multi objective genetic algorithm compared to a genetic algorithm using weighted sum objective functions. In [4] the problem's complexity was used to show that breaking the problem down into a very large number of smaller parts and then using another method to decide which of these smaller parts to solve, is a very effective way of solving large complex problems.

[4] uses reduced Variable Neighborhood Search (rVNS) [9] amongst other heuristics. rVNS is a faster form of Variable Neighbourhood Search. Variable Neighbourhood Search (VNS) [10] is based on the idea of systematically changing the

neighbourhood of a local search algorithm. Variable Neighbourhood Search enhances local search using a variety of neighbourhoods to "kick" the search into a new position after it reaches a local optimum. Several variants of VNS exist as extensions to the VNS framework [9] which have been shown to work well on various optimisation problems. In our experience, VNS has the advantage for complex, real-world problems, of requiring limited additional effort, once a basic local search framework is established.

3 The Variable Fitness Function

The Variable Fitness Function [1] describes how the weights of a weighted sum fitness function change over the iterations of a search process. The variable fitness function is piecewise linear, describing the relative importance of objectives at each iteration. There are two variations: the standard variable fitness function fixes the number of discontinuities and the number of iterations between them and the adaptive variable fitness function allows the points of discontinuity to evolve along with the variable fitness function objective weights (Figure 1). Work so far provides evidence that the adaptive version is more effective than the fixed version, as more complex functions can be evolved [1].

For the adaptive variable fitness function, we define a set of weights $\{W_{b,a}\}$ where a indexes the weight set ($a=1...A$) and b indexes the objective ($b=1...B$). We define I_a, the number of iterations between the weight sets. The variable fitness function is now defined as:

$$f(s,i) = \sum_{b=1}^{B} O_b(s)W_b(i)$$

where s is the solution to be evaluated and i is the iteration, $O_b(s)$ is the value of objective b for solution s and

$$W_b(i) = W_{b,c} + \frac{i-j}{k-j}(W_{b,c+1} - W_{b,c})$$

where iteration i occurs in the range from weight set c (starting at iteration j) to weight set $c+1$ (starting at iteration k) i.e. the linear interpolation of the weight of objective b for iteration i.

Figure 1 shows an example adaptive variable fitness function. This describes how the weights change over the iterations, for example, that the weight of objective 1 (W1) starts off at 2 (hence the objective is to be maximized) and then after iteration 200 its importance starts to decrease and objective 3 (that has weight W3) is to be minimized, and its importance is higher at the start and end of the search process.

3.1 Evolution

Little work has been done in encoding piecewise linear functions such as these into chromosomes. [11] uses a complex encoding for polynomial expressions. The

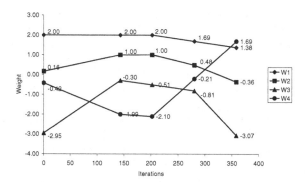

Fig. 1. An example adaptive variable fitness function. (the number of iterations between the weight sets and the number of weight sets may vary).

encoding is used to optimise a curve to fit a function described by a set of data points (and is not an appropriate method for the VFF). The evolution here is similar to work done on tuning of parameters for another algorithm using genetic algorithms [12].

When optimizing the weights of the variable fitness function, each weight in the variable fitness function appears as a gene in a GA chromosome. When the adaptive variable fitness function is used, the iterations between the weight sets are also included. Figure 2 shows how the weight sets are mapped to the genes of a chromosome.

| $W_{1,1}$ | ... | $W_{1,B}$ | I_1 | $W_{1,1}$ | ... | $W_{1,B}$ | I_1 | ... | $W_{A,1}$ | ... | $W_{A,B}$ |

Fig. 2. Mapping the weights to a chromosome for an adaptive variable fitness function

A modified version of 1 point crossover [13] will be used. It works the same way as normal 1-point crossover but the crossover point may only be on a weight set boundary. This method will keep mutually compatible weight sets together. The thick lines in Figure 2 show these crossover points. Each gene will have a chance to be mutated with a probability of p_{mut}, the mutation rate. Mutation will simply mutate the value of the gene by a random variable normally distributed around 0 and with a standard deviation as predefined for each weight. Hence $W_{a,b}$ is mutated by a value from the normal distribution $N(0, V_b)$ with probability p_{mut}. Where V_b is the standard deviation of mutation associated with objective b and p_{mut} is the probability of mutation which is the same for all alleles. This is similar to work done on mutation of artificial neural network weights evolved using GAs [14] where the network weight is mutated by a random number selected from a normal distribution.

The initial population of variable fitness functions is generated randomly where $W_{a,b}$ is picked uniformly at random out of the interval $[L_b, U_b]$ for objective b. There is a p_{adapt} probability that the chromosome will change length. If a chromosome is to change length there is an equal probability it will either shrink or grow by one weight set. If it is to shrink, a random weight set is chosen and removed from the chromosome. If it is to grow, a new weight set is inserted between two randomly chosen adjacent weight sets. The inserted weight set does not change the shape of the variable fitness function as it is inserted exactly half way between the two adjacent weight sets and has weight values that are the mean of the bordering weight sets. The new weight set is then mutated. Lastly, the I_a genes also have a p_{mut} probability of being mutated using the same method as the $W_{a,b}$ genes. This gives the chromosomes a chance to get more and less complex and to also expand to more or less iterations.

In our experiments, L_b and U_b will be -1 and 1 respectively (since objective values are normalized), $V_b = 0.05(U_b - L_b)$, and $p_{mut} = p_{adapt} = 0.05$ for all objectives b. These are known good values from previous work [1] and this previous work also shows that the sensitivity to parameters is low.

4   The Case Study Problem and Solution Heuristics

The problem we study is based on the workforce scheduling problem in [3]. The problem consists of assigning resources and time slots to geographically dispersed tasks. Tasks require various skills and resources possess these skills at different competencies. Resources are mobile and must travel between tasks and to and from their "home" location at the end and start of the day. Tasks have a priority associated with them which indicates their urgency or the reward for completing the task.

In the problem instances we study, 10 resources possess between 1 and 5 skills of which there are various bottlenecks in the availability. The resources travel at varying speeds. There are 300 tasks requiring between 0.5 and 1 hours to complete to be scheduled over 3 days and each has a 4 hour time window in which it must be completed. A task requires a resource to possess a certain skill, of which some skills are in more demand than others. Tasks are to be completed as early in the 4 hour time window as possible. Tasks have precedence constraints such that some task may not be started before another has completed. A chain of tasks is a subgraph of the precedence digraph of maximum indegree 1 where the indegree (outdegree) of a task in the subgraph is one if the indegree (outdegree) in the precedence digraph is greater than zero.

For this is a real world complex problem there are many objectives. Table 1 lists some of the principal objectives we have identified. The global fitness function is defined as $f = 5$ *(Scheduled High)* $+ 2$ *(Scheduled Low)* $+(Complete Chains) - 0.1$ *(Overrun)*, following reflection with our industrial collaborator. We use a constructive heuristic, *CON*, to build an initial schedule then an improvement metaheuristic, *IMP*, to improve it.

Table 1. Objectives used for the workforce scheduling problem

Objective	Description
Scheduled High	The number of high priority tasks scheduled.
Scheduled Low	The number of low priority tasks scheduled.
Complete Chains	The number of task chains that have been completed.
Travel Distance	The total distance traveled by the resources.
Travel Time	The total time spent traveling by the resources
Overrun	The total number of hours the task have overrun.

For the improvement heuristic we have decided to use Variable Neighbourhood Search (VNS). VNS is relatively simple to implement and we have seen that this kind of method can work well for scheduling problems [1]. We use a local search heuristic where we define the neighborhood as schedules which result from opti- mally reinserting a task, i.e. placing a task optimally in the schedule in terms of the current (variable) fitness function. If the task is not yet scheduled, this means allocat- ing the resources and time to it that yield the best improvement in fitness. If the task is already scheduled, this may mean moving the task in time, allocating new re- sources or a combination of the two (Figure 3). At each iteration of the local search, the entire neighbourhood is sampled and the best solution accepted. When the local search of the VNS reaches a local optimum, the search is kicked into a new area of the search space. We define these kicks as removing between 1 and 4 tasks and all dependent tasks. We remove dependent tasks so that precedence constraints are not broken.

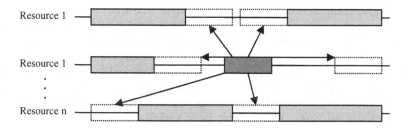

Fig. 3. Task reinsertion. The task is moved to the resource and time in the schedule which provides the best change in fitness according to the variable fitness function. The light grey boxes represent other tasks, the dark grey is the task being optimized and the dotted boxed are the positions being considered.

The construction method, *CON*, uses the local search operator of the VNS and ter- minates when a local optimum is found. The improvement metaheuristic, *IMP*, has a stopping criterion of 10,000 iterations. Table 2 lists the methods we will try. We can see the first two are normal heuristics and the last three are enhanced using VFF.

Table 2. Various methods to be used and their VFF enhance versions

Heuristics	Description
CON	Construction heuristic using the global fitness function.
CON + IMP	Construction heuristic and improvement heuristic using the global fitness function.
VFF(CON)	Construction heuristic using a variable fitness function.
VFF(CON + IMP)	Construction heuristic and improvement heuristic using a variable fitness function.
CON + VFF(IMP)	Construction heuristic using the global fitness function and improvement heuristic using a variable fitness function.

5 Computational Experiments

The five methods are used ten times on each of the five training instances and averages taken to reduce the effect of randomness. These five training instances are chosen to contain variations that a workforce would see on a day to day basis. By using multiple problem instances to evolve variable fitness functions we are trying to ensure that variable fitness functions learn characteristics of the problem through learning specific problem instances. For the methods enhanced with the variable fitness function, a test set of five problems instances will be solved using VFFs which were evolved using the training instances. Good performance on the test data will imply that a lot of CPU time could be used to train a "general purpose" variable fitness function, then that variable fitness function could be used very quickly in "real time".

The methods requiring the evolution of a variable fitness function will be given 50 generations with a population size of 10 (equivalent to 500 evaluations). Methods without a variable fitness function will also be given 500 evaluations and the best one taken to give them the same amount of CPU time. The *CON* heuristic takes approximately 30 seconds to construct the five schedules and the *IMP* heuristic takes approximately 25 minutes on a 3.0 GHz PC. As these experiments will take over 260 CPU days to complete they will be run in parallel on approximately 95 computers.

Table 3 shows the results of the individual methods and their standard deviations and Figure 4 graphs these with 90% confidence intervals. From the results we can see that the variable fitness functions were indeed able to enhance the standard methods significantly in all cases. We see very large variations in fitness when the variable fitness function is used on the constructive part of the search (*VFF(CON)* and *VFF(CON + IMP)*). Further investigation leads us to believe that this is because when the variable fitness function affects the constructive part of the search, it has the possibility to move a great distance in the search space from the "normal" constructive algorithm. The best variable fitness function enhancement for the *CON + IMP* method was to just enhance the improvement part. The variation of *CON + IMP* is extremely low and the solutions that it has found are far from optimal. This may be because the kick method of the VNS we have chosen was not sufficiently disruptive.

Table 3. Average fitness and standard deviation of ten runs of each method assessed using the global fitness function.

Method	Average Fitness	Standard Deviation
CON	3189.6774	N/A
VFF(CON)	3308.0253	150.3236
CON + IMP	3617.4517	8.4949
VFF(CON + IMP)	3689.9001	111.2555
CON + VFF(IMP)	3770.2434	35.4282

Figure 5 shows the breakdown of the individual objectives and Table 4 shows the difference in the average objective measures the enhanced methods have produced. This chart and table show that the VFF approach has found a way to obtain improvements to high priority objectives at the expense of low priority ones.

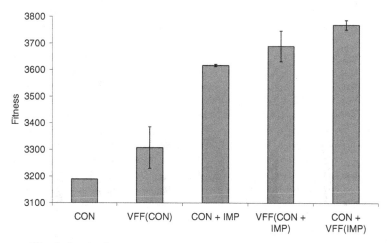

Fig. 4. Graph of the results in Table 3 with 90% confidence intervals

In all of the cases where the method is enhanced by the variable fitness function, the number of scheduled tasks, both low priority and high priority, has increased. This intuitively makes sense as these are the highest weighted objectives in the global fitness function. Travel time was decreased in both the cases where the variable fitness function was used to enhance the metaheuristic. The increase in travel time and other penalty objectives for the CON approach is not surprising as CON has no way to optimize these objectives by reinserting. Travel time is not included in the global fitness function, however, it would appear that when task reinsertion is permitted, the VFFs have learnt that less time spent traveling means more time can be spent doing tasks. In all cases, overrun increased, indicating that tasks were not scheduled as close to their start time as possible. This may be because tasks were shifted later in time so other tasks could be completed before them, enabling preceding tasks to also be completed.

Table 4. Average change in objectives as a result of variable fitness function enhancement

Base Method	CON	CON + IMP	
Improvement Using	VFF(CON)	VFF(CON + IMP)	CON + VFF(IMP)
Scheduled High (max)	20.20	4.30	17.10
Scheduled Low (max)	40.50	37.10	43.20
Travel Distance (min)	13.79	-24.03	-67.69
Travel Time (min)	14.91	-24.85	-64.34
Overrun (min)	603.52	220.51	203.08
Complete Chains (max)	-3.30	-1.20	1.20

Figure 6 shows the evolution process in action. Not only do these graphs show that the evolution process is working, and that the populations are evolving, but it shows the difference between randomly generated variable fitness functions (those in the

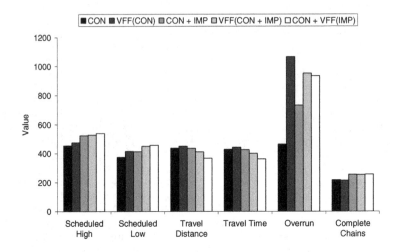

Fig. 5. Individual objective break down for each method

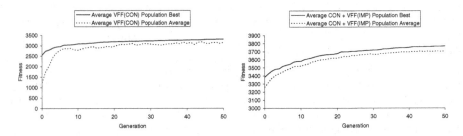

Fig. 6. Average population fitness and best of the population's fitness at each generation show-ing the evolution for *VFF (CON)* and *CON + VFF(IMP)* methods

initial population at generation 0) and evolved ones (those in the final population at generation 50). The plot showing the evolution of *VFF(CON)* method shows a greater increase in fitness from random variable fitness functions to evolved variable fitness functions than that of *CON + VFF(IMP)* (note the difference in "fitness" scale be-tween the graphs). This is because the *CON + IMP* is a better method than *CON*, and hence there is less room for improvement.

Figure 7 shows an example of how the variable fitness function is working. The top plot shows a typical evolved variable fitness function from the population and how the objective weights change over the iterations. Highlighted are the Travel Time and Overrun objective weights. Plotted below are the objective values obtained at each iteration from a single run using this variable fitness function. A quite obvious correlation can be seen between the weight of overrun and the average overrun observed. When the weight is positive, overrun increases and when the weight is negative it decreases. This variable fitness function has in fact learnt a type of right-left shift heuristic [15], which is frequently used in schedule repair.

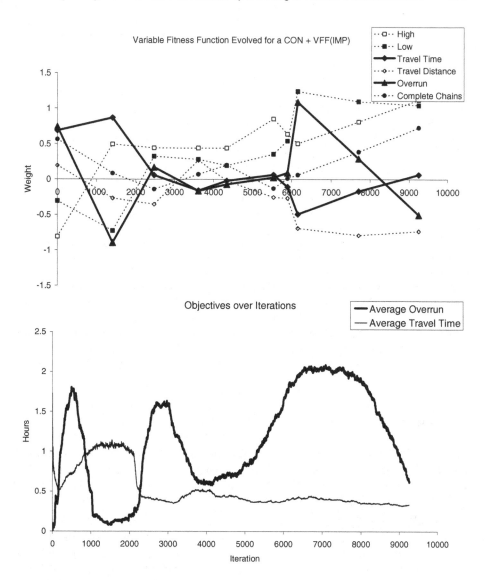

Fig. 7. A selected evolved VFF shown above and a plot below showing how two selected objective measures change over the course of a search

Figure 8 shows the improvement gained in the global fitness function from using the variable fitness function enhanced methods over the standard methods, for both the training data and the test data. Note that for test data instances, the amount of CPU time for the VFF and standard approaches are the same. As seen in the chart, the variable fitness functions enhanced methods are still significantly better than their standard versions on the test data (with the exception of the *VFF(CON + IMP)* method whose 90% confidence interval takes it below 0%). This is a good indication that

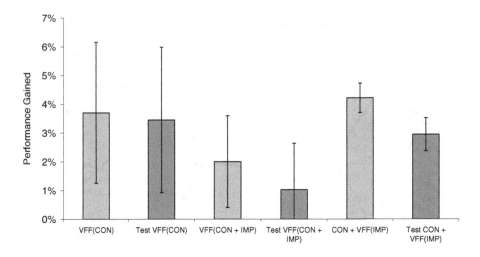

Fig. 8. Average method performance gained using variable fitness function on test data compared to training data

variable fitness functions trained for the *VFF(CON)* and *CON + VFF(IMP)* could be reused on different problem instances with good performance, and that they have "learned" generalisable information about the problem as well as specific information about the training instances.

6   Conclusions

In this paper we have demonstrated the application of an evolutionary variable fitness function to a constructive heuristic and a metaheuristic for a complex, real-world workforce scheduling problem. We have shown that statistically significant increases in heuristic and metaheuristic performance can be gained by using the variable fitness function. We have also seen that evolution plays a key role in getting these gains. To show the reusability of the evolved variable fitness functions they were used on another set of problem instances and showed gains of nearly equal magnitude. This is a strong indicator that a variable fitness function could be evolved offline and then the evolved variable fitness function be used in a real time situation. Arguably, the variable fitness function can be used for any optimization problem where multiple objectives can be defined. In future work we will investigate further how and why the VFF approach works its potential as a general problem-solving approach.

References

[1] Remde, S., Cowling, P., Dahal, K., Colledge, N.: Evolution of Fitness Functions to Improve Heuristic Performance. In: Proceedings of Learning and Intelligent Optimization (LION) II. LNCS, Springer, Heidelberg (to be published, 2008)

[2] Tsang, E., Voudouris, C.: Fast local search and guided local search and their application to British Telecom's workforce scheduling problem. Operations Research Letters 20(3), 119–127 (1997)

[3] Cowling, P., Colledge, N., Dahal, K., Remde, S.: The Trade Off between Diversity and Quality for Multi-objective Workforce Scheduling. In: Gottlieb, J., Raidl, G.R. (eds.) EvoCOP 2006. LNCS, vol. 3906, pp. 13–24. Springer, Heidelberg (2006)

[4] Remde, S., Cowling, P., Dahal, K., Colledge, N.: Exact/Heuristic Hybrids using rVNS and Hyperheuristics for Workforce Scheduling. In: Evolutionary Computation in Combinatorial Optimization Proc. LNCS, Springer, Heidelberg (2007)

[5] Hartmann, S.: Project Scheduling under Limited Resources: Model, methods and applications. Springer, Heidelberg (1999)

[6] Vanlaarhoven, P.J.M., Aarts, E.H.L., Lenstra, J.K.: Job Shop Scheduling by Simulated Annealing. Operations Research 40(1), 113–125 (1992)

[7] Toth, P., Vigo, D. (eds.): The Vehicle-Routing Problem. Siam, Philadelphia (2001)

[8] Lawler, E.L., Lenstra, J.K., Rinnooy Kan, A.H.G., Shmoys, D.B. (eds.): The Traveling Salesman Problem: A Guided Tour of Combinatorial Optimization. Wiley, Chichester (1991)

[9] Hansen, P., Mladenovic, N.: Variable neighborhood search: Principles and applications. European Journal of Oper. Res. 130(3), 449–467 (2001)

[10] Mladenovic, N., Hansen, P.: Variable neighborhood search. Computers & Operational Research 24(11), 1097–1100 (1997)

[11] Potgieter, G., Engelbrecht, A.P.: Genetic Algorithms for the Structure Optimisation of learned Polynomial Expressions. Applied Mathematics and Computation 186(2), 1441–1466 (2007)

[12] Shimojika, K., Fukuda, T., Hasehawa, Y.: Self-Tuning Fuzzy Modeling with adaptive membership function, rules, and hierarchical structure-based on Genetic Algorithm. Fuzzy Sets And Systems 71(3), 295–309 (1995)

[13] Reeves, C.R.: Genetic Algorithms and Combinatorial Optimization. In: Rayward-Smith, V.J. (ed.) Applications of Modern Heuristic Methods, Alfred Waller, Henley-on-Thames, pp. 111–125 (1995)

[14] Yao, X.: Evolving artificial neural networks. Proceedings Of The IEEE 87(9), 1423–1447 (1999)

[15] Valls, V., Ballestin, F., Quintanilla, S.: Justification and RCPSP: A technique that pays. European Journal of Operational Research 165(2) (2005)

Improving Query Expansion with Stemming Terms: A New Genetic Algorithm Approach*

Lourdes Araujo[1] and José R. Pérez-Agüera[2]

[1] Dpto. Lenguajes y Sistemas Informáticos. UNED, Madrid 28040, Spain
lurdes@lsi.uned.es
[2] Departamento de Ingeniería del Software e Inteligencia Artificial, UCM, Madrid
28040, Spain
jose.aguera@fdi.ucm.es

Abstract. Nowadays, searching information in the web or in any kind
of document collection has become one of the most frequent activities.
However, user queries can be formulated in a way that hinder the re-
covery of the requested information. The objective of automatic query
transformation is to improve the quality of the recovered information.
This paper describes a new genetic algorithm used to change the set
of terms that compose a user query without user supervision, by com-
plementing an expansion process based on the use of a morphological
thesaurus. We apply a stemming process to obtain the stem of a word,
for which the thesaurus provides its different forms. The set of candidate query terms is constructed by expanding each term in the original
query with the terms morphologically related. The genetic algorithm is
in charge of selecting the terms of the final query from the candidate
term set. The selection process is based on the retrieval results obtained
when searching with different combination of candidate terms. We have
obtained encouraging results, improving the performance of a standard
set of tests.

1   Introduction

Providing answers automatically to client's information needs has become a crucial task nowadays. The spectacular growth of the World-Wide-Web has called
for new solutions to access the information needed. An area of particular interest
is the reformulation of the queries that users present to the browsers or digital
libraries to access the information they need. The information contained in the
huge amount of available documents is characterized by a small set of representative terms, called index terms. These terms can be composed of one or more
words. Index terms are usually obtained by means of statistical techniques [1],
such as the construction of an inverted index[1] whose terms have associated a
list of pointers to the occurrences of the term in the text collection.

---

* Supported by projects TIN2007-68083-C02-01 and TIN2007-67581-C02-01.
[1] Given a set of documents containing words, the *inverted index* "inverts" that, so that
it consists of a set of words each listing all the documents containing that word.

J. van Hemert and C. Cotta (Eds.): EvoCOP 2008, LNCS 4972, pp. 182–193, 2008.

It very often happens that user queries are composed of terms which are different of the index terms of the collection, despite they refer the same concept, or they correspond to the same stem. Another reason that hinder the retrieval of the requested information is that very often users employ too small sets of search terms because they rely on implicit knowledge that the search engine lacks. An additional problem for retrieval is word and language ambiguity.

All these reasons have made query reformulation an interesting area of research. This operation amounts to changing the original query by adding, removing or replacing terms. In other cases, what is modified is the weights assigned to the terms in the search, to indicate their relevance. Let us consider an example. Suppose that a user, who is interested in learning about modifying the search queries to improve the results of his searches, has posed the following query:

modify query technique

As these three terms are very general, the list of documents that a Web searcher, such as Google or Yahoo, retrieves is extremely large (820,000 documents), and the order in which these documents are presented to the user is inappropriate. In fact, among the first ten documents presented to the user, there are only two more or less related to the concept of *query modification* or *query reformulation*.

One can easily think of expanding the search term by adding to the original query synonyms and related terms. In our example, the query could become:

modify query technique modification transform transformation method

Though the expansion reduces the amount of retrieved documents (now they are about 490,000), there are still many documents among the first ones which have nothing to do with *query reformulation*. Specifically, among the ten first retrieved documents only four are somehow related to the intended topic. Furthermore, a blind query expansion can worsen the result by lowering the precision. Searchers usually present first those documents which contain every term in the query, and this can lead to exclude many relevant documents. On the other hand, there exist other possible transformations which can provide a much better query for the user needs. Suppose that we have a method available that provides us with terms which are truly related to the information the user is looking for, not just to the terms of the query. In our example these terms could be:

query transformation
query modification
query expansion
information retrieval
web search

Now, an appropriate selection of terms among those of the original query and the related ones can lead to much more precise results. In the example, let us assume that the query is transformed to

"query expansion" "query transformation"

Now the amount of retrieved documents is 212 and all of them are relevant to the user needs.

Because of the extended use of information search in our society, query reformulation has become a very active research area and different approaches have been applied. Salton [16] and Robertson and Sparck Jones [15] use relevance feedback to reformulate the query. Systems based on relevance feedback modify the original query taking into account the user relevance judgements on the documents retrieved by this original query.

There is an underlying common background in many of the works done in query reformulation, namely the appropriate selection of a subset of search terms among a list of candidate terms. Because the number of possible subsets can be very large, it makes sense to apply heuristic search techniques to the problem, such as genetic algorithms (GAs).

GAs have been previously applied [4] to different aspects of information retrieval. Proposals devoted to the query expansion problem with GAs can be classified into relevance feedback techniques and Inductive Query by Example (IQBE) algorithms. In systems based on relevance feedback [21] the user gives feedback on the relevance of documents retrieved by his original query. IQBE [2] is a process in which the user does not provide a query, but document examples and the algorithms induce the key concepts in order to find other relevant documents.

Several works apply GAs to assigning weights to the query terms [14, 22, 18, 9, 8], while others are devoted to selecting the query terms. Let us review some proposals in the latter case, the one on which we focus our work. Chen et al. [2] apply a GA as an IQBE technique, i.e. to select the query terms from a set of relevant documents provided by the user. In this work, the authors propose an individual representation that has also been used in later works: chromosomes are binary vectors of fixed size in which each position is associated with one of the candidate terms. In [10], the authors propose a genetic programming (GP) algorithm to learn boolean queries encoded as trees whose operators are AND, OR and NOT. This work was later extended in [5] by incorporating multiobjective optimization techniques to the problem. Fernández-Villacañas and Shackleton [11] compared two evolutionary IQBE techniques for boolean query learning, one based on GP and the other on a classic binary GA, which obtained the best results. Kraft et al. [12] propose the use of GP to learn fuzzy queries. The queries are encoded as expression trees with boolean operators as inner nodes, and query terms with weights as terminal nodes. Cordón et al. [3] extend Kraft's proposal by applying a hybrid simulated annealing-genetic programming scheme, what allows them to use new operators to adjust the term weights. Tamine et al. [19] use knowledge-based operators instead of the classical blind operators, as well as niching techniques.

A common factor of the above mentioned works is that they relay on some kind of information provided by the user. In some cases, the user has to provide a set of documents that are used for the inductive learning of terms. In other cases, the user provides relevance judgements on the retrieved documents, that are use to compute the fitness.

In this work we propose a new application of GAs to the selection of query terms. The novelty is that our system does not require any user supervision: new candidate terms for the query are provided by a morphological thesaurus. Some of the new terms will allow to retrieve new relevant documents. However, including the whole set of candidate terms in the query will, in general, degrade the system performance because documents fetched by expanded terms can relegate documents relevant for the original query terms. Accordingly, we apply a GA to perform a selection of terms. The GA works with individuals which represent different combinations of candidate query terms or queries. The selection process is based on the relevance, with respect to the original query, of the first N (a parameter) documents retrieved when submitting the query to an automatic searcher[2]. To compute the relevance of a retrieved document we use the classical vector space model of information retrieval (IR) [6]. We have performed experiments to investigate the limit of the performance that can be reached by the GA, by assuming that we have available the relevance judgements provided by the user. After studying the GA parameters for this case, we have obtained an important improvement of the performance. Then, we have studied the GA which is applied without user supervision.

The rest of the paper proceeds as follows: section 2 describes the general scheme of the process to construct the set of candidate query terms; section 3 is devoted to describe the evolutionary algorithm used to select the terms of the final query, including different fitness functions tested in the experiments; section 4 presents and discusses the experimental results, and section 5 draws the main conclusions of this work.

2   The Query Expansion System

A classical operation in information retrieval to improve the performance of the systems is to use morphological variants. In most cases, morphological variants of words have similar semantic interpretations and can be considered as equivalent for the purpose of IR applications. For this reason, a number of so-called stemming algorithms, or stemmers, have been developed. They reduce a word to its stem or root form. One can view stemming as a form of global query expansion: we expand a term in the query with all the terms in the dictionary sharing the same stem. Our system is based on this idea of expanding the query with morphological variants.

Figure 1 shows a scheme of the process followed to transform a query. First of all, we obtain the stems which correspond to the original query terms. This is done with the well known Porter stemmer [20,13]. Then, we use the *morphological thesaurus* for Spanish available along with the Porter stemmer, which provides the different forms (plural and grammatical declinations) corresponding to each stem. All these terms are candidate for the final query. Finally, the GA is in

---

[2] The system does not require the user supervision.

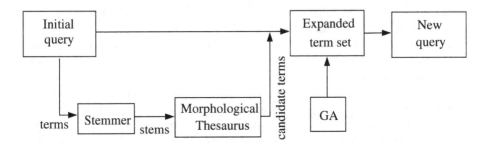

Fig. 1. Scheme of the process to select candidate terms to expand the query and the later selection of them by the Genetic Algorithm

charge of selecting, from the candidate term set, the final query terms, that are submitted to the searcher.

3   The Genetic Algorithm

Chromosomes [7] of our GA are fix-length binary strings where each position corresponds to a candidate query term. A position with value one indicates that the corresponding term is present in the query. Individuals of the initial population are randomly generated. Because of some preliminary experiments we have performed have shown that, in most cases, the elimination of the original query terms degrades the retrieval performance, we force to maintain them among the selected terms of every individual. The set of candidate terms is composed of the original query terms, along with related terms provided by the applied thesaurus.

The selection mechanism to choose individuals for the new population uses roulette wheel. We apply one-point crossover operator. Our algorithm uses random mutation which flips a bit randomly chosen. We also apply elitism.

3.1   Fitness Functions Tested

The fitness functions that we have proposed are different measures of the degree of similarity between a document belonging to the document collection and the submitted query. To compute this similarity, we apply the vector space model of information retrieval [6, 1]. In this model, a document d_j and a query q are represented as n-dimensional vectors. To construct the vectors, we have to assign weights to index terms in queries and in documents. The classic vector space model computes the term weight as:

$$w = tf \cdot log\frac{D}{d}$$

where tf stands for term frequency, $log\frac{D}{d}$ is the inverse document frequency, D is the total number of documents in the document set and d is the number of documents containing the term.

Then, as proposed by Salton and McGill [17], the degree of similarity of the document d_j with respect to the query q is evaluated as the distance between the vectors $\boldsymbol{d_j}$ and $\boldsymbol{q}$. This distance can be quantified by the cosine of the angle between these two vectors. Based on this similarity measure, we have considered three alternative fitness functions: $\sqrt{\cos\theta}$, $\cos\theta$, and $\cos^2\theta$. They differ on the distance among the values assigned to different individuals, which can affect the GA selection process.

4   Experimental Results

The system has been implemented in Java, using the JGAP library[3], on a Pentium 4 processor. We have used a set of tests provided by CLEF (Cross-Language Evaluation Forum) for the Spanish language. The collection and tests used come from EFE94. This document collection came from the international news agency EFE, from all the news received during 1994 and consists of 215.738 documents stored in files with SGML format.

In the first place, we have performed experiments to know the limit to the improvement we can reach according to the data provided by the collection used in the experiments, and in this way to have an idea of the quality of the fitness function that have been tested later.

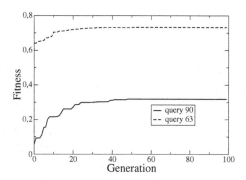

Fig. 2. Fitness evolution for two queries of the test set. The GA parameters have been a population size of 100 individuals, a crossover rate of 25% and a mutation rate of 1%.

To this purpose, we take the user relevance judgements as the best fitness function that we can use. Since for the CLEF collection used in the experiments we have the user relevance judgements, we have used them to guide the selection process. Specifically, we have used as fitness function the standard precision measure ($\frac{|Ra|}{|A|}$) defined as the fraction of retrieved documents (the set A) which are relevant (R_a). Table 1 shows the precision of the best individual obtained

---

[3] http://jgap.sourceforge.net/

by the GA. We have compared our system performance with the results of the original user query (*Baseline*) and with the results obtained expanding with the stems provided by the Porter stemming (*Porter Stemming*). The latter works by substituting the original query terms by their stems and performs the search in the document collection indexed by stems. We can observe that our system achieved an important improvement of the performance, greater than the one achieved with other stemming methods traditionally used in query expansion, such as Porter. We consider the improvement achieved a ceiling for the improvement of the unsupervised GA. Figure 2 shows the evolution of two queries of the test set. We can observe that both of them reach convergence very quickly.

Table 1. Global precision results for the whole set of tested queries. Each individual datum has been computed as the average over 5 different GA runs. Prec. stands for precision (all documents), Prec10 stands for the precision of the results for the first ten documents retrieved. Last column is the rate of precision (all documents) improvement.

	Prec.	Prec10	Improvement
Baseline	0.3567	0.4150	–
Porter Stem.	0.4072	0.45	+12.40%
Genetic Stem.	0.4682	0.54	+23.81%

4.1   Selecting the Fitness Function

Let us now investigate the proposed unsupervised GA. To select the fitness function to be used in the remaining experiments, we have studied the fitness evolution for different queries of our test set. Figure 3 compares the fitness evolution for the query which reaches the greatest improvement (*best_query*). The most relevant point in this figure is the generation at which each function reaches its optimum. The three functions converge to different numerical values that correspond to the same precision value (.68). We can observe that the square-root cosine function is the first one to converge to its optimum. Probably because this function emphasizes the distance between the values assigned to different individuals, thus improving the selection process of the GA. Accordingly, the square-root cosine has been the fitness function used in the remaining experiments.

4.2   Tuning the GA Parameters

The next step taken has consisted in tuning the parameters of the GA. Figure 4 shows the fitness evolution using different crossover rates for two queries, the best one (Figure 4(a)), and the worst one (Figure 4(b)). Results show that, in both cases, we can reach a quickly convergence with values around 25%.

Figure 5 presents the fitness evolution using different mutation rates for the best (a) and the worst (b) queries. Values around 1% are enough to produce a quick convergence.

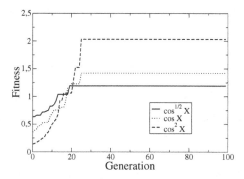

Fig. 3. Fitness functions comparison for the *best_query*, the one for which the greatest precision improvement is achieved. The GA parameters have been a population size of 100 individuals, a crossover rate of 25% and a mutation rate of 1%.

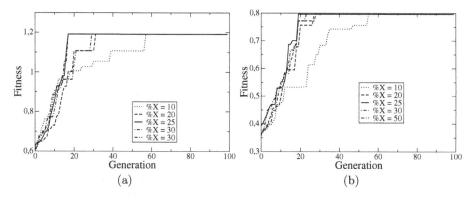

Fig. 4. Studying the best crossover rate for the best_query (a) and the worst one (b). The GA parameters are a population size of 100 individuals and a mutation rate of 1%.

Figure 6 show the fitness evolution for the best (a) and the worst (b) queries, with different population sizes. The plots indicate that small population sizes, such as one of 100 individuals, are enough to reach convergence very quickly.

4.3   Overall Performance

Table 2 presents precision and recall results obtained for the whole set of 40 test queries that we have considered. Recall (Recall $= \frac{|R_a|}{|R|}$), is a coverage measure defined as the fraction of the relevant documents (the set R) which has been retrieved (R_a). We have compared our system performance with the results obtained with the original user query (*Baseline*) and with the results obtained using the Porter stemming (*Porter Stemming*). Each individual datum has been computed as the average over 5 different GA runs. We can observe that our system is able to obtain an improvement in precision of 15.20% over the baseline,

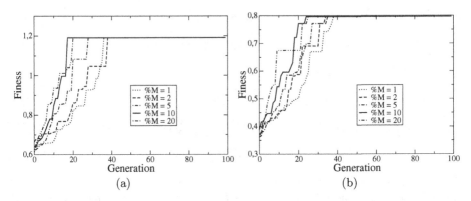

Fig. 5. Studying the best mutation rate for the best query (a) and the worst one (b). The GA parameters are a population size of 100 individuals and a crossover rate of 25%.

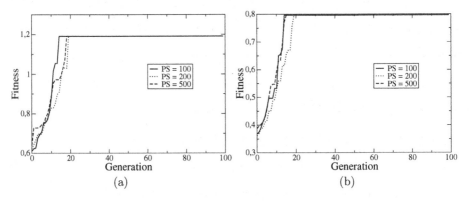

Fig. 6. Studying the population size for the best query (a) and the worst one (b). The GA parameters are a crossover rate of 25% and a mutation rate of 1%.

being this improvement larger that the one obtained by other methods, such as Porter stemming. Furthermore, our system also achieves a great improvement in the system recall. Table 2 also shows the recall results. Thus, our system is able to improve the coverage, improving precision at the same time. With the current implementation the mean execution time per query is 45 seconds.

4.4   Analyzing a Final Query

Apart from evaluating the numerical performance of our system, we have analyzed the queries resulting from the applied expansion process. Let us first consider the query 63 of the collection, the one with best performance (precision increases from .55 to .68, 19.11% of improvement). Table 3 shows the results for this query.

Table 2. Global precision and recall results for the whole set of tested queries. Prec. stands for precision (all documents), Prec5 stands for the precision of the results for the first five documents retrieved, and Prec10, stands for precision for the first ten documents retrieved. Last column is the rate of precision (all documents) improvement.

	Prec.	Prec5	Prec10	Improvement	Recall	Improvement
Baseline	0.3567	0.4750	0.4150	–	0.7035	–
Porter Stem.	0.4072	0.50	0.45	+12.40%	0.7228	+2.67%
Genetic Stem.	0.4206	0.5150	0.4575	+15.20%	0.7521	+6.46%

Table 3. Retrieval results of a query example

User query: *reserva de ballenas*
Set of candidate terms:
reserva, reservaba, reservaban, reservaciones, reservación, reservada, reservadamente, reservadas, reservado, reservados, reservamos, reservan, reservando, reservandose, reservar, reservara, reservaran, reservarlas, reservarle, reservarles, reservarlo, reservarlos, reservarnos, reservaron, reservarse, reservará, reservarán, reservaría, reservarían, reservas, reservase, reserve, reserven, reservista, reservistas, reservo, reservoir, reservándole, reservándolos, reservándose, reservó, ballenas, ballena, ballén, balléna
GA final query: *ballenas ballena reserva*

The original query is a compound term in Spanish, whose meaning is *whale reserve*. The first observation is the large size of the set of candidate terms, what makes clear the need for a selection. In the final query we can observe that the most representative terms related to the topic, such as *ballena* and *ballenas* (singular and plural Spanish words for whale), and the word *reserva* (reserve) have been added to the query. This query suggests investigating the application of a special treatment of compound terms, using "reserva de ballenas" as a single search term. This is a matter of future work.

Another query is *energía renovable* (Renewable Energy), for which the final query is *energías renovables energía renovable*. It achieves an improvement of 39.21%.

In other cases, such as for the query *Verduras, frutas y cáncer* (vegetables, fruits and cancer), the final query and the original query are the same. In this case, the whole point of the query is the relationship between the query terms. Because of this, searching independently for alternative forms of the query terms can only worsen the precision. However, the GA is capable of clearing the expanded query, recovering the original query and thus maintaining the system performance.

5 Conclusions

In this paper we have shown how an evolutionary algorithm can help to reformulate a user query to improve the results of the corresponding search. Our method does not require any user supervision. Specifically, we have obtained the candidate terms to reformulate the query from a *morphological thesaurus*, with provides, after applying stemming, the different forms (plural and grammatical declinations) that a word can adopt. The evolutionary algorithm is in charge of selecting the appropriate combination of terms for the new query. To do this, the algorithm uses as fitness function a measure of the proximity between the query terms selected in the considered individual and the top ranked documents retrieved with these terms.

We have carried out some experiments to have an idea of the possible improvement that the GA can achieve. In these experiments we have used the precision obtained from the user relevance judgements as fitness function. Results have shown that in this case the GA can reach a very high improvement.

We have investigated different proximity measures as fitness functions without user supervision, such as cosine, square cosine, and square-root cosine. Experiments have shown that the best results are obtained with square-root cosine. However, results obtained with this function do not reach the reference results obtained using the user relevance judgements. This suggests investigating other similarity measures as fitness functions.

A study of the queries resulting after the reformulation has shown that in many cases the GA is able to add terms which improve the system performance, and in some cases in which the query expansion spoils the results, the GA is able to recover the original query.

We are now working on reducing the execution time per query following two lines. On the one hand we are improving the GA implementation, and on the other hand we are investigating alternative fitness functions which do not require to retrieve documents at each evaluation.

References

1. Baeza-Yates, R.A., Ribeiro-Neto, B.A.: Modern Information Retrieval. ACM Press / Addison-Wesley (1999)
2. Chen, H., Shankaranarayanan, G., She, L., Iyer, A.: A machine learning approach to inductive query by examples: An experiment using relevance feedback, id3, genetic algorithms, and simulated annealing. JASIS 49(8), 693–705 (1998)
3. Cordón, O., de Moya Anegón, F., Zarco, C.: A new evolutionary algorithm combining simulated annealing and genetic programming for relevance feedback in fuzzy information retrieval systems. Soft Comput. 6(5), 308–319 (2002)
4. Cordón, O., Herrera-Viedma, E., López-Pujalte, C., Luque, M., Zarco, C.: A review on the application of evolutionary computation to information retrieval. Int. J. Approx. Reasoning 34(2-3), 241–264 (2003)
5. Cordón, O., Herrera-Viedma, E., Luque, M.: Improving the learning of boolean queries by means of a multiobjective iqbe evolutionary algorithm. Inf. Process. Manage. 42(3), 615–632 (2006)
6. Salton, G.: Automatic Information Organization and Retrieval. McGraw Hill Book Co, New York (1968)
7. Holland, J.J.: Adaptation in Natural and Artificial Systems. University of Michigan Press (1975)

8. Horng, J.-T., Yeh, C.-C.: Applying genetic algorithms to query optimization in document retrieval. Inf. Process. Manage. 36(5), 737–759 (2000)

9. Lopez-Pujalte, C., Bote, V.P.G., de Moya Anegón, F.: A test of genetic algorithms in relevance feedback. Inf. Process. Manage. 38(6), 793–805 (2002)

10. Smith, M., Smith, M.: The use of genetic programming to build boolean queries for text retrieval through relevance feedback. Journal of Information Science 23(6), 423–431 (1997)

11. Martín, J.L.F.-V., Shackleton, M.: Investigation of the importance of the genotype-phenotype mapping in information retrieval. Future Generation Comp. Syst. 19(1), 55–68 (2003)

12. Petry, F.E., Buckles, B.P., Sadasivan, T., Kraft, D.H.: The use of genetic programming to build queries for information retrieval. In: International Conference on Evolutionary Computation, pp. 468–473 (1994)

13. Porter, M.F.: An algorithm for suffix stripping. In: Readings in information retrieval, pp. 313–316. Morgan Kaufmann Publishers Inc., San Francisco (1997)

14. Robertson, A.M., Willet, P.: An upperbound to the performance of ranked-output searching: optimal weighting of query terms using a genetic algorithm. J. of Documentation 52(4), 405–420 (1996)

15. Robertson, S., Jones, K.S.: Relevance weighting of search terms. Journal of the American Society for Information Science 27(3), 129–146 (1996)

16. Salton, G.: The SMART Retrieval System—Experiments in Automatic Document Processing. Prentice-Hall, Inc., Upper Saddle River (1971)

17. Salton, G., McGill, M.: Introduction to Modern Information Retrieval. McGraw-Hill Book Company, New York (1983)

18. Sanchez, E., Miyano, H., Brachet, J.: Optimization of fuzzy queries with genetic algorithms. application to a data base of patents in biomedical engineering. In: VI IFSA Congress, vol. II, pp. 293–296 (1995)

19. Tamine, L., Chrisment, C., Boughanem, M.: Multiple query evaluation based on an enhanced genetic algorithm. Inf. Process. Manage. 39(2), 215–231 (2003)

20. van Rijsbergen, C., Robertson, S., Porter, M.: New models in probabilistic information retrieval. Technical Report 5587, British Library, London (1980)

21. van Rijsbergen, C.J.: Information retrieval, 2nd edn., Butterworths, London (1979)

22. Yang, J.-J., Korfhage, R.R.: Query modification using genetic algorithms in vector space models. Int. J. Expert Syst. 7(2), 165–191 (1994)

Inc*: An Incremental Approach for Improving Local Search Heuristics

Mohamed Bader-El-Den and Riccardo Poli

Department of Computing and Electronic Systems,
University of Essex, Wivenhoe Park, Colchester CO4 3SQ, United Kingdom
{mbbade,rpoli}@essex.ac.uk

Abstract. This paper presents Inc*, a general algorithm that can be used in conjunction with any local search heuristic and that has the potential to substantially improve the overall performance of the heuristic. Genetic programming is used to discover new strategies for the Inc* algorithm. We experimentally compare performance of local heuristics for SAT with and without the Inc* algorithm. Results show that Inc* consistently improves performance.

1 Introduction

Many NP-Complete problems, like scheduling, time tabling, satisfiability, graph colouring, etc., are routinely solved through the use of heuristics. A heuristic is effectively a rule of thumb or an educated guess that reduces the search required to find a solution. Heuristics make it possible to solve NP-Complete problems in practical situations which are beyond complete/exhaustive solvers. However, they provide no guarantee of success. So, finding ways to improve the performance of heuristics could have important and far-reaching ramifications.

We present Inc*, a general algorithm that can be used in conjunction with any local search heuristic to substantially improve the overall performance of the heuristic. Genetic programming is used to discover new strategies for the Inc* algorithm. We demonstrate the approach with Boolean satisfiability problems (SAT).

The paper is organised as follows. In Section 2 we introduce SAT problems and describe some of the best-known local-search heuristics used to solve them. We also review recent evolutionary systems developed for learning and evolving SAT heuristics. In Section 3, we introduce the Inc* algorithm along with the GP system used to evolve strategies for Inc*. A description of the experiments we performed with the GP Inc* framework is given in Section 4, Finally, we draw some conclusions in Section 5.

2 SAT Problem

SAT is a classical combinatorial optimisation problem. It was the first problem to be proved to be NP-Complete [3]. Many heuristics have been proposed and successfully used for solving the SAT problem (e.g., [8,18,20,19]). SAT has many different practical applications. Also, many other problems can be transformed into SAT problems.

J. van Hemert and C. Cotta (Eds.): EvoCOP 2008, LNCS 4972, pp. 194–205, 2008.
© Springer-Verlag Berlin Heidelberg 2008

The target in SAT is to determine whether it is possible to set the variables of a given Boolean expression in such a way to make the expression true. The expression is said to be satisfiable if such an assignment exists. If the expression is satisfiable, we often want to know the assignment that satisfies it. The expression is typically represented in Conjunctive Normal Form (CNF), i.e., as a conjunction of clauses, where each clause is a disjunction of variables or negated variables.

There are many algorithms for solving SAT. Incomplete algorithms attempt to guess an assignment that satisfies a formula. So, if they fail, one cannot know whether that's because the formula is unsatisfiable or simply because the algorithm did not run for long enough. Complete algorithms, instead, effectively *prove* whether a formula is satisfiable or not. So, their response is conclusive. They are in most cases based on backtracking. That is, they select a variable, assign a value to it, simplify the formula based on this value, then recursively check if the simplified formula is satisfiable. If this is the case, the original formula is satisfiable and the problem is solved. Otherwise, the same recursive check is done using the opposite truth value for the variable originally selected.

The best complete SAT solvers are instantiations of the Davis Putnam Logemann Loveland procedure [4]. Incomplete algorithms are often based on local search heuristics (Section 2.1). These algorithms can be extremely fast, but success cannot be guaranteed. On the contrary, complete algorithms guarantee success, but their computational load can be considerable, and, so, they can be unacceptably slow on large SAT instances.

2.1   Stochastic Local-Search Heuristics for SAT

Stochastic local-search heuristics have been widely used since the early 90s for solving the SAT problem following the successes of GSAT [20]. The main idea behind these heuristics is to try to get an educated guess as to which variable will most likely, when flipped, give us a solution or will move us one step closer to a solution. Normally the heuristic starts by randomly initialising all the variables in a CNF formula. It then flips one variable at a time until either a solution is reached or the maximum number of flips allowed has been exceeded. Algorithm 1 shows the general structure of a typical local-search heuristic for the SAT problem. The algorithm is normally repeatedly restarted for a certain number of times if it is not successful.

Some of the best-known heuristics of this type include:

GSAT   [20] which, at each iteration, flips the variable with the highest gain score, where the gain of a variable is the difference between the total number of satisfied clauses after flipping the variable and the current number of satisfied clauses. The gain is negative if flipping the variable reduces the total number of satisfied clauses.

HSAT   [8] In GSAT more than one variable may present the maximum gain. GSAT chooses among such variables randomly. HSAT, instead, it selects the variable with the maximum age, where the age of a variable is the number of flips since it is was last flipped. So, the most recently flipped variable has an age of zero.

WalkSat   [19] starts by selecting one of the unsatisfied clauses C. Then it flips randomly one of the variables that will not break any of the currently satisfied clauses (leading to a "zero-damage" flip). If none of the variables in C has a "zero-damage" characteristic, it selects with probability p the variable with the maximum score gain, and with probability $(1 - p)$ a random variable in C.

Algorithm 1. General algorithm for SAT stochastic local search heuristics

---
1: L = initialise the list of variables randomly
2: **for** $i = 0$ to MaxFlips **do**
3:     **if** L satisfies formula F **then**
4:         **return** L
5:     **end if**
6:     select variable V from L using some selection heuristic
7:     flip V
8: **end for**
9: **return** n o assignment satisfying F found

As one case see, SAT heuristics use one of two main strategies for choosing the next variable to flip. The first strategy is to make a greedy move. In other words, the SAT solver chooses to flip the variable which transforms the current solution state to a state which is closest to a solution. The gain of the variable is typically the most important factor is selecting such a move, although also the age of the variable is sometimes used to avoid looping. The second strategy is to perform a random walk. Mainly this is done to avoid (or escape from) local optima. This is done by selecting a random variable to flip from a designated set variables. There are different ways of choosing this set. For example, the set can include all the variables in the CNF formula, as in the GSAT, or just the variables in unsatisfied clauses, as in WalkSat.

2.2   Incremental SAT

In a standard SAT algorithm the input is a problem instance and the target is to state whether this instance could be satisfied or not, and what are the variable assignment that satisfies it. In some cases it is also important to know if the instance could be still satisfied if further (arbitrary) Boolean clauses were added to the current set. This is known in the literature as incremental or dynamic SAT [16]. In incremental SAT the solver normally starts with a certain number of clauses and determines whether this set can be satisfied or not. In case it is satisfied, the solver gives the user the opportunity of adding more clauses to the existing set. The solver then checks whether the solution is still valid. If not, it attempts to repair it.

Most incremental SAT solvers are based on exact algorithms as in [10], although some researchers have also used incomplete or heuristic-based solvers to deal with incremental SAT problems [11]. The main problem with the latter is that heuristics give no grantee that a solution can be found. Their main advantage is speed.

In this paper we will introduce an approach that presents some similarity with incremental SAT, but where the objective is to solve SAT problems, not incremental SAT problems. In particular, we will use Genetic Programming (GP) [13,14] to investigate the benefits of dynamically changing the number of active clauses during the course of solving SAT problems. So, the solver is given a CNF formula including *all* the clauses from the beginning, but we give the solver the ability to decide which clauses to start with and in which order to tackle them. We will explain this in more detail later.

2.3   Evolutionary Algorithms and SAT Problem

There have been a number of proposals of using evolutionary algorithms for SAT. An example is FlipGA which was introduced by Marchiori and Rossi in [15], a genetic algorithm was used to generate offspring solutions to SAT using standard genetic operators. However, offspring were then improved by means of local search methods. The same authors later proposed ASAP, a variant of FlipGA [17]. A good overview of other algorithms of this type is provided in [9].

GP has evolved competitive SAT solvers. For example, Fukunaga evolved local search heuristics [6,7]. Also, GP has been used to enhance the performance of exact algorithms for SAT by helping the algorithm decide which variables to start the backtracking process with or to evolve heuristics for initialising dynamic decisions [12]. Furthermore, a general framework for evolving local-search 3-SAT heuristics, called GP-HH, has recently been proposed [1,2]. The aim there is to obtain "disposable" heuristics which are evolved and used for a specific subset of instances of a problem. Results were promising with GP-HH evolving very competitive heuristics.

3   The Inc* Framework

3.1   Principles Behind Inc*

As mentioned above, Inc* is a general algorithm that can be used in conjunction with any local search heuristic to improve its performance. The general idea of the algorithm is the following: rather than attempting to directly solve a difficult problem, let us first derive a sequence of progressively simpler and simpler instances of the problem; then let us give the solver these instances one by one starting from the simplest, and progressing in the sequence only after all previous simplified instances are solved. The search is not restarted when a new instance is presented to the solver. In this way, it is hoped that the solver will effectively and progressively be biased towards areas of the search space where there is a higher chance of finding a solution to the original problem.

While this is the fundamental idea, the Inc* framework goes one step further and makes the choice of the simplified problems dynamic. The objective of this is to limit the chances of the algorithm getting stuck in local optima. Whenever the system detects that one of the simplified instances in the chain leading to the original problem is too difficult, it backtracks and creates a new simplified instance (of the same size as the previous one, but radically different from it) in the attempt to continue the progression towards the goal problem instance.

The Inc* framework is particularly applicable to the SAT problem, where one can easily and dynamically create the necessary set of simplified problems. Effectively the algorithm starts by selecting a subset of the clauses in the formula. It then use one of the SAT heuristics to tests the satisfiability of this portion of the formula, which we will call the *clauses active list*. Depending on the result of the heuristic on this portion of the formula, the algorithm then increases or decreases the number of clauses in the active list. In some cases adding a clause has no effect on the satisfiability of the active list with the current variable assignment, so no additional flips are necessary. In other cases, more work is needed to find a new valid assignment.

To illustrate the benefits of the main ideas behind Inc*, in Figure 1 we show the results of two simple experiments using GSAT. In both experiments, we added clauses to the active list one by one and we used GSAT after each addition to find an assignment that satisfies the clauses currently active. The graph on the left of the figure shows the case of a formula with 20 variables and 91 clauses. The graph reports the number of flips GSAT used to find a new satisfying assignment after the addition of each new clause and the number of variables in use in the active list. The plot on the right of the figure shows the number of flips required by GSAT to find assignments that satisfy the active list for a SAT instance with 50 variable and 218 clauses. The plots show that adding a new clause to the currently active clauses requires no flips or a very small number of flips most of the time. We even find that in both instances the full formula is actually satisfied by an assignment found before all clauses had been added.[1] This is really the reason why the use of a progression of problems may make a problem easier. On rare occasions, however, finding a new assignment after the addition of a clause may require hundreds or even thousands of flips. It is precisely to avoid these high peaks that Inc* backtracks.[2]

We have turned these ideas into a detailed algorithm (Algorithm 2). The algorithm starts by initialising all the variables in the formula F randomly and by activating an initial set of clauses by adding them to active clause list AC. The algorithm then runs one of the SAT local search heuristics. The heuristic is given relatively a small number of flips to run with at the beginning. The number of allowed flips is incremented gradually if the SAT solver fails to satisfy the AC, until, of course, the total number of flips used exceeds a predefined maximum ($MaxFlips$). A weight is assigned to each clause, which indicates how many flips have been necessary to satisfy the active list after the addition of this clause. So, after each run of the SAT heuristic, clause weights are updated. If the heuristic found a variable assignment L that satisfies the current AC, then the size of AC is increased by adding new classes to it. Otherwise, the algorithm removes from AC a small set of clauses, giving preference to those with the lowest clause weight, and the number of allowed flips is increased, as previously mentioned.

Two key elements in the effectiveness of Inc* are the decisions taken at Steps 2 and 2 in Algorithm 2 as to how many clauses to add or remove from the active list after a success or failure, respectively. In this paper, we have used GP to find optimal strategies to make these decisions. In the next section we describe the GP system used and the evolved strategies.

---

[1] This is not entirely surprising, since it is well-known that in most hard SAT instances there are (sometimes numerous) redundant clauses. A redundant clause is a clause that has no effect on the SAT formula [21]. There are algorithms for finding and removing redundant clauses [5], but the process is complex and very time consuming, especially in large SAT instances.

[2] Recently, some researchers have attempted to detect possible hidden structural information inside real-world SAT instances (e.g., backbones, backdoors, equivalences and functional dependencies) in an effort to improve the efficiency of SAT solvers on hard instances. Inc* does not explicitly attempt to detect the hidden SAT structure. However, it effectively finds assignments for the variables that minimise the chance of violating the backbone of a SAT problem as early as possible in the construction of a solution. Furthermore, it does this quickly and without ever requiring complex operations, simply acting as a wrapper for standard SAT heuristics.

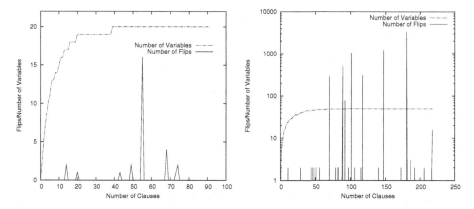

Fig. 1. Behaviour of Inc* with GSAT on two SAT problems (see text) when adding clauses one by one (and no backtracking)

3.2 Inc* Optimisation Via GP

In this section we will describe the GP system used to evolve strategies for the Inc* algorithm. As we mentioned above, an evolved strategy (which takes the form of a computer program) needs to decide how many clauses the algorithm should add/remove to/from the active list after each success or failure at finding a valid variable assignment that satisfies the current active clauses of the full SAT formula.

We use a tree representation for programs. The function and terminal sets are shown Table 1. We constrain the representation requiring that the root node of each individual in the population be the binary function $ifSuccess(d_1, d_2)$, where d_1 and d_2 assumed to be reals. This function returns the integer part of its first argument, $\lfloor d_1 \rfloor$, if the last run of the SAT heuristic was successful at satisfying the current AC. If this is the case the value $\lfloor d_1 \rfloor$ is taken to represent how many clauses should be added to AC.[3] If, instead, the SAT heuristic failed to satisfy AC, then $ifSuccess$ returns the value $\lfloor d_2 \rfloor$, which is taken to represent how many clauses should be removed from AC.

The other elements of the primitive set behave as follows: the function $add(d_1, d_2)$ returns the sum of d_1 and d_2, $sub(d_1, d_2)$ subtracts d_2 from d_1, $mul(d_1, d_2)$ returns the product of d_1 by d_2, $div(d_1, d_2)$ safely divides d_1 by d_2, and $neg(d_1)$ inverts the sign of d_1. The terminals vNo and cNo return the total number of variables and the total number of clauses in the full SAT formula, respectively. The terminals $used\_cNo$ and $used\_vNo$, instead, return the number of unique variables and the number of clauses currently loaded in the active list, respectively. Finally, $constX$ represent random integers between 0 and 9.

To evolve general Inc* strategies, we used a training set including many SAT problems with different numbers of variables. The problems were taken from the widely used SATLIB benchmark library. All problems were randomly generated satisfiable instances of 3-SAT. In total we used 50 instances: 10 with 100 variables, 15 with 150

---

[3] Note that, to give complete freedom to evolution, negative return values are allowed. If $\lfloor d_1 \rfloor$ is negative clauses are removed, rather than added, from AC.

Algorithm 2. Inc* approach to solving SAT problems

1: L = random variable assignment
2: AC = small set of random clauses from the original problem
3: $Flips$ = number of allowed flips at each stage
4: $Flips\_Total = 0$ {This keeps track of the overall number of flips used}
5: $Flips\_Used = 0$ {This keeps track of the flips used to test the active list}
6: Inc_Flip_Rate = rate of increment in the number of flips after each fail
7: **repeat**
8:    **for** $i = 0$ to Flips **do**
9:       **if** L satisfies formula F **then**
10:          **return** L
11:       **end if**
12:       select variable V from AC using some selection heuristic
13:       flip V in L
14:    **end for**
15:    Flips_Total = Flips_Total + Flips_Used
16:    update clause weights
17:    **if** L satisfies AC **then**
18:       **if** AC contains all clauses in F **then**
19:          **return** L
20:       **end if**
21:       AC = add more clauses to the active list
22:    **else**
23:       sort AC
24:       AC = remove some clauses from the active list
25:       $Flips$ = increment allowed flips
26:    **end if**
27: **until** $Flips\_Total < MaxFlips$
28: **return** n o assignment satisfying F found

variables and 25 with 250 variables. The fitness $f(s)$ of an evolved strategy s was measured by running the Inc* algorithm under the control of s on all the 50 fitness cases. More precisely

$$f(s) = \sum_i \left(inc_s(i) * \frac{v(i)}{10} \right) + \frac{1}{flips(s)}$$

where $v(i)$ is the number of variables in fitness case i, $inc_s(i)$ is a flag representing whether or not running the Inc* algorithm with strategy s on fitness case i led to success (i.e., $inc_s(i) = 1$ if fitness case i is satisfied and 0 otherwise), and $flips(s)$ is the number of flips used by strategy s averaged over all fitness cases. The factor $v(i)/10$ is used to emphasise the importance of fitness cases with a larger number of variables, while the term $1/flips(s)$ is added to give a slight advantage to strategies which use fewer flips (this is very small and typically plays a role only to break symmetries in the presence of individuals that solve the same fitness cases, but with different degrees of efficiency).

There is only one exception to this fitness calculation. In the system we keep a count of the number of attempts the SAT solver made at solving the AC list. If a maximum number of tries is reached, fitness is computed differently. Imagine, for example, what

Table 1. GP function and terminal sets

Function Set	
$ifSuccess(d_1,d_2)$	: returns d_1 if the last attempt to solve the CNF formula was successful
$add(d_1,d_2)$	: returns the sum of d_1 and d_2 as doubles
$sub(d_1,d_2)$	: subtracts d_2 from d_1
$mul(d_1,d_2)$	: returns the multiplication of d_1 by d_2
$div(d_1,d_2)$	: safe division of d_1 by d_2
$abs(d_1)$	: returns the absolute value of d_1
$neg(d_1)$	: multiplies d_1 by -1
$sqrt(d_1)$	: returns the a safe square root of d_1
Terminal Set	
vNo	: total number of variables in the CNF SAT formula
cNo	: total number of clauses in the CNF SAT formula
$used\_cNo$	: number the currently active variables
$used\_vNo$	: number the currently active clauses
$constX$	: constant integer number form 0 to 9

would happen if an evolved strategy added zero clauses after each successful attempt and removed zero clauses after each unsuccessful one. After a small number of flips have been expended to satisfy the initial active clauses, since no clauses are added or removed, no further flips would ever be necessary. So, the total number of flips used would never reach the maximum number of flips allowed, leading to an infinite loop. By using a maximum number of tries, we can avoid this and we can signal to the system that this individual (strategy) went into an infinite loop on the current fitness case. The system reacts by setting the fitness of this strategy to zero and stopping the evaluation of any remaining fitness cases.

The GP system initialises the population by randomly drawing nodes from the function and terminal sets. This is done uniformly at random using the GROW method, except that the selection of the function $ifSuccess$ is forced for the root node and is not allowed elsewhere. After initialisation, the population is manipulated by the following operators:

- Roulette wheel selection (proportionate selection) is used. Reselection is permitted.
- The reproduction rate is 0.1. Individuals that have not been affected by any genetic operator are not evaluated again to reduce the computation cost.
- The crossover rate is 0.8. Offspring are created by extracting a random subtree from the first parent and inserting it at a random point (excluding the root of the tree) in a copy of the second parent.
- Mutation is applied with a rate of 0.1. This is done by selecting a random node from the parent (including the root of the tree), deleting the sub-tree rooted there, and then regenerating it randomly as in the initialisation phase.

4   Experimental Results

In our experiments we used a population of 1000 individuals, run for 51 generations. While strategies are evolved using 50 fitness cases, the generality of best of run individuals is then evaluated on an independent test set including more 500 SAT instances.

Table 2. Comparison between average performance of GSAT and GSAT with Inc* SR=success rate, AT = average tries, AF=average number of flips (out of a maximum of 100,000)

Instance Test Set	variables	clauses	GSAT		GSAT+Inc*		
			SR	AF	SR	AF	AT
uf20-(901-1000)	20	91	0.632	43853	1	1336.52	3.37
uf50-(901-1000)	50	218	0.368	75623	0.935	14949.3	12.29
uf75-(51-100)	75	325	0.348	78423.1	0.780	31614.6	18.47
uf100-(901-1000)	100	430	0.297	86599	0.723	39856	22.59

This section will show a comparison between the performance of standard handcrafted heuristics (GSAT and WalkSat) and the same heuristics when combined with Inc* controlled by strategies evolved by GP. We have used the following parameters values for the Inc* algorithm.[4] We allow 100 flips to start with. Upon failure, the number of flips is incremented by 20%. We allow a maximum total number of flips of 100,000. The maximum number of tries is 1000 (including successful and unsuccessful attempts).

The GP system has managed to evolve a number of successful strategies. Most of these can be categorised into three groups. In the first group, strategies start by activating a relatively small number of clauses w.r.t. the total, after which they then rapidly increase the number of active clauses. This was almost always the best performing group. In the second group, strategies start by activating a very large number clauses at the beginning, then they remove some clauses after each fail and try to go forward again until a solution for all clauses is found. Strategies in this category perform slightly worse than those in the first category. Strategies in the third group were generally outperformed by those in the other groups. Strategies in this group acted in an unexpected manner. Namely, these strategies kept moving forward by adding clauses after both successful and unsuccessful tries. In the testing phase this kind of strategies performed well on instances with fewer than 100 variables in terms of number of flips used to solve the instance. However, they had a lower success rate than other strategies on larger instances. The reason of this will be explained after showing detailed results of the strategies. The following is some evolved heuristics after they have been normalizef representing the diffrent groups:

- First group: $ifSuccess(abs(vno),neg(sroot(used\_vNo)))$
- First group: $ifSuccess(mul(used\_vno,div(used\_cno,2)),neg(div(used\_vno,9)))$
- Second group: $ifSuccess(cno,neg(sub(cno,div(used\_cno,3))))$
- Third group: $ifSuccess(mul(vno,2),abs(add(vno,sroot(used\_cno))))$

Table 2 shows a first set of experimental results. In particular, it shows the difference between the average performance of the GSAT and the average performance of GSAT combined with the best evolved Inc* strategies, which we will call IncGSAT. Both heuristics in this experiment are allowed a maximum of 100,000 total flips. The performance of the heuristics on an instance is the average of 10 different runs, to ensure

---

[4] Many different combinations of parameter values have been tested, but this particular combination gave almost invariably the best results.

Table 3. Comparison between average performance of WalkSat and WalkSat with Inc* SR=success rate, AT = average tries, AF=average number of flips (out of a maximum of 100,000)

Instance Test Set	variables	clauses	WalkSat		WalkSat+Inc*		
			SR	AF	SR	AF	AT
uf20-(901-1000)	20	91	1	104.43	1	116.239	1.18
uf50-(901-1000)	50	218	1	673.17	1	696.174	4.95
uf75-(51-100)	75	325	1	1896.74	1	2000.59	8.07
uf100-(901-1000)	100	430	1	3747.32	1	3825.82	11.51
uf150-(51-100)	150	645	0.974	15021.3	0.987	14275	16.45
uf200-(51-100)	200	860	0.9	26639.2	0.936	28526.2	21.39
uf225-(51-100)	225	960	0.87	29868.5	0.91	31258.8	22.17
uf250-(51-100)	250	1065	0.816	38972.4	0.875	38304.2	24.09

the results are statistically meaningful. The AF column shows the average number of flips used by each heuristic in successful attempts only. As one case see IncGSAT has a better performance than GSAT in terms of success rate (as also shown in Figure 2) and average number of flips used to solve the test instances. So, IncGSAT managed to solve many more instance in less time. We believe this is due to the lack of random moves in the GSAT, which makes GSAT easy pray of local optima. IncGSAT improves GSAT by going forward and backward adding and removing clauses through the course of a run thereby avoiding the problem. The AT column shows the number of tries IncGSAT used. This corresponds to the number of times IncGSAT modified the search space to escape local optima and find a better path for satisfying the instance.

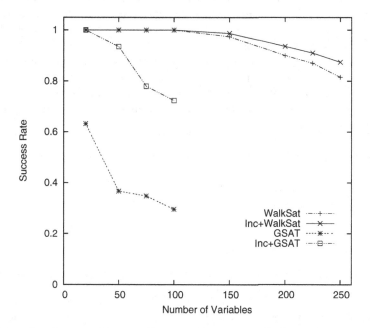

Fig. 2. Comparison between GSAT, WalkSAT, IncGSAT and IncWalk average success rate performance

Table 3 shows the results of another set of experiment using WalkSat and a combination of Sat* and WalkSat which we will call IncWalkSat. Also in this experiment, both heuristics were given a maximum of 100,000 total flips. Again, the performance of the heuristics on an instance is the average of 10 different runs. WalkSat is among the best performing local search heuristics for SAT.

We categorize the results in this table to two groups. The first group includes instances with no more than 100 variables. The second group includes instances with more than 100 variables. In the first group of problems both heuristics have a perfect success rate of 100%. However, WalkSat used a slightly smaller number of flips, thereby running marginally faster than IncWalkSat on this group of problems. In the second group, which contains larger instances, however, IncWalkSat has a higher success rate than WalkSat, and the difference in the performance increases as the size of the instance increases. This means that Inc* can solve complex instances where local heuristics alone fail. This explains why when training the GP system on small instances some evolved strategies (the strategies in group three) always tried to go foreword by adding more clauses after both successful and unsuccessful tries as we mentioned above. Effectively, these strategies tried to imitate the standard heuristics behaviour, and, indeed, they were slightly faster on small instances.

5   Conclusion

In this paper we provided a proof of concept, supported by results, of the ideas behind our Inc* algorithm, which tries to solve problems incrementally. Results on the SAT problem showed that combining local search heuristics with Inc* improved their performance especially on instance which standard local search alone failed to solve.

In future work, we will try to generalise the algorithm to other problem domains, including scheduling, time tabling, TSP, etc. Also we will test the algorithm on diffident types of SAT benchmarks (e.g., structured and handcrafted SAT problems). Also we would like to embed Inc* within a hyperheuristic framework where multiple agents perform the search in parallel. Each agent might, for example, use a different heuristic and would search for solutions to a part of the original problem (e.g., a subset of the clauses in a SAT formula).

Acknowledgements

The authors acknowledge financial support from EPSRC (grants EP/C523377/1 and EP/C523385/1).

References

1. Bader-El-Den, M.B., Poli, R.: A GP-based hyper-heuristic framework for evolving 3-SAT heuristics. In: GECCO 2007: Proceedings of the 9th annual conference on Genetic and evolutionary computation, July 7-11, 2007, vol. 2, pp. 1749–1749. ACM Press, London (2007)
2. Bader-El-Din, M.B., Poli, R.: Generating SAT local-search heuristics using a GP hyper-heuristic framework. In: Proceedings of the 8th International Conference on Artificial Evolution, vol. 36(1), pp. 141–152 (2007)

3. Cook, S.A.: The complexity of theorem-proving procedures. In: STOC 1971: Proceedings of the third annual ACM symposium on Theory of computing, pp. 151–158. ACM Press, New York (1971)
4. Davis, M., Logemann, G., Loveland, D.: A machine program for theorem-proving. Commun. ACM 5(7), 394–397 (1962)
5. Fourdrinoy, O., Gregoire, E., Mazure, B., Sais, L.: Eliminating redundant clauses in sat instances. In: Van Hentenryck, P., Wolsey, L.A. (eds.) CPAIOR 2007. LNCS, vol. 4510, pp. 71–83. Springer, Heidelberg (2007)
6. Fukunaga, A.: Automated discovery of composite SAT variable selection heuristics. In: Proceedings of the National Conference on Artificial Intelligence (AAAI), pp. 641–648 (2002)
7. Fukunaga, A.S.: Evolving local search heuristics for SAT using genetic programming. In: Deb, K., et al. (eds.) GECCO 2004. LNCS, vol. 3103, pp. 483–494. Springer, Heidelberg (2004)
8. Gent, I.P., Walsh, T.: Towards an understanding of hill-climbing procedures for SAT. In: Proc. of AAAI 1993, Washington, DC, pp. 28–33 (1993)
9. Gottlieb, J., Marchiori, E., Rossi, C.: Evolutionary algorithms for the satisfiability problem. Evol. Comput. 10(1), 35–50 (2002)
10. Han, H., Somenzi, F.: Alembic: an efficient algorithm for cnf preprocessing. In: DAC 2007: Proceedings of the 44th annual conference on Design automation, pp. 582–587. ACM, New York (2007)
11. Hoos, H.H., O'Neill, K.: Stochastic local search methods for dynamic SAT- an initial investigation. Technical Report TR-00-01, 1 (2000)
12. Kibria, R.H., Li, Y.: Optimizing the initialization of dynamic decision heuristics in DPLL SAT solvers using genetic programming. In: Collet, P., Tomassini, M., Ebner, M., Gustafson, S., Ekárt, A. (eds.) EuroGP 2006. LNCS, vol. 3905, pp. 331–340. Springer, Heidelberg (2006)
13. Koza, J.R.: Genetic Programming: On the Programming of Computers by Means of Natural Selection. MIT Press, Cambridge (1992)
14. Langdon, W.B., Poli, R.: Foundations of Genetic Programming. Springer, Heidelberg (2002)
15. Marchiori, E., Rossi, C.: A flipping genetic algorithm for hard 3-SAT problems. In: Banzhaf, W., Daida, J., Eiben, A.E., Garzon, M.H., Honavar, V., Jakiela, M., Smith, R.E. (eds.) Proceedings of the Genetic and Evolutionary Computation Conference, Orlando, Florida, USA, 13-17, 1999, vol. 1, pp. 393–400. Morgan Kaufmann, San Francisco (1999)
16. Minton, S., Johnston, M.D., Philips, A.B., Laird, P.: Minimizing conflicts: A heuristic repair method for constraint satisfaction and scheduling problems. Artificial Intelligence 58(1-3), 161–205 (1992)
17. Rossi, C., Marchiori, E., Kok, J.N.: An adaptive evolutionary algorithm for the satisfiability problem. In: SAC 2000, vol. 1, pp. 463–469 (2000)
18. Selman, B., Kautz, H.: Domain-independent extensions to GSAT: solving large structured satisfiability problems. In: Proceedings of theInternational Joint Conference on Artificial Intelligence(IJCAI 1993), Chambry, France (1993)
19. Selman, B., Kautz, H.A., Cohen, B.: Noise strategies for improving local search. In: Proceedings of the Twelfth National Conference on Artificial Intelligence (AAAI 1994), Seattle, pp. 337–343 (1994)
20. Selman, B., Levesque, H.J., Mitchell, D.: A new method for solving hard satisfiability problems. In: Rosenbloom, P., Szolovits, P. (eds.) Proceedings of the Tenth National Conference on Artificial Intelligence, pp. 440–446. AAAI Press, Menlo Park (1992)
21. Zeng, H., McIlraith, S.A.: The role of redundant clauses in solving satisfiability problems. In: van Beek, P. (ed.) CP 2005. LNCS, vol. 3709, p. 873. Springer, Heidelberg (2005)

Metaheuristics for the
Bi-objective Ring Star Problem

Arnaud Liefooghe[1], Laetitia Jourdan[1], Matthieu Basseur[2],
El-Ghazali Talbi[1], and Edmund K. Burke[2]

[1] Laboratoire d'Informatique Fondamentale de Lille (LIFL), INRIA, CNRS
Université des Sciences et Technologies de Lille
Parc scientifique de la Haute Borne, 40 avenue Halley, Bât. A, Park Plaza
59650 Villeneuve d'Ascq, France
{Arnaud.Liefooghe,Laetitia.Jourdan,El-Ghazali.Talbi}@lifl.fr
[2] School of Computer Science,
University of Nottingham, Jubilee Campus, Wollaton Road,
Nottingham, NG8 1BB, United Kingdom
Matthieu.Basseur@info.univ-angers.fr,ekb@cs.nott.ac.uk

Abstract. The bi-objective ring star problem aims to locate a cycle through a subset of nodes of a graph while optimizing two types of cost. The first criterion is to minimize a ring cost, related to the length of the cycle, whereas the second one is to minimize an assignment cost, from non-visited nodes to visited ones. In spite of its natural multi-objective formulation, this problem has never been investigated in such a way. In this paper, three metaheuristics are designed to approximate the whole set of efficient solutions for the problem under consideration. Computational experiments are performed on well-known benchmark test instances, and the proposed methods are rigorously compared to each other using different performance metrics.

1 Introduction

The purpose of the bi-objective ring star problem is to find a cycle (the ring) which visits a subset of nodes of a graph. The two objectives are the minimization of a cost associated to the ring itself and the minimization of a cost associated to the arcs directed from non-visited nodes to visited ones. Although this problem is clearly bi-objective, it has always been investigated in a single-objective way. It was introduced by Labbé et al. [12], where the goal was to minimize the sum of both costs. Another mono-criterion formulation of the problem, where one of the objectives is regarded as a constraint, has been investigated, for instance, by Renaud et al. [16]. These two formulations are commonly used to convert a multi-objective problem into a single-objective one by using scalar approaches [14].

The ring star problem has a wide range of industrial applications, including telecommunication networks design, school bus routing, routing of essential medical care services, circular-shaped transportation, and post-box location. However, in spite of its real-world applications, this is the first time that such a

J. van Hemert and C. Cotta (Eds.): EvoCOP 2008, LNCS 4972, pp. 206–217, 2008.
© Springer-Verlag Berlin Heidelberg 2008

problem is studied in a bi-objective way, perhaps because of its complexity. Indeed, it is particularly challenging because, once it is decided which nodes have to be visited or not, a classical travelling salesman problem still remains to be solved. Nevertheless, Current and Schilling [5] investigated two variants of a similar problem: the *median tour problem* and the *maximal covering tour problem*. In both, one criterion is the minimization of the total length of the tour, while another one is the maximization of the access to the tour for non-visited nodes. To tackle these problems, the authors used a kind of lexicographic method, where one objective function is optimized after another. Furthermore, Dorner et al. [8] recently formulated a problem of tour planning for mobile healthcare facilities in Senegal. A mobile facility has to visit a subset of nodes. Non-visited nodes are then assigned to their closest tour stop or are regarded as unable to reach a tour stop. The obectives are the minimization of the ratio between medical working time and total working time, the minimization of the average distance to the nearest tour stops, and the maximization of a coverage criterion. The authors designed a Pareto ant colony optimization algorithm and two multi-objective genetic algorithms to solve real-world instances.

In this paper, we investigate metaheuristic solution methods for the problem under consideration. Three metaheuristics are designed to approximate the whole set of efficient solutions. A population-based local search and two evolutionary algorithms are compared on state-of-the-art instances involving up to 300 nodes. The reminder of the paper is organized as follows. In Section 2, we provide the basic definitions for multi-objective optimization and the formulation of the bi-objective ring star problem. Section 3 presents three resolution methods designed to tackle the problem under consideration. Some experimental results and a comparative study are provided in Section 4, while the last section concludes the paper and discusses perspectives about this work.

2   Preliminaries

This section presents some basic concepts related to multi-objective optimization and provides the formulation of the bi-objective ring star problem.

2.1   Multi-objective Optimization

A *Multi-objective Optimization Problem* (MOP) aims to optimize a set of $n \geq 2$ objective functions $f_1, f_2, \ldots, f_n$ simultaneously. Without loss of generality, we assume that all n objective functions have to be minimized. Let X denote the set of feasible solutions in the *decision space*, and Z the set of feasible points in the *objective space*. To each decision vector $x \in X$ is assigned exactly one objective vector $z \in Z$ on the basis of a vector function $f : X \to Z$ with $z = f(x) = (f_1(x), f_2(x), \ldots, f_n(x))$. In the case of a *Multi-objective Combinatorial Optimization Problem* (MCOP), note that a decision vector $x \in X$ has a finite set of possible values.

Definition 1. *An objective vector $z \in Z$ weakly dominates another objective vector $z' \in Z$ if and only if $\forall i \in [1..n]$, $z_i \leq z_i'$.*

Definition 2. *An objective vector $z \in Z$ dominates another objective vector $z' \in Z$ if and only if $\forall i \in [1..n]$, $z_i \leq z_i'$ and $\exists j \in [1..n]$ such as $z_j < z_j'$.*

Definition 3. *An objective vector $z \in Z$ is non-dominated if and only if there does not exist another objective vector $z' \in Z$ such that z' dominates z.*

A solution $x \in X$ is said to be *efficient* (or *Pareto optimal*) if its mapping in the objective space results in a non-dominated point. The set of all efficient solutions is the *efficient* (or *Pareto optimal*) *set*, denoted by X_E. The set of all non-dominated vectors is the *non-dominated front* (or the *trade-off surface*), denoted by Z_N. A common approach in solving MOPs is to find or to approximate the set of efficient solutions; or at least a solution $x \in X_E$ for each non-dominated vector $z \in Z_N$ such as $f(x) = z$. A reasonable basic introduction to multi-objective optimization can be found in [6].

Note that we will assume, throughout the paper, that objective values are normalized. To achieve this, the minimum and the maximum value of each objective function are used in order to adaptively replace each objective function by its corresponding normalized function, so that its values lie in the interval $[0, 1]$.

2.2 The Bi-objective Ring Star Problem

The *Ring Star Problem* (RSP) can be described as follows. Let $G = (V, E, A)$ be a complete mixed graph where $V = \{v_1, v_2, \ldots, v_n\}$ is a set of vertices, $E = \{[v_i, v_j] | v_i, v_j \in V, i < j\}$ is a set of edges, and $A = \{(v_i, v_j) | v_i, v_j \in V\}$ is a set of arcs. Vertex v_1 is the depot. To each edge $[v_i, v_j]$ we assign a non-negative *ring cost* c_{ij}, and to each arc (v_i, v_j) we assign a non-negative *assignment cost* d_{ij}. The RSP consists of locating a simple cycle through a subset of nodes $V' \subset V$ (with $v_1 \in V'$) while (i) minimizing the sum of the ring costs related to all edges that belong to the cycle, and (ii) minimizing the sum of the assignment costs of arcs directed from every non-visited node to a visited one so that the associated cost is minimum. An example of solution is given in Figure 1, where solid lines represent edges that belong to the ring and dashed lines represent arcs of the assignments. The first objective is called the *ring cost* and is defined as:

$$\sum_{[v_i,v_j]\in E} c_{ij}x_{ij} \, , \tag{1}$$

where x_{ij} is a binary variable equal to 1 if and only if the edge $[v_i, v_j]$ belongs to the cycle. The second objective, the *assignment cost*, can be computed as follows:

$$\sum_{v_i\in V\setminus V'} \min_{v_j\in V'} d_{ij} \, . \tag{2}$$

The RSP is an NP-hard combinatorial problem since the particular case of visiting the whole set of nodes is equivalent to a traditional travelling salesman problem.

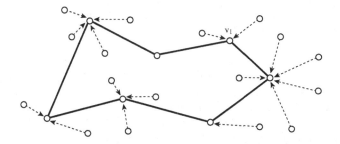

Fig. 1. An example of a solution for the ring star problem

3   Metaheuristics for the Bi-objective Ring Star Problem

Three metaheuristics are proposed to tackle the bi-objective RSP: a variable neighbourhood iterative Local Search (LS) and two Evolutionary Algorithms (EAs). These algorithms are respectively steady-state variations of IBMOLS [1], IBEA [17] and NSGA-II [7]. IBMOLS and IBEA are both recent indicator-based metaheuristics, whereas NSGA-II is one of the most often used Pareto-based resolution methods. In this section, RSP-specific components are described after we have presented the main characteristics of the LS and of the EAs.

3.1   A Multi-objective Local Search

Since they are easily adaptable to the multi-objective context, many of the search algorithms proposed to tackle MOPs are EAs. However, LS algorithms are known to be effective metaheuristics for solving real-world applications [4,9]. Several multi-objective LS approaches have been proposed in the literature. In particular, the Indicator-Based Multi-Objective Local Search (IBMOLS for short) has recently been presented in [1]. IBMOLS is a generic population-based multi-objective LS dealing with a fixed population size. This allows to obtain a set of efficient solutions in a single run without specifying any mechanism to control the number of solutions during the search process. Moreover, IBMOLS represents an alternative to aggregation- and Pareto-based multi-objective metaheuristics. Indeed, as proposed in [17], it is assumed that the optimization goal is given in terms of a binary quality indicator I [19] which can be regarded as an extension of the Pareto dominance relation. A value $I(A, B)$ quantifies the difference in quality between two non-dominated sets A and B. So, if Z_N denotes the optimal non-dominated front, the overall optimization goal can be formulated as:

$$\arg \min_{A \in \Omega} I(A, Z_N) , \qquad (3)$$

where Ω denotes the space of all non-dominated set approximations. As noted in [17], Z_N does not have to be known, it is just required in the formalization of the optimization goal. Since Z_N is fixed, I actually represents a unary function that assigns a real number reflecting the quality of each approximation set.

One of the main advantages of indicator-based optimization is that no specific diversity preservation mechanism is generally required, according to the indicator being used.

The IBMOLS algorithm maintains a population P. Then, it generates the neighbourhood of a solution contained in P until a good solution is found (*i.e.* one that is better than at least one solution of P in terms of the indicator being used). By iterating this simple principle to every solution of P, we obtain a local search step. The whole local search stops when the archive of potentially efficient solutions has not received any new item during a complete local search step. Moreover, as local search methods are usually performed in an iterative way, a population re-initialization scheme has to be designed after each local search. Several strategies can be used within an iterative IBMOLS [1]. Solutions can be re-initialized randomly, and crossover or random noise can be applied to solutions of the efficient set approximation. The interested reader could refer to [1] for more details about IBMOLS.

A beneficial feature of this LS is the low number of parameters that are required. In addition to the population size, the binary quality indicator to be used and the population re-initialization strategy (between each local search) are the two only other problem-independent parameters. Indeed, several quality indicators can be used within IBMOLS. The binary additive ϵ-indicator [17] is particularly well-adapted to indicator-based search and seems to be efficient on different kinds of problems (see, for instance, [1,17]). It is capable of obtaining both a well-converged and a well-diversified Pareto set approximation. This indicator computes the minimum value by which a solution $x_1 \in X$ can be translated in the objective space to weakly dominate another solution $x_2 \in X$. For a minimization problem, it is defined as follows:

$$I_{\epsilon+}(x_1, x_2) = \max_{i \in \{1,\dots,n\}} (f_i(x_1) - f_i(x_2)) .$$ (4)

Furthermore, to evaluate the quality of a solution according to a whole population P and a binary quality indicator I, different approaches exist. As proposed in [17], we will here consider an additive technique that amplifies the influence of solutions mapping to dominating points over solutions mapping to dominated ones which can be outlined as follows:

$$I(P \setminus \{x\}, x) = \sum_{x^* \in P \setminus \{x\}} -e^{-I(x^*, x)/\kappa} ,$$ (5)

where $\kappa > 0$ is a scaling factor. However, the initial experiments were not satisfactory because the algorithm was not able to find the extreme points of the trade-off surface. This is known to be one of the drawbacks of the ϵ-indicator, apparently due to the high convexity of the front. To tackle this problem, we add a condition preventing the deletion of solutions corresponding to the extreme non-dominated vectors during the replacement step of IBMOLS. Additionally, the population re-initialization scheme used between each local search is based on random noise, such as in the basic simulated annealing algorithm [4]. Random noise consists of multiple mutations applied to N different randomly chosen solutions contained in the archive of potentially efficient solutions. If the size of the archive is less than N, the population is filled with random solutions.

3.2   Multi-objective Evolutionary Algorithms

The multi-objective EAs designed for the RSP are variations of two state-of-the-art methods, namely IBEA [17] and NSGA-II [7]. Some minor modifications have been carried out to improve the algorithms for the particular case of the addressed problem, for which the set of non-dominated points is, in general, very large.

IBEA. Introduced by Zitzler and Künzli [17], the *Indicator-Based Evolutionary Algorithm* (IBEA) is, like IBMOLS, an indicator-based metaheuristic. The fitness assignment scheme of this EA is based on a pairwise comparison of solutions contained in a population by using a binary quality indicator. As within IBMOLS, no diversity preservation technique is required, according to the indicator being used. The selection scheme for reproduction is a binary tournament between randomly chosen individuals. The replacement strategy is an environmental one that consists of deleting, one-by-one, the worst individuals, and in updating the fitness values of the remaining solutions each time there is a deletion; this is continued until the required population size is reached. Moreover, an archive stores solutions mapping to potentially non-dominated points, in order to prevent their loss during the stochastic search process. However, in our case, and in contrast to the IBEA defined in [17], this archive is updated at each generation since the beginning of the EA, so that the output size is not necessarily less than or equal to the population size. Just like for the IBMOLS algorithm, the indicator used within IBEA is the additive ϵ-indicator; and the same mechanism has been used to prevent the loss of the extreme points on the trade-off surface.

NSGA-II. At each generation of NSGA-II (*Non-dominated Sorting Genetic Algorithm II* [7]), the solutions contained in the population are ranked into several classes. Individuals mapping to vectors from the first front all belong to the best efficient set; individuals mapping to vectors from the second front all belong to the second best efficient set; and so on. Two values are then computed for every solution of the population. The first one corresponds to the *rank* the corresponding solution belongs to, and represents the quality of the solution in terms of convergence. The second one, the *crowding distance*, consists of estimating the density of solutions surrounding a particular point of the objective space, and represents the quality of the solution in terms of diversity. A solution is said to be better than another if it has the best rank, or in the case of a tie, if it has the best crowding distance. The selection strategy is a deterministic tournament between two random solutions. At the replacement step, only the best individuals survive, with respect to the population size. Likewise, an external population is added to the steady-state NSGA-II in order to store every potentially efficient solution found during the search.

Vertex	v_1	v_2	v_3	v_4	v_5	v_6	v_7	v_8	v_9	v_{10}
Random key	0	0.7	-	0.3	-	0.8	0.2	-	0.5	-

Fig. 2. A RSP solution represented by random keys

3.3   Application to the Bi-objective Ring Star Problem

This section provides the problem-specific steps of the metaheuristics introduced earlier. Components designed for the particular case of the bi-objective RSP, such as the encoding mechanism, the population initialization as well as the neighbourhood, mutation and crossover operators, are described below.

Solution Encoding. The encoding mechanism used to represent a RSP solution, for both the LS and the EAs, is based on the random keys concept proposed by Bean [2]. This implementation has already been successfully applied for a single-objective version of the RSP in [16]. To each node v_i belonging to the ring we assign exactly one *random key* $x_i \in [0, 1[$, where $x_1 = 0$. A special value is assigned to unvisited nodes. Thus, the ring route associated to a solution corresponds to the nodes ordered according to their random keys in an increasing way; *i.e.* if $x_i < x_j$, then v_j comes after v_i. As an example, a possible representation for the cycle $(v_1, v_7, v_4, v_9, v_2, v_6)$ is given in Figure 2. Vertices v_3, v_5, v_8 and v_{10} are assigned to a visited node in such a way that the associated assignment cost is minimum.

Population Initialization. For every optimization method, the initial population has been generated randomly. Each node has a probability $p = 0.5$ that it will be visited or not, and to each visited vertex we associate a key randomly generated between 0 and 1.

Neighbourhood and Mutation Operators. As the RSP is both a routing problem and an assignment problem, different neighbourhood and mutation operators have to be designed. Here, we consider the following:

- *insert operator*: adds an unvisited node v_i in the cycle, the position where to insert v_i is chosen in order to minimize the ring cost
- *remove operator*: removes a vertex v_j of the ring
- *2-opt operator*: applies a 2-opt operator between two nodes of the cycle v_i and v_j, *i.e.* it reverses the sequence of visited nodes between v_i and v_j.

For the LS, the neighbours of a solution are randomly explored, without considering any order between these three operators; and each neighbour is at most visited once. Moreover, note that it is not necessary to completely re-evaluate a solution each time a neighbourhood operator is applied. Thus, after an *insert* neighbourhood operator, we just have to re-assign unvisited nodes in order to minimize the assignment cost. After a *remove* neighbourhood operator, we just have to re-assign the nodes that were previously assigned to the one that has been removed. And, after a *2-opt* neighbourhood operator, we just have to recompute the ring cost, the assignment cost being unchanged. In the case of mutations, the operators are applied to randomly chosen vertices.

Crossover Operator. The crossover operator is a quadratic crossover closely related to the one proposed in [16]. Two randomly selected solutions s_1 and s_2 are divided in a particular position. Then, the first part of s_1 is combined with the second part of s_2 to build a first offspring, and the first part of s_2 is combined with the second part of s_1 to build a second offspring. Every node retains its random key so that it enables an easy reconstruction of the new individuals. Thanks to the random keys encoding mechanism, solutions having a different ring size can easily be recombined, even if the initial ring structures are generally broken in the offspring solutions.

4   Experiments

The metaheuristics described in the previous section have all been implemented using the ParadisEO-MOEO library[1] [13]. ParadisEO-MOEO is a C++ white-box object-oriented framework dedicated to the reusable design of metaheuristics for multi-objective optimization. All the algorithms share the same base components for a fair comparison between them. Computational runs were performed on an Intel Core 2 Duo 6600 (2×2.40 GHz) machine, with 2 GB RAM.

4.1   Experimental Protocol

Benchmarks. The performance of the metaheuristics has been tested on different instances taken from the TSPLIB[2] [15]. These instances involve between 50 and 300 nodes. The number at the end of an instance's name represents the number of nodes for the instance under consideration. Let l_{ij} denote the distance between two nodes v_i and v_j of a TSPLIB file. Then, the ring cost c_{ij} and the assignment cost d_{ij} have both been set to l_{ij} for every pair of nodes v_i and v_j.

Parameter Setting. For each one of the metaheuristics proposed to tackle the bi-objective RSP, the search process stops after a certain ammount of run time. As shown in Table 1, this run time is defined according to the size of the instance under consideration. Likewise, the population size depends on the number of vertices involved in the instance (see Table 1). For each instance, a small (S), a medium (M), a large (L) and an extra-large (XL) population size have been tested. The noise rate for the population re-initialization in the iterated version of IBMOLS is set to a fixed percentage of the instance's size. We investigate three different values for this noise rate: 5%, 10% and 20%. Then, $0.05 \times n$, $0.1 \times n$ and $0.2 \times n$ random mutations are applied respectively for a problem with n nodes. For both IBMOLS and IBEA, the scaling factor κ is set to 0.05. Finally, for the EAs, the crossover probability is set to 0.25, and the mutation probability to 1.00, with a probability of 0.25, 0.25 and 0.50 for the *remove*, the *insert* and the *2-opt* operator, respectively.

Performance Assessment. For each TSPLIB instance and each metaheuristic proposed in Section 3, a set of 20 runs, with different initial populations,

---

[1] ParadisEO is available at http://paradiseo.gforge.inria.fr.
[2] Benchmarks are available at http://elib.zib.de/pub/mp-testdata/tsp/tsplib.

Table 1. Instance-dependant parameters setting

Instance	Population size				Running
	S	M	L	XL	time
eil51	5	10	15	100	20"
st70	5	10	15	100	1'
kroA100	10	15	20	100	2'
bier127	10	15	20	100	5'
kroA150	15	20	30	100	10'
kroA200	15	20	30	100	20'
pr264	15	20	30	100	50'
pr299	15	20	30	100	100'

has been performed. In order to evaluate the quality of the non-dominated front approximations obtained for a specific test instance, we follow the protocol given in [11]. First, we compute a reference set $Z_N^\star$ of non-dominated points extracted from the union of all these fronts. Second, we define $z^{max} = (z_1^{max}, z_2^{max})$, where z_1^{max} (respectively z_2^{max}) denotes the upper bound of the first (respectively second) objective in the whole non-dominated front approximations. Then, to measure the quality of an output set A in comparison to $Z_N^\star$, we compute the difference between these two sets by using the unary hypervolume metric [18], $(1.05 \times z_1^{max}, 1.05 \times z_2^{max})$ being the reference point. The hypervolume difference indicator (I_H^-) computes the portion of the objective space that is weakly dominated by $Z_N^\star$ and not by A. Furthermore, we also consider the R2 indicator proposed in [10] with a Chebycheff utility function defined by $z^\star = (1,1)$, $\rho = 0.01$ and a set Λ of 500 uniformly distributed normalized weighted vectors. As a consequence, for each test instance, we obtain 20 hypervolume differences and 20 R2 measures, corresponding to the 20 runs, per algorithm. As suggested by Knowles et al. [11], once all these values are computed, we perform a statistical analysis on pairs of optimization methods for a comparison on a specific test instance. To this end, we use the Mann-Whitney statistical test as described in [11], with a p-value lower than 5%. Note that all the performance assessment procedures have been achieved using the performance assessment tool suite provided in PISA[3] [3].

4.2   Computational Results and Discussion

Table 2 presents the results obtained by the metaheuristics on eight different test instances. Due to space limitation and in order to simplify the reading of the table, only the results obtained by a large population size and by a noise rate of 5% for IBMOLS and by an extra-large population size for NSGA-II and IBEA are reported in the paper. These parameters have respectively been chosen as they were globally more efficient for each one of the algorithms. Overall, with respect to the metrics we used, we can see that IBMOLS performs significantly

---

[3] The package is available at http://www.tik.ee.ethz.ch/pisa/assessment.html.

Table 2. Comparison of different metaheuristics for the I_H^- and the R2 metrics by using a Mann-Whitney statistical test with a p-value of 5%. According to the metric under consideration, either the results of the algorithm located at a specific row are significantly better than those of the algorithm located at a specific column ($\succ$), either they are worse ($\prec$), or there is no significant difference between both ($\equiv$).

		I_H^-			R2		
		IBMOLS	IBEA	NSGA-II	IBMOLS	IBEA	NSGA-II
eil51	IBMOLS	-	$\succ$	$\succ$	-	$\succ$	$\succ$
	IBEA	$\prec$	-	$\succ$	$\prec$	-	$\succ$
	NSGA-II	$\prec$	$\prec$	-	$\prec$	$\prec$	-
st70	IBMOLS	-	$\equiv$	$\succ$	-	$\succ$	$\succ$
	IBEA	$\equiv$	-	$\succ$	$\prec$	-	$\succ$
	NSGA-II	$\prec$	$\prec$	-	$\prec$	$\prec$	-
kroA100	IBMOLS	-	$\equiv$	$\succ$	-	$\equiv$	$\succ$
	IBEA	$\equiv$	-	$\succ$	$\equiv$	-	$\succ$
	NSGA-II	$\prec$	$\prec$	-	$\prec$	$\prec$	-
bier127	IBMOLS	-	$\succ$	$\succ$	-	$\succ$	$\succ$
	IBEA	$\prec$	-	$\succ$	$\prec$	-	$\succ$
	NSGA-II	$\prec$	$\prec$	-	$\prec$	$\prec$	-
kroA150	IBMOLS	-	$\succ$	$\succ$	-	$\succ$	$\succ$
	IBEA	$\prec$	-	$\succ$	$\prec$	-	$\succ$
	NSGA-II	$\prec$	$\prec$	-	$\prec$	$\prec$	-
kroA200	IBMOLS	-	$\succ$	$\succ$	-	$\succ$	$\succ$
	IBEA	$\prec$	-	$\equiv$	$\prec$	-	$\equiv$
	NSGA-II	$\prec$	$\equiv$	-	$\prec$	$\equiv$	-
pr264	IBMOLS	-	$\equiv$	$\succ$	-	$\prec$	$\prec$
	IBEA	$\equiv$	-	$\succ$	$\succ$	-	$\succ$
	NSGA-II	$\prec$	$\prec$	-	$\succ$	$\prec$	-
pr299	IBMOLS	-	$\equiv$	$\succ$	-	$\prec$	$\prec$
	IBEA	$\equiv$	-	$\succ$	$\succ$	-	$\succ$
	NSGA-II	$\prec$	$\prec$	-	$\succ$	$\prec$	-

better than IBEA and NSGA-II on most test instances. Nevertheless, it is not the case on large problems (*pr264* and *pr299*), where IBMOLS is outperformed by both algorithms according to the R2 metric. Additionally, although IBEA is in general statistically outperformed by IBMOLS, it performs significantly better than NSGA-II on a large number of the tested instances, and never performs significantly worse on each one of them. Furthermore, we can see that the overall efficiency of NSGA-II is very poor since it is statistically outperformed on most problems, except occasionally where it performs better than the IBMOLS algorithm, as pointed out above. To summarise, IBMOLS performs well on small-size RSP instances, but seems to have more trouble in dealing with large ones. Moreover, we also compared our results to the ones given in [12] for a mono-objective version of the problem. For each test instance, the error ratio between the point belonging to $Z_N^\star$ that minimizes the single objective function investigated in [12] and the exact optimal value is is always under 2% and is averagely under 0.5%.

One of the main characteristics of the problem under consideration seems to be the high number of points located in the trade-off surface. Then, after a certain number of iterations, a large part of the population involved in all the algorithms might map to potentially non-dominated points. This could explain the low efficiency of NSGA-II. Since the same rank is assigned to the major part of the population, only the crowding distance is used to compare solutions. However, the indicator-based fitness assignment scheme is obviously much more suited to determine potentially efficient solutions than the single crowding distance. Moreover, the high performance of IBMOLS in comparison to IBEA might depends on how close are the solutions which map to non-dominated points in the decision space. If these solutions are close to each other according to the neighbourhood operators, a LS is known to be particularly well-suited to find additional interesting solutions by exploring the neighbourhood of a potentially efficient solution. On the contrary, an EA usually explores the decision space in a more random way. Thus, a landscape analysis could be interesting to study the bi-objective RSP in more depth.

5   Conclusion

In this paper, a multi-objective routing problem, the bi-objective ring star problem, has been investigated. It has already been studied in a single-objective form where either both objectives have been combined [12] or one objective has been treated as a constraint [16]. Here, for the first time, this problem is formulated in such a way that multiple conflicting criteria have to be optimized simultaneously. Three metaheuristics have been proposed to approximate the minimal complete set of efficient solutions: a population-based local search with variable neighbourhood and two evolutionary algorithms. Experiments were conducted using various test instances. We concluded that the local search method was significantly more efficient than the evolutionary algorithms on a large majority of instances, with respect to the performance metrics we used. The only instances for which the local search was outperformed were large ones. As a next step, we will try to solve ring star problem instances involving an even bigger number of nodes, to verify if our observations are still valid. If it is the case, it could be interesting to design a cooperation scheme between two different methods (*i.e.* the local search procedure and an evolutionary algorithm) in order to benefit from the respective quality of each one of them. The resulting hybrid metaheuristic could be particularly effective for solving large size problems.

References

1. Basseur, M., Burke, E.K.: Indicator-based multi-objective local search. In: CEC 2007, Singapore, pp. 3100–3107 (2007)
2. Bean, J.: Genetic algorithms and random keys for sequencing and optimization. ORSA Journal on Computing 6(2), 154–160 (1994)

3. Bleuler, S., Laumanns, M., Thiele, L., Zitzler, E.: PISA — a platform and programming language independent interface for search algorithms. In: Fonseca, C.M., Fleming, P.J., Zitzler, E., Deb, K., Thiele, L. (eds.) EMO 2003. LNCS, vol. 2632, pp. 494–508. Springer, Heidelberg (2003)
4. Burke, E.K., Kendall, G. (eds.): Search Methodologies: Introductory Tutorials in Optimization and Decision Support Techniques. Springer, Heidelberg (2005)
5. Current, J.R., Schilling, D.A.: The median tour and maximal covering tour problems: Formulations and heuristics. European Journal of Operational Research 73, 114–126 (1994)
6. Deb, K.: Multi-objective optimization. In: [4], ch. 10, pp. 273–316 (2005)
7. Deb, K., Agrawal, S., Pratap, A., Meyarivan, T.: A fast and elitist multiobjective genetic algorithm: NSGA-II. IEEE Transactions on Evolutionary Computation 6(2), 182–197 (2002)
8. Doerner, K., Focke, A., Gutjahr, W.J.: Multicriteria tour planning for mobile healthcare facilities in a developing country. European Journal of Operational Research 179(3), 1078–1096 (2007)
9. Glover, F.W., Kochenberger, G.A. (eds.): Handbook of Metaheuristics. International Series in Operations Research & Management Science, vol. 57. Kluwer Academic Publishers, Boston, USA (2003)
10. Hansen, M.P., Jaszkiewicz, A.: Evaluating the quality of approximations of the non-dominated set. Technical Report IMM-REP-1998-7, Institute of Mathematical Modeling, Technical University of Denmark (1998)
11. Knowles, J., Thiele, L., Zitzler, E.: A tutorial on the performance assessment of stochastic multiobjective optimizers. Technical report, Computer Engineering and Networks Laboratory (TIK), ETH Zurich, Switzerland (revised version) (2006)
12. Labbé, M., Laporte, G., Rodríguez Martín, I., Salazar González, J.J.: The ring star problem: Polyhedral analysis and exact algorithm. Networks 43, 177–189 (2004)
13. Liefooghe, A., Basseur, M., Jourdan, L., Talbi, E.-G.: ParadisEO-MOEO: A framework for evolutionary multi-objective optimization. In: Obayashi, S., Deb, K., Poloni, C., Hiroyasu, T., Murata, T. (eds.) EMO 2007. LNCS, vol. 4403, pp. 386–400. Springer, Heidelberg (2007)
14. Miettinen, K.: Nonlinear Multiobjective Optimization, vol. 12. Kluwer Academic Publishers, Boston, USA (1999)
15. Reinelt, G.: TSPLIB – A traveling salesman problem library. ORSA Journal on Computing 3(4), 376–384 (1991)
16. Renaud, J., Boctor, F.F., Laporte, G.: Efficient heuristics for median cycle problems. Journal of the Operational Research Society 55(2), 179–186 (2004)
17. Zitzler, E., Künzli, S.: Indicator-based selection in multiobjective search. In: Yao, X., Burke, E.K., Lozano, J.A., Smith, J., Merelo-Guervós, J.J., Bullinaria, J.A., Rowe, J.E., Tiño, P., Kabán, A., Schwefel, H.-P. (eds.) PPSN 2004. LNCS, vol. 3242, pp. 832–842. Springer, Heidelberg (2004)
18. Zitzler, E., Thiele, L.: Multiobjective evolutionary algorithms: A comparative case study and the strength pareto approach. IEEE Transactions on Evolutionary Computation 3(4), 257–271 (1999)
19. Zitzler, E., Thiele, L., Laumanns, M., Foneseca, C.M., da Fonseca, V.G.: Performance assessment of multiobjective optimizers: An analysis and review. IEEE Transactions on Evolutionary Computation 7(2), 117–132 (2003)

Multiobjective Prototype Optimization with Evolved Improvement Steps

Jiri Kubalik[1], Richard Mordinyi[2], and Stefan Biffl[3]

[1] Department of Cybernetics
Czech Technical University in Prague
Technicka 2, 166 27 Prague 6, Czech Republic
kubalik@labe.felk.cvut.cz
[2] Space-Based Computing Group
Institute of Computer Languages
Vienna University of Technology
Argentinierstr. 8, A-1040 Vienna, Austria
richard@complang.tuwien.ac.at
[3] Institute of Software Technology and Interactive Systems
Vienna University of Technology
Favoritenstr. 9/188, A-1040 Vienna, Austria
Stefan.Biffl@tuwien.ac.at

Abstract. Recently, a new iterative optimization framework utilizing an evolutionary algorithm called "Prototype Optimization with Evolved iMprovement Steps" (POEMS) was introduced, which showed good performance on hard optimization problems - large instances of TSP and real-valued optimization problems. Especially, on discrete optimization problems such as the TSP the algorithm exhibited much better search capabilities than the standard evolutionary approaches. In many real-world optimization problems a solution is sought for multiple (conflicting) optimization criteria. This paper proposes a multiobjective version of the POEMS algorithm (mPOEMS), which was experimentally evaluated on the multiobjective 0/1 knapsack problem with alternative multiobjective evolutionary algorithms. Major result of the experiments was that the proposed algorithm performed comparable to or better than the alternative algorithms.

Keywords: multiobjective optimization, evolutionary algorithms, multiobjective 0/1 knapsack problem.

1   Introduction

In many real-world optimization problems a solution is sought that is optimal with respect to multiple (often conflicting) optimization criteria. Multiple objectives specify quality measures of solutions that typically do not result in a single optimal solution. Instead there is a set of alternative solutions that are optimal in a sense that (*i*) none of them is superior to the others and (*ii*) there is no superior solution in the search space that to these optimal solutions considering

J. van Hemert and C. Cotta (Eds.): EvoCOP 2008, LNCS 4972, pp. 218–229, 2008.

all objectives. Thus, a good multi-objective optimization technique must be able to search for a set of optimal solutions concurrently in a single run.

For this purpose, evolutionary algorithms seem to be well suited because they evolve a population of diverse solutions in parallel. Many evolutionary-based approaches for solving multiobjective optimization problems have been proposed in the last 25 years.

The recently introduced POEMS optimization framework proved to be efficient for solving hard optimization problems - the Traveling Salesman Problem (TSP)[4], a binary string optimization problem [4], a real-valued parameter optimization problem [5], and a network flow optimization problem [6]. This paper introduces an extension of the basic POEMS algorithm for solving multiobjective optimizations.

For an experimental evaluation of the presented approach a multiobjective 0/1 knapsack problem was used, which is a well-known NP hard combinatorial optimization problem, the particular formulation of the problem is given in Section 5. Results achieved by our approach were analyzed and compared to several evolutionary-based approaches presented in [9]. First results indicate that the proposed multiobjective POEMS algorithm performs very well on the test problem. It also scales well as it outperforms the alternative algorithms even on the largest instances of the problem.

This paper is structured as follows. First, a short overview of multiobjective optimization techniques is given with a focus on evolutionary-based approaches. Section 3 briefly describes the single-objective POEMS algorithm. In section 4, the multiobjective version mPOEMS is introduced. Section 5 describes the test problem, test datasets, and the configuration of the multiobjective POEMS algorithm used in the experiments. Results achieved with our approach are analyzed in section 6. Section 7 concludes and suggests directions for analyzing and improving the proposed approach.

2   Multiobjective Optimization Techniques

There are many evolutionary approaches for solving multiobjective optimization problems. The most distinguishing features are (i) the fitness assignment strategy for evaluating the potential solutions, (ii) the evolutionary model with a specific selection and replacement strategy, and (iii) how the diversity of the evolved population is preserved. Note the last issue is extremely important as the desired outcome of the algorithm is a set of optimal solutions that is as diverse as possible. One of the early approaches is Schaffer's Vector Evaluated Genetic Algorithm (VEGA) [7] that does not make the use of a single fitness value when selecting solutions to a mating pool. Instead, it carries out selection for each objective separately. Then, crossover and mutation are used in a standard way. Another approaches make the use of a weighted-sum aggregation of objectives in order to assign a scalar fitness value to solutions, see [1]. However, such methods are highly sensitive to the weight vector used in the scalarization process.

Perhaps the most widespread and successful are multiobjective evolutionary algorithms that use a concept of *dominance* for ranking of solutions. By definition [1], a solution x dominates the other solution y, if the solution x is no worse than y in all objectives and the solution x is strictly better than y in at least one objective. Naturally, if solution x dominates solution y then x is considered better than y in the context of multiobjective optimization. However, many times there are two different solutions such that neither of them can be said to be better than the other with respect to all objectives. When this happens between two solutions, they are called *non-dominated* solutions.

The concept of dominance can be used to divide a finite set S of solutions chosen from the search space into two non-overlapping sets, the *non-dominated set* S_1 and the *dominated set* S_2. The set S_1 contains all solutions that do not dominate each other. The set S_2, which is a complement of S_1, contains solutions that are dominated by at least one solution of S_1. If the set S is the whole feasible search space then the set S_1 is a set of optimal solutions called *Pareto-optimal* solutions and the curve formed by joining these solutions is called a Pareto-optimal front. Note that in the absence of any higher-level information, all Pareto-optimal solutions are equally important [1]. That is why the goal in a multiobjective optimization is to find a set of solutions that is (i) as close as possible to the Pareto-optimal front and (ii) as diverse as possible so that the solutions are uniformly distributed along the whole Pareto-optimal front.

Of the Pareto-based approaches, perhaps the most well-known are Pareto Archived Evolution Strategy (PAES) [3], Non-dominated Sorting GA (NSGA and NSGA-II) and Strength Pareto Evolutionary Algorithm (SPEA and SPEA2). We just briefly describe the NSGA-II [2] and SPEA2 [9] algorithms, because we chose them for as alternative approaches for the empirical comparison with our approach.

SPEA2 uses a regular population and an archive (a set of constant size of best solutions found so far). An archive truncation method guarantees that the boundary solutions are preserved. Fitness assignment scheme takes for each individual into account how many individuals it dominates and it is dominated by which is further refined by the incorporation of density information. In order to maintain a good spread of solutions, NSGA-II uses a density estimation metric called *crowding distance*. The crowding distance of a given solution is defined as the largest cuboid enclosing the solution without including any other solution in the population. Then, so called *crowding comparison operator* guides the selection process towards solutions of the best non-domination rank and with crowding distance. In each generation, a population Q_t of offspring solutions is generated from the current population of solutions P_t. The two populations are merged together resulting in the temporary population R_t of size $2 \cdot N$, where N is the population size. From this population a better half of solutions is chosen in the following way to constitute a new population P_{t+1}. First, the population R_t is sorted according to non-domination. Then the solutions are taken starting from the best non-domination level and are put to the new population P_{t+1}. If a set of solutions of currently processed non-domination level is bigger than the remaining empty space in the population P_{t+1}, then the best solutions in terms of the crowding distance are used only.

3   Singleobjective POEMS

Standard evolutionary algorithms (EAs) typically evolve a population of candidate solutions to a given problem. Each of the candidate solutions encodes a complete solution, e.g., a complete set of the problem parameters in parameter optimizations, a complete schedule in the case of scheduling problems, or a complete tour for the traveling salesman problem. This implies, especially for large instances of the solved problem, that the EA operates with very big and complex structures.

In POEMS [4], the evolutionary algorithm does not operate on a population of complete solutions to the problem to be solved. Instead, one candidate solution, called the *prototype*, is generated at the beginning and then it is iteratively improved with the best-performing modification of the current prototype provided by an EA, see Figure 1.

The prototype modifications are represented as a sequence of primitive actions/operations, defined specifically for the problem at hand. The evaluation of action sequences is based on how well/badly they modify the current prototype, which is passed as an input parameter to the EA. Moreover, sequences that do not change the prototype at all are penalized in order to avoid generating useless trivial solutions. After the EA finishes, it is checked whether the best evolved sequence improves the current prototype or not. If an improvement is achieved, then the sequence is applied to the current prototype and resulting in the new prototype. Otherwise the current prototype remains unchanged for the next iteration. The process of iterative prototype improvement stops when the termination condition is fulfilled. A common termination condition is the number of fitness evaluations performed in the run.

The following paragraphs briefly discuss POEMS implementation issues.

Representation of action sequences. The EA evolves linear chromosomes of length *MaxGenes*, where each gene represents an instance of a certain action cho-

---

```
1 generate(Prototype)
2 repeat
3    BestSequence ← run_EA(Prototype)
4    Candidate ← apply(BestSequence, Prototype)
5    if(Candidate is_better_than Prototype)
6        Prototype ← Candidate
7 until(POEMS termination condition)
8 return Prototype
```

Fig. 1. An outline of the single-objective POEMS algorithm

sen from a set of elementary actions defined for the given problem. Each action is represented by a record, with an attribute *action_type* followed by parameters of the action. Besides actions that truly modify the prototype, there is also a special type of action called *nop* (no operation). Any action with *action_type* = *nop* is interpreted as a void action with no effect on the prototype, regardless of the values of its parameters. A chromosome can contain one or more instances of the *nop* operation. This way a variable effective length of chromosomes is implemented.

Operators. The representation of action sequences allows to use a variety of possible recombination and mutation operators such as standard 1-point, 2-point or uniform crossover and a simple gene-modifying mutation. In [4] a generalized uniform crossover was used, that forms a valid offspring as an arbitrary combination of parental genes. Both parents have the same probability of contributing their genes to the child, and each gene can be used only once. The mutation operator changes either the *action_type* (activates or inactivates the action) or the parameters of the action.

Evolutionary model. In general, the EA is expected to be executed many times during the POEMS run. Thus, it must be designed and configured to converge fast in order to get the output in short time. As the EA is evolving sequences of actions to improve the solution prototype, not the complete solution, the maximal length of chromosomes *MaxGenes* can be short compared to the size of the problem. For example *MaxGenes* would be much smaller than the size of the chromosome in case of binary string optimization or much smaller than the number of cities in case of the TSP problem [4]. The relaxed requirement on the expected EA output and the small size of evolved chromosomes enables to setup the EA so that it converges within a few generations. Examples of typical configurations can be found in [4], [5] and [6].

It is important to note, that the evolved improving alterations of the prototype do not represent just local moves around the prototype. In fact, long phenotypical as well as genotypical distances between the prototype and its modification can be observed if the algorithm possesses a sufficient explorative ability. The space of possible modifications of the current prototype is determined by the set of elementary actions and the maximum allowed length of evolved action sequences *MaxGenes*, see [4]. If the actions are less explorative and the sequences are shorter, the system searches in a prototype neighborhood only and is more prone to get stuck in a local optimum early. And vice versa, if larger alterations of the prototype can be evolved using the primitive actions, the exploration capability the algorithm allows to converge to better and hopefully globally optimal solutions.

4   Multiobjective POEMS

The multiobjective "Prototype Optimization with Evolved iMprovement Steps" (mPOEMS) belongs to a class of multiobjective optimization algorithms that

```
1 generate(SolutionBase)
2 repeat
3    Prototype ← choose_prototype(SolutionBase)
4    ActionSequences ← MOEA(Prototype, SolutionBase)
5    NewSolutions ← apply_to(ActionSequences, Prototype)
6    SolutionBase ← merge(NewSolutions, SolutionBase)
7 until(termination condition is fulfilled)
8 return SolutionBase
```

Fig. 2. An outline of the mPOEMS algorithm

uses the concept of dominance. In this section we describe the way the set of non-dominated solutions progressing towards the Pareto-optimal set is evolved in mPOEMS. Note that in multiobjective optimization the goal is to find a set of optimal solutions (as close as possible to the Pareto-optimal set) that are as diverse in both the variable space and the objective space as possible. Thus, the main differences between mPOEMS and POEMS are that

- mPOEMS maintains a set of best solutions found so far, called a *solution base*, not just one prototype solution that is maintained in POEMS. In each iteration of mPOEMS one solution from the set of non-dominated solutions in the solution base is chosen as the prototype for which the action sequences will be evolved by the EA.
- mPOEMS uses a kind of a multiobjective EA (MOEA) based on the dominance concept, not just a simple EA. The output of the MOEA is a set of action sequences (not just one action sequence) generating new solutions that are merged with the current solution base resulting in a new version of the solution base.

Figure 2 shows the main steps of the mPOEMS algorithm. It starts with generating the initial solutions of the solution base. The size of the solution base is denoted as *SBSize* and stays constant through the whole mPOEMS run.

The first step of the main body of the iterative process is the selection of the prototype for the current iteration. The prototype is chosen among non-dominated solutions of the solution base in a way that guarantees that all parts of the non-dominated front of the evolved solution base are processed equally, see paragraph "prototype selection" below. The prototype is passed as an input parameter to the multiobjective EA, where the action sequences possibly altering the prototype towards the Pareto-optimal set are evolved. The other input parameter of MOEA is the current solution base that is used for evaluation purposes, see below. MOEA returns the final population of action sequences, which are then applied to the current prototype resulting in a set of new solutions.

Prototype selection. In each iteration a new prototype is chosen among non-dominated solutions of the solution base. The selection scheme is designed so that all partitions of the non-domination set have as equal sampling rate as possible. In the first iteration a set S of n *candidate prototype* solutions is chosen according to the following procedure:

1. $S = \{\}$, $i = 1$, Choose a solution s_i by random
2. $i + +$, choose a solution s_i so that its normalized Euclidean distance to the nearest solution in S is maximal,
   $S = S + s_i$,
3. Repeat Step 2 until $|S| = n$.

The Euclidean distance between two solutions i and j is calculated with the objective function values according to the following formula

$$d_{ij} = \sqrt{\sum_{k=1}^{m}(\frac{o_k^{(i)} - o_k^{(j)}}{u_k - l_k})^2},$$

where $o_k^{(i)}$ and $o_k^{(j)}$ are k-th objective values of solutions i and j, u_k and l_k are the upper and lower bounds for the k-th objective and m is the number of objectives. Each time a new prototype is to be chosen it is selected from the set S by random and removed from the set. Also if any solution in S becomes dominated by any solution in S it is removed from S. If the set is empty a new sample S of non-dominated solutions is selected according to the above described procedure.

The outline of the multiobjective EA used in mPOEMS is shown in Figure 3. First, it generates a starting population of action sequences of size *PopSize*. The action sequences are evaluated based on the quality of the solution that is produced by applying the given action sequences to the prototype. Then, the population of action sequences is evolved within a loop until some stopping condition is fulfilled. In the first step of the loop, a new population of action sequences is generated using standard operations of selection, crossover and mutation. The action sequences are evaluated and assigned fitness values. Finally, the new population and the old one are merged and *PopSize* solutions of the best non-dominated fronts of that joint population are used to constitute the resulting population.

Fitness assignment schema. Since we are dealing with multiobjective optimization problems, each solution is assigned multiple objective values. The evaluation procedure uses a concept of dominance between solutions in order to find a single fitness value specifying the solution quality in terms of its non-domination level. In order to have more levels of non-domination that better distinguishes solutions the evaluated solutions are merged with solutions from the solution base resulting in a temporary set of solutions S (the prototype solution is included in the set S as well). The process of calculating the level of non-dominance starts with finding the non-dominated solutions among the whole set S. These solutions belong to the first level of non-domination front and are assigned a non-domination level $ND_{level} = 1$. Then they are temporarily disregarded from the set S and the set of non-dominated solutions is sought

input: *Prototype, SolutionBase*
output: Population of evolved action sequences

```
1 generate(OldPop)
2 evaluate(OldPop)
3 repeat
4     NewPop ← evolutionary_cycle(OldPop)
5     evaluate(NewPop)
6     OldPop ← merge(OldPop, NewPop)
7 until(EA termination condition is fulfilled)
8 return OldPop
```

Fig. 3. An outline of the multiobjective evolutionary algorithm used in mPOEMS

among the remaining solutions. These are the solutions of the second level of non-domination and are assigned a non-domination level $ND_{level} = 2$. The process goes on until there is no solution left in S, i.e. every solution has assigned its ND_{level} value. In the second phase of the evaluation procedure, the evaluated solutions are assigned their fitness value. Solutions that belong to a better than or the same level of non-domination as the prototype solution are assigned a fitness value equal to their ND_{level} value. Solutions with the ND_{level} higher than the prototype solution are assigned a fitness value equal to $ND_{level} + 0.5 * P_D$, where P_D is 1 if the given solution is dominated by the prototype, and 0 otherwise. Note that the smaller fitness the better solution. So, the selection pressure is towards the solutions that

1. belong to a better non-domination front than the prototype, if possible, and
2. are not dominated by the prototype solution.

Evolutionary model. New solutions produced by action sequences evolved by the MOEA are merged with the current solution base resulting in a temporary population of size $PopSize + SBSize$. From this population a new solution base of size $SBSize$ is selected according to the schema used in NSGA-II. First, the joint set is sorted based on the non-domination. Then the non-dominated fronts are added to the new solution base one by one, starting from the best non-dominated front. The non-dominated front that can not fit the whole into the remaining space in the new solution base is ranked according to the *crowding distance* value introduced in [2], and only the best solutions are added to the new solution base. This strategy together with the prototype selection scheme ensures that (i) the boundary solutions of the non-dominated front of the solution base will not get lost and (ii) the most unique solutions will retain in the solution base and (iii) the non-dominated front will be sampled uniformly.

5   Test Data and Experimental Setup

Test problem. For experimental evaluation we chose a well-known NP-hard Multiobjective 0/1 Knapsack Problem. We used the same formulation of the problem as was used in the comparative study by Zitzler and Thiele [8] and we compared results achieved by mPOEMS with alternative approaches presented there. As stated in [8], the multiobjective 0/1 knapsack problem is a good test problem, because its description is simple, yet the problem itself is difficult to solve and the problem is important in practice.

The multiobjective 0/1 knapsack problem considered in [8] is defined in the following way: Given a set of m items and a set of n knapsacks, with $p_{i,j}$ being profit of item j according to knapsack i, $w_{i,j}$ being weight of item j according to knapsack i, and c_i being capacity of knapsack i, find a vector $\mathbf{x} = (x_1, x_2, \ldots, x_m) \in \{0, 1\}^m$, such that

$$\forall i \in \{1, 2, \ldots, n\} : \sum_{j=1}^{m} w_{i,j} \cdot x_j \leq c_i$$

and for which $f(\mathbf{x}) = (f_1(\mathbf{x}), f_2(\mathbf{x}), \ldots, f_n(\mathbf{x}))$ is maximum, where

$$f_i(\mathbf{x}) = \sum_{j=1}^{m} p_{i,j} \cdot x_j$$

and $x_j = 1$ iff item j is selected.

Since the solution of the problem is encoded as a binary string of length m, many codings do not represent a feasible solution (i.e. the capacity constraint of one or more knapsacks is violated). Thus, a greedy *heuristic repair algorithm*, which was also used in [8], was applied to every illegal solution. It removes items from the solution until all capacity constraints are fulfilled. Items with the least *profit/weight ratio* are removed first.

Compared algorithms. Out of the algorithms tested in [9] we chose the following two – SPEA2 and NSGA-II. We used the largest datasets with 750 items that were used in [9]. Both the datasets and the data of results presented in [9] are available on the web [10]. Thirty independent runs were carried out with mPOEMS on each test problem resulting in a similar set of values as in [9].

Performance measures and indicators. For all datasets the algorithms were compared on the basis of the following performance measures

- *Coverage of two sets* $C(X, Y)$, proposed in [8]. The measure is defined in the following way: Given the two sets of non-dominated solutions found by the compared algorithms, the measure $C(X, Y)$ returns a ratio of a number of solutions of Y that are dominated by or equal to any solution of X to the whole set Y. Thus, it returns values from the interval [0, 1]. The value $C(X, Y) = 1$ means that all solutions in Y are covered by solutions of the set X. And vice versa, the value $C(X, Y) = 0$ means that none of the solutions in Y are covered by the set X. For the problems with 2 knapsacks we also present plots showing the tradeoff fronts constituted of non-dominated solutions of a union set of ten sets of non-dominated solutions obtained by each algorithm on the given dataset. The plots also show the Pareto-optimal front for the respective dataset if available.

- *Size of the space covered* $S(X)$, proposed in [8] and modified in [9]. A reference volume between the origin and an utopian objective vector (defined by the profit sums of all items in each objective) is taken into account. This measure is defined as a fraction of that volume that is not dominated by the final non-dominated solutions. So, the smaller the value of this measure the better the spread of solutions is, and vice versa.

Configuration of mPOEMS. In [9], the total number of solutions sampled (and evaluated) through the whole EA was 480.000 for 2 knapsacks, 576.000 for 3 knapsacks and 672 for 4 knapsacks. We used the same number of solution evaluations $N_{Evaluations}$ for each dataset. Parameters of mPOEMS were constant for all datasets as follows

- $PopulationSize = 70$. Size of the population of evolved action sequences.
- $SolutionBaseSize = 100$. The size of the solution base.
- $MaxGenes = 50$. The length of the evolved action sequences. Note, that it is much smaller than the solution size m (which is 750 in this study).
- $N_{Iterations}$ was 274, 330, and 384 for 2, 3, 4 knapsacks. The number of iterations in mPOEMS algorithm, see *repeat-until* cycle in Fig. 2.
- $N_{Generations=25}$. The number of generations carried out in the MOEA.
- $P_{Cross} = 0.8$, $P_{Mutate} = 0.2$. Probability of crossover and mutation.
- $Tournament = 3$. Parameter of the tournament selection used in MOEA.
- $n = 20$. A size of the set S of candidates for the prototype.
- Both the solution base as well as the starting population of action sequences in each iteration were initialized by random.

6   Results

Table 1 provides a comparison of mPOEMS and the other approaches on the basis of the coverage of two sets performance measure $C(X, Y)$. Each cell of the table is interpreted so that it indicates a proportion of non-dominated solutions obtained by the approach given in the corresponding column covered by the set of non-dominated solutions obtained by the approach given in the corresponding row. For example, for $n = 4$ we see that non-dominated solutions found by mPOEMS dominate 96.1% of non-dominated solutions found by NSGA-II and 97.1% of non-dominated solutions found by SPEA2, while only 0.1% of mPOEMS non-dominated solutions are dominated by solutions found by SPEA2 and none of them is dominated by solutions found by NSGA-II.

Table 2 shows the average values of the *size of the space covered* measure achieved by the compared algorithms. It shows that mPOEMS is no worse on any dataset and is significantly better (proved by t-test) on datasets with 2 and 4 knapsacks than the other two algorithms. This indicates, that mPOEMS finds a better spread of non-dominated solutions than NSGA-II and SPEA2.

Table 1. Comparison of mPOEMS, NSGA-II and SPEA2 using the *coverage of two sets* measure on datasets with 2, 3 and 4 knapsacks and 750 items. n is the number of knapsacks. Numbers in each cell show the fraction of non-dominated solutions obtained by the algorithm in the given column that are dominated by non-dominated solutions obtained by the algorithm in the given row.

		NSGA-II	SPEA2	mPOEMS
	NSGA-II	-	0.604	0.178
n=2	SPEA2	0.283	-	0.047
	mPOEMS	0.708	0.966	-
	NSGA-II	-	0.02	0.0
n=3	SPEA2	0.887	-	0.323
	mPOEMS	0.955	0.431	-
	NSGA-II	-	0.006	0.0
n=4	SPEA2	0.844	-	0.001
	mPOEMS	0.961	0.971	-

Table 2. Comparison of mPOEMS, NSGA-II and SPEA2 using the *size of the space covered* measure on datasets with 2, 3 and 4 knapsacks and 750 items

	NSGA-II	SPEA2	mPOEMS
$n = 2$	0.497	0.492	**0.490**
$n = 3$	0.689	0.689	0.688
$n = 4$	0.8195	0.822	**0.8180**

7   Conclusions and Future Work

This paper proposed a new multiobjective optimization algorithm mPOEMS based on the single-objective POEMS, recently introduced in [4]. mPOEMS extends the single-objective version of POEMS so that it (i) maintains a set of best solutions found so far called *solution base*, (ii) uses multiobjective EA (MOEA) instead of a simple EA to evolve a population alterations of the current prototype, and (iii) employs a strategy for proper selecting the prototype in each iteration.

mPOEMS was evaluated on a multiobjective 0/1 knapsack problem with 2 and 4 knapsacks (objectives) and 750 items. Results obtained by mPOEMS were compared to the results achieved by two state-of-the-art multiobjective evolutionary algorithms - NSGA-II and SPEA2, presented in [9]. The approaches were compared using a performance measure checking the mutual dominance of their outcomes and the size of the volume covered by the found non-dominated solutions. Performance of mPOEMS is at least as good or even better as the compared algorithms on all datasets. This is a very promising observation because the NSGA-II and SPEA2 were the best performing algorithms among the algorithms analyzed in [9].

As this is the first study on the mPOEMS algorithm, there are many open issues that should be investigated in the future:

- Computational complexity of the algorithm and its sensitivity to parameters' setting should be investigated.
- The performance of mPOEMS should be evaluated on other test problems with different characteristics, such as the problems with discontinuous Pareto-optimal front.
- Since the mPOEMS is in fact a local search approach which uses an EA to choose the next move to perform so it should be compared with other local search approaches as well.

Acknowledgement

Research described in the paper has been supported by the research program No. MSM 6840770012 "Transdisciplinary Research in Biomedical Engineering II" of the CTU in Prague.

References

1. Deb, K.: Multi-Objective Optimization using Evolutionary Algorithms. John Wiley & Sons, Ltd., New York (2002)
2. Deb, K., Pratap, A., Agarwal, S., Meyarivan, T.: A fast and elitist multiobjective genetic algorithm: NSGA-II. IEEE Trans. on Evolutionary Computation 6(2), 182–197 (2002)
3. Knowles, J.D., Corne, D.W.: Approximating the nondominated front using the Pareto archived evolution strategy. Evolutionary Computation 8(2), 149–172 (2000)
4. Kubalik, J., Faigl, J.: Iterative Prototype Optimisation with Evolved Improvement Steps. In: Collet, P., Tomassini, M., Ebner, M., Gustafson, S., Ekárt, A. (eds.) EuroGP 2006. LNCS, vol. 3905, pp. 154–165. Springer, Heidelberg (2006)
5. Kubalik, J.: Real-Parameter Optimization by Iterative Prototype Optimization with Evolved Improvement Steps. In: 2006 IEEE Congress on Evolutionary Computation, pp. 6823–6829. IEEE Computer Society, Los Alamitos (2006) [CD-ROM]
6. Kubalik, J., Mordinyi, R.: Optimizing Events Traffic in Event-based Systems by means of Evolutionary Algorithms. In: Event-Based IT Systems (EBITS 2007) organized in conjunction with the Second International Conference on Availability, Reliability and Security (ARES 2007), Vienna, April 10-13 (2007)
7. Schaffer, J.D.: Multiple Objective Optimization with Vector Evaluated Genetic Algorithms. Genetic Algorithms and Their Applications. In: Proceedings of the First International Conference on Genetic Algorithms, pp. 93–100 (1985)
8. Zitzler, E., Thiele, L.: Multiobjective Evolutionary Algorithms: A Comparative Case Study and the Strength Pareto Approach. IEEE Transactions on Evolutionary Computation 3(4), 257–271 (1999)
9. Zitzler, E., Laumanns, M., Thiele, L.: SPEA2: Improving the Strength Pareto Evolutionary Algorithm For Multiobjective Optimization. In: Evolutionary Methods for Design, Optimisation, and Control, Barcelona, Spain, pp. 19–26 (2002)
10. Zitzler, E., Laumanns, M.: Test Problem Suite: Test Problems and Test Data for Multiobjective Optimizers,
    http://www.tik.ee.ethz.ch/~zitzler/testdata.html

Optimising Multiple Kernels for SVM by Genetic Programming

Laura Dioşan[1,2], Alexandrina Rogozan[1], and Jean-Pierre Pecuchet[1]

[1] Babeş-Bolyai University, Cluj-Napoca, Romania
[2] LITIS, EA 4108, INSA, Rouen, France
lauras@cs.ubbcluj.ro, {arogozan,pecuchet}@insa-rouen.fr

Abstract. Kernel-based methods have shown significant performances in solving supervised classification problems. However, there is no rigorous methodology capable to learn or to evolve the kernel function together with its parameters. In fact, most of the classic kernel-based classifiers use only a single kernel, whereas the real-world applications have emphasized the need to consider a combination of kernels - also known as a multiple kernel (MK) - in order to boost the classification accuracy by adapting better to the characteristics of the data. Our aim is to propose an approach capable to automatically design a complex multiple kernel (CMK) and to optimise its parameters by evolutionary means. In order to achieve this purpose we propose a hybrid model that combines a Genetic Programming (GP) algorithm and a kernel-based Support Vector Machine (SVM) classifier. Each GP chromosome is a tree that encodes the mathematical expression of a MK function. Numerical experiments show that the SVM involving our evolved complex multiple kernel (eCMK) perform better than the classical simple kernels. Moreover, on the considered data sets, our eCMK outperform both a state of the art convex linear MK (cLMK) and an evolutionary linear MK (eLMK). These results emphasize the fact that the SVM algorithm requires a combination of kernels more complex than a linear one.

1 Introduction

Various classification techniques have been used in order to detect correctly the labels associated to some items. Kernel-based techniques (such as Support Vector Machine (SVM) [1]) are an example of such intensively explored classifiers. These methods represent the data by means of a kernel function, which defines similarities between pairs of data [2]. One reason for the success of kernel-based methods is that the kernel function takes relationships that are implicit in the data and makes them explicit, the result being that the detection of patterns takes place more easily.

The selection of an appropriate kernel K is the most important design decision in SVM since it implicitly defines the feature space $\mathcal{F}$ and the map ϕ. An SVM will work correctly even if we do not know the exact form of the features that are used in $\mathcal{F}$. The performance of an SVM algorithm depends also on several parameters. One of them, denoted C, controls the trade-off between maximizing the margin and classifying without error. The other parameters regard the kernel function. For simplicity, Chapelle in [3] has proposed to denote all these parameters as hyper parameters. All that hyper

J. van Hemert and C. Cotta (Eds.): EvoCOP 2008, LNCS 4972, pp. 230–241, 2008.
© Springer-Verlag Berlin Heidelberg 2008

parameters have to be tuned. This is a difficult problem, since the estimate of the error on a validation set is not an explicit function of these parameters.

The selection of an optimal kernel function and the values of the hyper parameters is known in literature as *model selection* [3]. This task is usually performed by training the classifier with different functions picked up from a range of kernels and several parameter values from a discrete set, which is fixed *a priori*. The optimal model corresponds to the configuration that generates the best classification performance by using a cross-validation technique [3]. Nevertheless, a simple kernel may not be always suitable especially for very complex classification problems like those related to multi-modal heterogeneous data. The real-world applications have emphasized the need to consider a combination of kernels, denoted by multiple kernel (MK) [4,5]. Recent research works have already shown that the MKs improve the performance of the SVM classifiers due to their flexibility, allowing for a better learning of complex and heterogeneous data. In addition, the optimisation of the hyper parameters plays a very important part.

The automatic MK designing is more than a simple kernel selection: in the MK framework the best expression of the kernel function is learnt as a more or less complex combination of simple kernels. In the same time, the optimal values of the hyper parameters are found. One has to answer several important questions concerning the design of a MK: *It is possible to learn the MK function by using some training annotated data?* And, in the case of a positive answer, *What kernels have to be used within an MK for a given classification problem? How to find the optimal parameters of the simple kernels involved in this combination?* and *What allows for better classification performances: a linear MK or a complex MK ?* If the answer for the first question can be found in the literature [4,5,6,7,8,9], the answer of the last ones will be given in this paper.

Therefore we choose to use the evolutionary framework in order to discover the optimal expression and its parameters of an MK function for several given problems. We combine the Genetic Programming (GP) [10] and the SVM algorithms [1] within a two-level hybrid model. The aim of the model we propose is to find the best MK function and to optimise its parameters, but also to adapt the regularisation kernel parameter C. These three objectives are achieved simultaneously because each GP chromosome encodes the expression of a complex multiple kernel (CMK) and its parameters. The GP-kernel is involved into a standard SVM algorithm to be trained in order to solve a particular classification problem. After an iterative process which runs more generations, an optimal evolved complex multiple kernel ($eCMK^*$) is provided. The proposed combination of kernels could be learnt from thousands of examples while combining hundreds of kernels within reasonable time. The $eCMK$ we introduce is compared not only to several well-known simple kernels, but also to a convex linear MK (cLMK) [4] and to an evolved linear MK (eLMK) [7]. We will show that our model is able to find more efficient complex MKs on the considered data sets.

The paper is organized as follows: The related work is presented in Section 2. Section 3 outlines the theory behind SVM classifiers giving a particular emphasis to the kernel functions. Section 4 describes our technique for evolving CMKs. This is followed by Section 5 where the results of the experiments are presented and discussed. Finally, Section 6 concludes our paper.

2   Related Work

An MK is in fact a combination of several simple kernels. This combination could be either linear or complex. Regarding the linear combination, each simple kernel is involved with a weight that represents its relative influence/importance in the *LMK*. The optimal weights of the simple kernels included in an *LMK* have been found by convex [4,5,6,3,11] or evolutionary methods [7,12].

About the shape of the complex combinations of kernels, to the best of our known, the genetic algorithms (GAs) [8,9] have only been used in order to learn the expression of an MK function.

As regards the optimisation of the hyper-parameters, extensive exploration such as performing line search for one hyper-parameter or grid search for two hyper-parameters is frequently applied. However, this search processes usually require training the model several times with different hyper-parameter values and hence is computationally prohibitive especially when the number of candidate values is large. Because of the computational complexity, grid search is only suitable for the adjustment of very few parameters. More elaborated techniques for optimising hyper-parameters are the gradient-based approaches [3]. Different optimisation criteria have been used: minimize the leave-one-out error [1], minimax (maximize the radius margin bound and minimize the validation or the leave-one-out errors) [3] or minimize the CV error [1]. Several promising recent approaches [13,14,15] are based on regularisation path algorithms that can trace the entire solution path as a function of the hyper-parameter without having to train the model multiple times. Evolutionary algorithms have also been used in order to optimise the hyper parameters of an SVM classifier [16,17].

There are very few approaches that deal with both the problem of hyper parameter optimisation and of MK function learning. Ong et al. [11] have shown that the MK function is a linear combination of a finite number of pre-specified hyper-kernel evaluations. The semi definite programming (SDP) approach [4,5,13,18] is applied for learning MK seen as a linear combination of positive semi definite matrices. Similar to this idea, Bousquet and Herrmann [19] further restricts the class of kernels to the convex hull of the kernel matrices normalized by their trace. The genetic algorithms have also been used in order to optimise both the MK shape and its hyper parameters [8,9]. In fact, other than a combination of kernels through different operations, these GA-based approaches could optimised the hyper parameters of the SVM algorithm.

3   Support Vector Machines

Initially, the *SVM* algorithm has been proposed for solving binary classification problems [1,20]. Later, these algorithms have been generalized for multi-classes problems. Consequently, we will explain the theory behind *SVM* only on binary-labelled data. Suppose the training data has the following form: $D = (x_i, y_i)_{i=\overline{1,m}}$, where $x_i \in \Re^d$ represents an input vector and each y_i, $y_i \in \{-1, 1\}$, the output label associated to the item x_i. The *SVM* algorithm maps the input vectors to a higher dimensional space where a maximal separating hyper-plane is constructed [21]. Learning the *SVM* means to minimize the norm of the weight vector w under the constraint that the training items of different classes belong to opposite sides of the separating hyper-plane. Since

$y_i \in \{-1, +1\}$ we can formulate this constraint as: $y_i(w^T x_i + b) \geq 1$, $i = 1, \ldots, m$. The items that satisfy this equation in case of equality are called support vectors since they define the resulting maximum-margin hyper-planes. To account for misclassification, the soft margin formulation of *SVM* has introduced some slack variables $\xi_i \in \Re -$ see Eq. (1). Moreover, the separation surface has to be nonlinear in many classification problems. The *SVM* algorithm can be extended to handle nonlinear separation surfaces by using a feature function $\phi(x)$. The *SVM* extension to nonlinear data sets is based on mapping the input variables into a feature space $\mathcal{F}$ of a higher dimension and then performing a linear classification in that higher dimensional space. The important property of this new space is that the data set mapped by ϕ becomes linearly separable if an appropriate feature function is used, even when that data set is not linearly separable in the original space. Hence, to construct a maximal margin classifier one has to solve the convex quadratic programming problem encoded by Eq. (1), which is the primal formulation of it:

$$\text{minimise}_{w,b,\xi} \tfrac{1}{2} w^T w + C \sum_{i=1}^{m} \xi_i$$
$$\text{subject to: } y_i(w^T \phi(x_i) + b) \geq 1 - \xi_i, \qquad (1)$$
$$\xi_i \geq 0, \forall i \in \{1, 2, \ldots, m\}.$$

The coefficient C is a tuning parameter that controls the trade off between maximizing the margin and classifying without error. The primal decision variables w and b define the separating hyper-planes. Instead of solving Eq. (1) directly, it is a common practice to solve its dual formulation described by Eq. (2), where a_i denotes the Lagrange variable for the ith constraint of Eq. (1):

$$\text{maximise}_{a \in \Re^m} \sum_{i=1}^{m} a_i - \tfrac{1}{2} \sum_{i,j=1}^{m} a_i a_j y_i y_j \phi(x_i)\phi(x_j)$$
$$\text{subject to } \sum_{i=1}^{m} a_i y_i = 0, \qquad (2)$$
$$0 \leq a_i \leq C, \forall i \in \{1, 2, \ldots, m\}.$$

The optimal separating hyper-plane $f(x) = w^T \phi(x) + b$, where w and b are determined by Eqs. (1) or (2) could be used in order to classify the un-labelled input data:

$$y_k = sign \left(\sum_{x_i \in S} a_i y_i \phi(x_i)\phi(x_k) + b \right) \qquad (3)$$

where S represents the set of support vector items x_i. Because not all input data-points are linear separable, it is suitable to use a kernel function. Cf. [2] a kernel is a function K, such that $K(x, z) = \langle \Phi(x), \Phi(z) \rangle$ for all $x, z \in \Re^d$. Note that all we required are the results of such an inner product. Therefore we do not even need to have an explicit representation of the mapping ϕ, nor to know the nature of the feature space. The only requirement is to be able to evaluate the kernel function on all the pairs of data items, which is much easier than computing the coordinates of those items in the feature space. Evaluating the kernel yields a symmetric, positive semi definite matrix known as the kernel or Gram matrix [22]. In order to obtain an *SVM* classifier with kernels, one has to solve the following optimization problem:

$$\text{maximise}_{a \in \Re^m} \sum_{i=1}^{m} a_i - \tfrac{1}{2} \sum_{i,j=1}^{m} a_i a_j y_i y_j K(x_i, x_j)$$
$$\text{subject to } \sum_{i=1}^{m} a_i y_i = 0, \qquad (4)$$
$$0 \leq a_i \leq C, \forall i \in \{1, 2, \ldots, m\}.$$

In this case, Eq. (3) becomes: $y_k = sign \left(\sum_{x_i \in S} a_i y_i K(x_i, x_k) + b \right)$.

Table 1. The expression of several classical kernels

Kernel name	Kernel expression
Polynomial	$K_{Pol}\ (x,z) = (x^T \cdot z + coef)^d$
Radial basis function	$K_{RBF}(x,z) = exp(-\sigma|x - z|^2)$
Sigmoid	$K_{Sig}\ (x,z) = \tanh(\sigma x^T \cdot z + r)$

4   The Model for Evolving Complex MKs

The model we propose can be used in order to discover the optimal expression of an eCMK. This model involves a hybrid approach, which combines a *GP* algorithm and an *SVM* classifier. Each *GP* chromosome is a tree that encodes the mathematical expression of an eCMK to be used by the *SVM* algorithm. The quality of a *GP* individual is given by the classification accuracy computed through running the *SVM* on the validation set in order to solve a particular classification problem. The hybrid approach is structured on two levels: a macro level and a micro level (see Figure 1(a)).

The macro level algorithm is a standard *GP* [10], which is used in order to evolve the mathematical expression of a CMK. We use steady-state evolutionary model [23] as an underlying mechanism for our *GP* implementation. The evolutionary algorithm starts by an initialisation step for creating a random population of individuals. The following steps are repeated until a given number of generations/iterations are reached: two parents are selected by using a binary selection procedure; the parents are recombined in order to obtain an offspring O; the offspring is than considered for mutation; the new individual O^* (obtained after mutation) replaces the worst individual W in the current population if O^* is better than W.

The micro level algorithm is an *SVM* classifier. The original implementation of the *SVM* algorithm proposed in *libsvm* [24] allows the use of several well-known kernels (linear, polynomial, RBF and sigmoid, respectively, kernels) – see Table 1. In numerical experiments, we also use a modified version of this algorithm, which is based on our evolutionary CMK. The quality of each *GP* individual is computed by running the *SVM* algorithm embedding the eCMK encoded in the current chromosome. The accuracy rate computed by the classifier (on the validation set) represents the fitness of the *GP* tree.

4.1   The GP Representation of an MK

In our model, the *GP* chromosome is a tree encoding the mathematical expression of an *eCMK* and its parameters. The tree-based representation of an *MK* allows for a larger search space of kernel combinations than an array-based representation. However, we constrained the *GP* individual representation to satisfy the kernel algebra [2] (regarding the positiveness and the symmetry of the Gram matrix required by valid kernels).

The approach we propose is based on a particular type of *GP* tree: its leaves contain either a simple kernel or a constant, where a kernel is a function with two arguments (x and z) that represent the input vectors. Note that a *GP* tree must contain at least one kernel in its leaves, otherwise the obtained expression will perform any dot

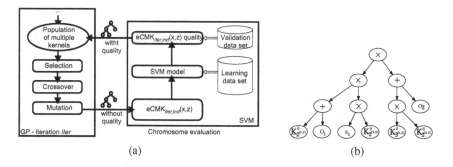

(a)   (b)

Fig. 1. a) Sketch of our hybrid approach; b) A GP chromosome that encodes the CMK expression: $(K_2^\theta(x,z) + o_1) \times s_1 \times K_2^\theta(x,z) \times (K_3^\theta(x,z) \times K_2^\theta(x,z) + o_2)$

product between the input vectors x and z. The leaves of the tree contain elements from the terminal set (*TS*), while the internal nodes contain elements from the function set (*FS*).

For a better adaptation to the classification problem, our terminal set contains not only the classic simple kernels, but also some ephemeral random constants [10]: $TS = KTS \cup \{o_i, s_j\}$, where *KTS* represents the terminal set that corresponds to those simple kernels (such as linear kernel, polynomial kernel, radial basis function (*RBF*) kernel, sigmoid kernel – see Table 1), o_i are offset (shifting) coefficients that control the threshold of the mapping from the original space into the feature space $\mathcal{F}$ and s_i are scaling coefficients that control the relative influence of the simple kernels in the *eCMK* expression. Both types of coefficients must be represented by positive real values.

Each simple kernel has associated a set of parameters θ that can affect the performance of the SVM algorithm. Therefore, we will consider more kernels for the TS, but with different parameters. The *RBF* kernel has only a parameter – the bandwidth σ (in this case $\theta = \{\sigma\}$). The sigmoid kernel has two parameters: the bandwidth σ and the shifting coefficients r that controls the threshold of the mapping ($\theta = \{\sigma, r\}$). The polynomial kernel has only a parameter: the degree d ($\theta = \{d\}$) – see Table 1.

In order to deal with shape and parameter optimisation, our solution is to consider in the *TS* different simple kernels with different parameters θ . We will denote these kernels as parametrised kernels K^θ. Thus, the *GP* algorithm will be able to discover the best *eCMK* expression by combining the best parametrised simple kernels.

The function set contains three operations ($FS = \{+, \times, \exp\}$) that preserve the key properties of a kernel function. The theory of kernel algebra also specifies the *power* function, but this operation (with a natural exponent) can be easily obtained as a repeated multiplication. An example of a *GP* chromosome is depicted in Figure 1(b). Although we have used $FS = \{+, \times, \exp\}$ and $TS = \{K_1^\theta, K_2^\theta, K_3^\theta, o, s\}$ for our *GP* chromosome representation, only two functions (+ and ×), two kernels (K_2^θ and K_3^θ) and three constants are actually involved in the current chromosome that represents the expression of the $eCMK_{iter,ind}(x,z)$.

4.2   Genetic Operations

Initialization. We have used the *grow method*, which is a recursive procedure, in order to initialize a *GP* individual. We have chosen this initialisation method, which is well known in the literature, for its robustness. The root of each *GP* tree must be a function from the *FS*. If a node contains a function, then its children are initialized either with another function or with a terminal (a kernel or a constant). The initialization process is stopped at the maximal depth of the *GP* tree. The leaves of the *GP* tree are initialised with terminals taken from the *TS*. At least one leaf of the *GP* tree has to contain a kernel in order to obtain a valid expression of the *MK*. The maximal kernel depth has to be large enough in order to ensure an important search space for the optimal expression of a CMK.

Crossover. We use the crossover operator in order to assure an important diversity of the *eCMK*s. The crossover is performed in a tree-structure preserving way in order to guarantee the syntactical validity of the offspring. Our model uses a one-cutting point crossover with the particularity that the offspring has to contain at least one kernel in its leaves.

Mutation. The purpose of the mutation operator is to produce a local small perturbation of the current chromosome. A cutting point is randomly chosen: the sub-tree belonging to that point is deleted and a new sub-tree is grown there by applying the same random growth process that was used in order to generate the initial population. Note that the maximal depth allowed for the *GP* trees limits the growth process of the sub-tree. The mutation operator may generate new constants at any point in a run, like in Koza's implementation [10]. In our model, these ephemeral random constants are represented by the scaling and offset coefficients. Note that in our model the initialization, recombination and mutation operators always generate valid *eCMK*s.

4.3   Fitness Assignment

The evaluation of the chromosome quality is based on a validation process. We must therefore provide some information about the data set partitioning before describing the fitness assignment process. Each data sample was randomly divided into two sets: a training set (80%) - for model building - and a testing set (20%) - for performance assignment. The training set was then randomly partitioned into learning (2/3) and validation (1/3) parts.

Each *eCMK* encoded by a *GP* tree is taken into consideration in order to learn the corresponding *SVM* model on the learning subset and for its classification performance assignment on the validation subset. Therefore, the quality of an *eCMK*, which is the current *GP* chromosome, can be measured by the classification accuracy rate (the number of correctly classified items over the total number of items) computed on the validation data set. Note that we are dealing with a maximization problem: the quality of the MK is proven by the accuracy of the *SVM* algorithm that uses the respective kernel. At the end of all *GP* generations (iterations), the optimal CMK, provided by the best *GP* chromosome, denoted as *eCMK**, is involved again into the *SVM* classification algorithm on the test data set.

4.4   Comparison to the Previous Models

The model we propose combines a *GP* technique with an *SVM* classifier. The idea of a hybrid approach *GP* + *SVM* is not new, but the representation that we proposed is novel.

Our approach is more general than the previous ones based on *GP* used in order to evolve the expression of a kernel function [25,26,27]. This time our purpose is to discover a new complex *MK* function and not only a simple one. The *GP* trees that encode the *MK*s are more elaborated since they could contain, in their leaves, other standard kernels whose good performance has been already proven.

Several important remarks can be also made regarding our previous *eLMK* model presented in [7] and the *eCMK* model we propose in this paper. The *eLMK* is a combination of kernels and it can be only linear, while the *eCMK* is a more complex one. The objective function is also different in these models: the *GP* algorithm used in the *eCMK* model optimises the shape and the parameters of an MK function, while the *GA* has optimised only the weights of a linear combination of kernels [7].

Moreover, the expression of the CMK obtained by a *GA*-based model in [8,9] could be actually less complex than that we are evolving now. Our approach is able to find a more adapted MK expression due to: a larger set of operations involved in the expression of the *eCMK* ($+, \times, \exp$). The power function with an integer exponent involved in the *GA*-based *eCMK* model [8,9] appears implicitly in our *GP*-based approach due to the tree-based representation of the MK; this representation is able to generate it by itself as a repeated multiplication. A more complex form of the MK expression is due to both the *GP* tree-based representation and the coefficients in the expression of a CMK. A better adaptability of the CMK to the data is the third difference - the previous *GA*-based *eCMK* model forces at least the polynomial and the ANOVA kernels to appear in the expression of the *MK*. Our *GP*-based approach for *eCMK*s allows, according to the data and their characteristics, to chose those simple kernels to be involved in the combination (either all the kernels, or just a few of them).

5   Experiments and Discussion

We have evaluated our *eCMK* learning approach on several data sets taken from Machine Learning Repository UCI and Statlog data sets. These data sets were chosen in order to allow comparisons with a state-of-the-art *LMK*s proposed in [4] and with an evolutionary *LMK* proposed in [7]. These data sets are also widely used in the classification community. All the data sets relate to binary-classification problems, but they have different sizes (the number of items and the number of characteristics) and they belong to different fields: medical, economical, and geographic: classification of radar returns from the ionosphere (P_1), breast cancer classification (P_2), heart disease diagnosis (P_3), classifications of personal income (P_4) and (P_5).

A population of 50 individuals is evolved during 50 generations, which is a reasonable limit in order to assure the diversity of our evolutionary complex multiple kernels (*eCMK*s). We have also limited the maximal depth of a *GP* tree to 10 levels, which allows us to consider complex combinations of maximum 2^{10} kernels in reasonable time. We have worked with the binary tournament mechanism for the chromosome selection. The crossover and mutation operations are performed by 0.8 and 0.3, respectively, probabilities.

Research works have shown that the optimisation of the kernel parameters together with the optimisation of the kernel expression (shape) are the most important steps to be considered when building an *SVM* classifier. Therefore, we use a method proposed in [3] in order to initialise the regularisation coefficient. A good value for the C parameter could be the inverse of the empirical variance s^2 of the data in the feature space [3]. We use this value also in the numerical experiments performed in order to evolve the expression of an eCMK function.

5.1 Evolving the Complex Multiple Kernel Function

In this first experiment, our aim is to evolve a complex multiple kernel *eCMK*. Two different terminal sets are used in order to evolve different combinations: a terminal set that contains only several simple kernels *KTS* and a mixed terminal set that contains not only simple kernels, but also some constants $MTS = KTS \cup \{c_1, c_2, \ldots c_n\}$. Note that in our experiments, these constants could be either scaling or shifting coefficients. Therefore, the *TS*s used in our numerical experiments are:

1. a TS composed by several simple kernels with different parameters $KTS = \{K^\theta_{Pol},$ $K^\theta_{RBF}, K^\theta_{Sig}\}$ where the parameters θ of each simple kernel have been considered in some discrete ranges: for the degree d of the Polynomial kernel 15 values (from 1 to 15) are considered, for the bandwidth σ of the RBF kernel the following values: $\sigma_{qt} = q \cdot 10^t$, $q = \{1, 2, \ldots, 9\}$, $t = \{-5, -4, \ldots, -1\}$ are considered and for the sigmoid kernel all the combination between σ and r, where the σ_{qt} and the $r = 10^u$, $u \in \{-1, 0, 1\}$ are taken into account.
2. a TS with different standard kernels and constants $MTS = KTS \cup \{c_1, c_2, \ldots, c_n\}$.

We have to note several things about the constant values c_i. Mercer conditions [22] impose these constants to be positive. The $[0, 1]$ range was suggested in [4,7] for all the constants as the authors have represented the relative weights of the simple kernels (*SK*s) involved into a linear MK (*LMK*). In our case, we have to deal with some scaling and shifting coefficients that can appear or not in the expression of the *eCMK*s. Therefore, several positive intervals have been considered for these coefficients in our experiments $[0, 1]$, $[0, 10]$ and $[0, 100]$, the best seems to be the $[0, 1]$ range. During different runs, various expressions of the eCMK function are obtained, all of them with about the same complexity.

The performances of the eCMKs based on various *TS*s are presented in Table 2: the first two rows contain the accuracy rates (for each problem) computed by the *SVM* algorithm involving our *eCMK* on the test set (unseen data). This *eCMK* is the best *MK* (*eCMK**) obtained at the end of the evolutionary process on the validation set (the best *GP* chromosome from the last generation). In Table 2 we also present the performances of three classic kernels for all the test problems (the last three rows). We emphasize the fact that the value of the penalty error C is adapted to the data and other parameters involved in each simple kernel were optimised in order to achieve the best classification performances. This allows us to verify if our *eCMK*s outperform these

Table 2. The accuracy rate of various kernels. The first two rows present the accuracy rates computed by the SVM algorithm embedding the eMK. The last three rows contain the performances of the simple kernels for each test problem.

		P_1	P_2	P_3	P_4	P_5
Multiple	KTS	86.11±1.13	97.81±0.13	86.98±0.51	84.27±0.14	86.93±0.19
kernels	MTS	91.67±0.90	98.03±0.13	86.98±0.51	84.38±0.14	88.99±0.18
Simple	Kpol	77.77±1.36	97.58±0.14	85.79±0.53	84.26±0.14	86.24±0.20
kernels	Krbf	80.55±1.29	97.81±0.13	85.21±0.54	83.65±0.14	83.49±0.21
	Ksig	66.67±1.54	97.81±0.13	77.91±0.63	82.73±0.14	84.52±0.21

optimised simple kernels and then to measure the improvements. In addition, Table 2 displays the corresponding confidence intervals (on the test set of each problem).

The values from Table 2 indicate that the $eCMK$s always perform better than the optimised simple kernels. This is a very important result, if we take into account the fact that GP algorithm has the possibility to choose among the simple kernels and their parameters. In addition, by taking into account different TS compositions, we can remark that the $eCMK$s based on a complex expression that contains simple kernels and coefficients seem to perform slightly better than the $eCMK$s based only on simple kernels ($MTS \gg KTS$). Therefore, it seems to be efficient to combine the kernels by the coefficients. Thus, we are tempted to promote the $eCMK$ based on a mixed TS to the detriment of the $eCMK$ based only on kernels.

5.2   Comparison between the Complex Evolved MKs and the Linear MKs

We aim to compare the improvements obtained by the SVM classifier which involves our most promising $eCMK^*$s based on MTS with both the state of the art convex LMK [4] and the evolutionary LMK already proposed in [7].

In order to emphasize the improvements obtained by involving an MK in the SVM algorithm, an average performance improvement ($\overline{\Delta}$) is computed for each MK as the mean of the improvements δ_i for all the problems. Note that δ_i is the difference between the accuracy rate computed by the SVM algorithm with an MK (Acc_{MK}) and the accuracy rate computed by the same SVM algorithm, but with a simple human-designed kernel (SK) for the ith problem: $\delta_i = \frac{Acc^i_{MK} - Acc^i_{SK}}{Acc^i_{SK}}, i = \overline{1,5}$, and $\overline{\Delta} = \frac{\sum_{i=1}^{5} \delta_i}{5}$, where SK could be one of the considered simple kernels: K_{Pol}, K_{RBF} and K_{Sig} and MK could be one of the MKs: $eCMK$ - the evolutionary complex multiple kernel proposed in this paper based on MTS, $eLMK$ - the evolutionary linear multiple kernel [7] and $cLMK$ - the convex linear multiple kernel [4].

The values of the performance improvements are given in Table 3 and they show that our $eCMK$s generally perform better than both the linear MKs (convex or evolutionary). This may be because our combination of kernels, being more complex and involving, is better adapted to each classification problem than the linear combinations.

In conclusion, the $eCMK$ model based on the MTS seems to be the most promising one. However, more experiments have to be performed in order to verify the advantages of such an $eCMK$.

Table 3. The average performance improvements for the MKs vs. SKs

Δ	K_{Pol}	K_{RBF}	K_{Sig}
MTS	**7.45%**	**6.84%**	**22.10%**
$eLMK$	2.00%	3.66%	9.33%
cMK	8.00%	3.00%	17.66%

6   Conclusion

A hybrid framework has been proposed in order to solve classification problems: a *GP* algorithm combined with an *SVM* classifier for evolving complex *MK*s. We have performed several numerical experiments in order to compare our evolutionary complex multiple kernel to other kernels (human designed or not, simple or multiple). The numerical results have shown that our evolutionary complex *MK*s perform better not only than the simple kernels, but also than the linear *MK*s (convex or evolutionary *LMK*s). Although our approach has a higher computational cost during the learning stage, once the *eCMK** is constructed, the classification stage is as fast as the previous *MK* approaches.

Moreover, the complex kernel functions must include not only different combination of operators (+, * exp) and kernels (sigmoid, polynomial, *RBF*), but also some scaling and shifting coefficients. The *eCMK*s based on efficient kernels, whose parameters are optimised for the evolved combination of kernels seems to be the most promising approach. The regularisation parameter C of the SVM-classifier allows also for performance improvement. We emphases the fact that, to our knowledge, the approach we propose is the only capable to achieve these three objectives in the same framework.

We will focus our further work on the validation of the approach proposed in this paper for large data sets and using multiple data sets for the training stage; this could help to evolve kernels that are more generic.

References

1. Vapnik, V.: The Nature of Statistical Learning Theory. Springer, Heidelberg (1995)
2. Schoelkopf, B., Smola, A.J.: Learning with Kernels. The MIT Press, Cambridge (2002)
3. Chapelle, O., Vapnik, V., Bousquet, O., Mukherjee, S.: Choosing multiple parameters for support vector machines. Machine Learning 46(1/3), 131–159 (2002)
4. Lanckriet, G.R.G., Cristianini, N., Bartlett, P., Ghaoui, L.E., Jordan, M.I.: Learning the kernel matrix with semidefinite programming. JMLR 5, 27–72 (2004)
5. Sonnenburg, S., Rätsch, G., Schafer, C., Scholkopf, B.: Large scale multiple kernel learning. Journal of Machine Learning Research 7, 1531–1565 (2006)
6. Rakotomamonjy, A., Bach, F.R., Canu, S., Grandvalet, Y.: More efficiency in multiple kernel learning. In: The Twenty-fourth ICML (accepted, 2007)
7. Diosan, L., Oltean, M., Rogozan, A., Pecuchet, J.P.: Improving SVM performance using a linear combination of kernels. In: Beliczynski, B., Dzielinski, A., Iwanowski, M., Ribeiro, B. (eds.) ICANNGA 2007. LNCS, vol. 4432, pp. 218–227. Springer, Heidelberg (2007)

8. Nguyen, H.N., Ohn, S.Y., Choi, W.J.: Combined kernel function for SVM and learning method based on evolutionary algorithm. In: Pal, N.R., Kasabov, N., Mudi, R.K., Pal, S., Parui, S.K. (eds.) ICONIP 2004. LNCS, vol. 3316, pp. 1273–1278. Springer, Heidelberg (2004)

9. Lessmann, S., Crone, S.R.S.: Genetically constructed kernels for support vector machines. In: Proc. of German Operations Research, pp. 257–262. Springer, Heidelberg (2005)

10. Koza, J.R.: Genetic Programming: On the Programming of Computers by Means of Natural Selection. MIT Press, Cambridge (1992)

11. Ong, C.S., Smola, A., Williamson, B.: Learning the kernel with hyperkernels. Journal of Machine Learning Research 6, 1043–1071 (2005)

12. Majid, A., Khan, A., Mirza, A.M.: Combination of support vector machines using genetic programming. International Journal of Hybrid Intelligent Systems 3(2), 109–125 (2006)

13. Bach, F.R., Thibaux, R., Jordan, M.I.: Computing regularization paths for learning multiple kernels. In: NIPS (2004)

14. Hastie, T., Rosset, S., Tibshirani, R., Zhu, J.: The entire regularization path for the support vector machine. J. Mach. Learn. Res. 5, 1391–1415 (2003/2004) (electronic)

15. Wang, G., Yeung, D.Y., Lochovsky, F.H.: A kernel path algorithm for support vector machines. In: ICML 2007, pp. 951–958. ACM Press, New York (2007)

16. Friedrichs, F., Igel, C.: Evolutionary tuning of multiple SVM parameters. Neurocomputing 64, 107–117 (2005)

17. Igel, C.: Multi-objective Model Selection for Support Vector Machines. In: Coello Coello, C.A., Hernández Aguirre, A., Zitzler, E. (eds.) EMO 2005. LNCS, vol. 3410, pp. 534–546. Springer, Heidelberg (2005)

18. Zhang, Z., Kwok, J.T., Yeung, D.Y.: Model-based transductive learning of the kernel matrix. Machine Learning 63(1), 69–101 (2006)

19. Bousquet, O., Herrmann, D.J.L.: On the complexity of learning the kernel matrix. In: Becker, S., Thrun, S., Obermayer, K. (eds.) NIPS, pp. 399–406. MIT Press, Cambridge (2002)

20. Burges, J.C.: A tutorial on support vector machines for pattern recognition. In: Knowledge Discovery and Data Mining, vol. 2, pp. 121–167. Kluwer_Academic (1998)

21. Vapnik, V.: Statistical Learning Theory. Wiley, Chichester (1998)

22. Cortes, C., Vapnik, V.: Support-vector networks. Machine Learning 20, 273–297 (1995)

23. Syswerda, G.: A study of reproduction in generational and steady state genetic algorithms. In: Rawlins, G.J.E. (ed.) Proceedings of FOGA Conference, pp. 94–101. Morgan Kaufmann, San Francisco (1991)

24. Chang, C.C., Lin, C.J.: LIBSVM: a library for support vector machines (2001), Software available at: http://www.csie.ntu.edu.tw/~cjlin/libsvm

25. Howley, T., Madden, M.G.: The genetic kernel support vector machine: Description and evaluation. Artif. Intell. Rev. 24(3-4), 379–395 (2005)

26. Dioşan, L., Rogozan, A., Pécuchet, J.P.: Evolving kernel functions for svms by genetic programming. In: ICMLA 2007 (accepted, 2007)

27. Gagne, C., Schoenauer, M., Sebag, M., Tomassini, M.: Genetic programming for kernel-based learning with co-evolving subsets selection. In: Runarsson, T.P., Beyer, H.-G., Burke, E.K., Merelo-Guervós, J.J., Whitley, L.D., Yao, X. (eds.) PPSN 2006. LNCS, vol. 4193, pp. 1008–1017. Springer, Heidelberg (2006)

Optimization of Menu Layouts by Means of Genetic Algorithms

Luigi Troiano[1], Cosimo Birtolo[2], Roberto Armenise[2], and Gennaro Cirillo[2]

[1] RCOST – University of Sannio
Viale Traiano – 82100 Benevento, Italy
troiano@unisannio.it
[2] Poste Italiane s.p.a. – Chief Information Office
Centro Sviluppo Servizi Innovativi Napoli
Piazza Matteotti 3 – 80133 Naples, Italy
{birtoloc,armenis5,ciril127}@posteitaliane.it

Abstract. Menu systems are key components in modern graphical user interfaces (GUIs), either for traditional desktop applications, or for the latest web applications. The design of interface layout must consider different aspects resulting in a trade-off between often conflicting requirements. This trade-off is aimed at making effective use of interfaces in order to meet user preferences and to conform to standard guidelines at the same time. Assuming we are able to quantify such a trade-off, the problem of finding a menu system able to maximize it figures as a combinatorial optimization problem. In this paper we investigate the application of genetic algorithms as a viable approach to identifying solutions that can be used as a starting point for further fine-tuning.

Keywords: GUI design, menu layout, optimization, search based software engineering.

1 Introduction

At the time of their introduction, menu systems represented a major shift from command-line interfaces. The early systems were very simple and not hierarchical in structure. Since then, they have been innovating continuously. Menu systems have been evolved not only in structure, functionality and purpose. Today, the menu system is a component of fundamental importance for making GUIs attractive and usable, and special care is paid to their design and implementation. This is not only related to traditional desktop applications. In modern web applications, the menu systems still play a key role in helping the user to navigate functionalities, especially after the advent of AJAX and Rich Internet Applications (RIA) more recently. The menu system layout is a basic ingredient for increasing productivity.

In designing a menu layout of good quality, engineers have to consider many aspects including how effectively functionalities are retrieved and activated, what standard guidelines suggest, and what are the preferences of users. These aspects

J. van Hemert and C. Cotta (Eds.): EvoCOP 2008, LNCS 4972, pp. 242–253, 2008.

are translated into several design requirements, that often are conflicting. For instance, although having flat hierarchical structures improves accessibility, a limitation to the number of items is necessary in order not to have long lists. At the same time, users could have preferences for the item order. A trade-off between these different requirements must be found in order to maximize the menu system quality.

The problem of finding the layout that maximizes quality is combinatorial in nature, as it depends on the arrangement of each item in different positions onto the menu structure, with no construction rules for building an optimal solution. This suggests that the problem is NP-hard. Nowadays, this task is not yet supported by search techniques, and it is left to the experience of engineers. In this paper we investigate the application of genetic algorithms in exploring the space of menu systems at the search of solutions that can maximize the requirements trade-off. The remainder of this paper is organized as follows: in Section 2 we introduce the issues related to menu system design and we formally define the optimization problem; in Section 3 we describe the characteristics of the genetic algorithm used in our experimentation; Section 4 is aimed to present some experimental results; Section 5 outlines conclusions and future work.

2   Designing and Optimizing a Menu System

Although menu systems are components each user has widely experienced, for the sake of clarity, it can be useful to define terms in order to disambiguate definitions. In the remainder of this paper we will refer to *menu layout* as the hierarchical structure by which the user gains access to application functionalities. A menu layout is made of menus; each *menu* is made of a list of *items* referring to *submenus* or to *actions*. The first are menus at lower level, the latter are aimed to activate functionalities, thus they represent the menu system leaves (i.e. terminals).

In designing a menu layout we have to take into the account:

− *accessibility*, as the ease of reaching desired actions.
− *guidelines*, as a set of best practices in organizing the menu layout
− *preferences*, as a wish list made explicit or implicit by the end user

As the menu system aims to quickly activate functionalities, we will consider accessibility as the optimization driver, whilst we will refer to guidelines and preferences as optimization preferences (constraints, when mandatory). Therefore, the problem solution relies on finding a menu layout that maximizes accessibility and compliance to guidelines and user preferences.

Menu selection involves most aspects of human information processing. Indeed, the user is asked for visually inspect the menu, reading and comprehending items in order to find a path that will lead to the desired functionality; choosing the best option until the action is not reached and task accomplished. The menu system layout is expected to better support the user in this task.

So the main issue in designing a good menu system is about how to organize menu items in the hierarchy in order to make actions and to easy accessible.

In early stages, researchers focused on functional features a menu system is demanded to have in order to improve accessibility. For instance, Walker and Smelcer [1] investigated the relationship between the structure made of walking menus or cascading menus and the time required by an user to reach the target action. In our work, we assumed this issue solved by the current standard implementations, so our focus is mostly on the relationship between the menu hierarchical structure (layout) and accessibility.

Various models for predicting the selection time have been proposed, and research aimed to find the structure that minimizes it. If items are sorted, e.g. alphabetically, search time can be predicted by Hick's Law [2], which states that the time to locate an item is a logarithmic function of the menu size. When menus are not alphabetically ordered, users have to scan them in a linear fashion to locate an item. However, if the user has memorized the position of items in a menu, search time becomes constant. Thus, selection times is reduced to the time needed to reach the item position. Fitts' Law [3,4] predicts the time required to move the cursor to a particular item. It describes the movement time taken to acquire, or point to, a visual target, stating that the movement time needed to acquire a target is a logarithmic function of the ratio between the target distance d and the target width w, known as the task's Index of Difficulty (ID). According to Fitts' law, menu items that appear further down the menu have a greater ID. As this model does not consider constraints in the motion trajectory, Fitts' law cannot accurately predict the movement time in cascading pull-down menus. If the cursor has to be steered along a tunnel, movement time is better modeled by Steering Law [5]. According to this law, movement time is determined by the ratio between the tunnel distance td and the tunnel width tw.

Hollink and van Someren [6] reviewed the assumptions underlying prediction models for the selection time, and proposed a method to validate these assumptions off-line. In their method, after the relationship between the path followed through the menu system and the navigation time, this last is determined by two structural properties of the path: the number of menu items a user has to open and for each navigation step the number of menu items the user has to read. Furthermore the prediction is based on the users' choice strategy, the node opening function and the node choice function.

Recently, Bernard [7] has presented a further model for predicting the selection time. The Hypertext Accessibility Index measure (H_{HAI}) is defined as

$$H_{HAI(x)} = \sqrt{\sum_{i=1}^{L} \sum_{j \in N_i} \log_2(b_j + 1) \log_2(d_j + 1)} \qquad (1)$$

where

- x is the menu structure
- L is the maximum number of levels of x
- N_i is the set of menus and menu items at level i
- b_j is the number of children of j
- d_j is the depth of j, assuming for the root and all menu item $d = 1$

It can be easily verified that $H_{HAI} \in [1, +\infty)$: the lower H_{HAI} is, the more the menu items are accessible; when all menu items are assigned to the root menu (i.e. no submenu is considered) $H_{HAI} = 1$. An interesting characteristic of this model is that H_{HAI} index predicts the expected navigation time on the basis only of the menu system layout. Bernard's model shows that though broader trees in general tend to have better search efficiency than deeper trees, topological shape has also an important effect. The H_{HAI} metric has been validated by comparing predictions with the empirical results found by others and Bernard himself.

Bernard's findings are in accordance with results of other researchers. For instance, Botafogo et al. [8] found that imbalance might indicate a poorly designed hypertext hierarchy, though this is sometimes unavoidable in some domains. They propose two metrics for imbalance, namely the "depth imbalance" and "child imbalance". The depth imbalance metric measures the variance in depth of a node's children; the child imbalance measures the variance in the number of descendants (i.e. sections, subsections, pages, etc.) of a node's children.

In the last years, other techniques have been introduced in order to improve the selection time in cascading pull-down menus, focusing on the selection of first-level items. Shorter selection times have been reached by either decreasing the distance to the menu items, or by increasing the size of the menu item. A split menu adapts to user behavior and relocates the menu items according to usage. Frequently selected items are moved into the top split of the menu and seldom selected items are pushed downward, i.e. the distance to an item depends on selection probability [9]. Ahlström [10] modeled and improved cascading menu selection times through the use of 'force-fields', a variant of sticky widgets, that attracts the cursor towards the cascading menu. The evaluation did not investigate whether the technique caused an adverse effect on selecting non-cascading items.

Designers usually use guidelines to organize the menu structure. They provide a collection of best practices in organizing and structuring the menu layout. Examples are Apple's Human Interface Guidelines [11] and Sun's Java Look and Feel Guidelines [12]. Guidelines are either too specific or too vague, so they do not always apply to the problem at hand [13]. For instance an Apple's Human Interface Guidelines suggests putting on menu bar some particular menus that an user expects to find such as "File", "View" and "Help". Guidelines say, as a general rule, to avoid creating long menus, in fact they are difficult for the user to scan and can be overwhelming, from other side it has not to put many items in a single menu and it needs to regrouping them in other menus. In most guidelines, it is suggested not to go further two levels of cascading menus, although in some cases it is convenient to violate this rule.

According to the current literature, building of menu hierarchy and optimization is a challenging task, whose applications go further the desktop and web applications. However, building a quality menu system requires a large group of users (e.g. focus groups) and a large number of trials in order to find the best way or structuring the menu layout. Search techniques, can provide a valuable support in screening alternatives and in providing starting point that can be refined more efficiently.

3   Algorithm

The algorithm is inspired to the Simple GA given by Goldberg [14]. The structure is outlined in Fig.1.

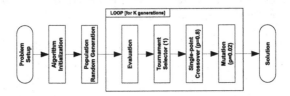

Fig. 1. Algorithm structure

After the problem is setup in terms of menu items and preferences, the algorithm is instanced and the initial population is randomly generated. The algorithm body is made of following stages:

- *evaluation*: a fitness score is assigned to each population individual.
- *genetic processing*: here individuals are genetically processed by selection, crossover and mutation

After K generations the best individual is obtained. Valid (i.e. legal) individuals are compliant with mandatory preferences (constraints), invalid not.

Preferences are given as a set of relational and structural properties, each with an assigned priority. In our case, we assumed priorities on a scale of five: 1- very important, 2- important, 3- medium, 4- not important, 5- not very important. Preferences are facultative. Besides them, we assumed mandatory preferences (i.e. constraints), with priority 0- mandatory.

3.1   Chromosome Structure and Genetic Operators

Among the different ways of representing a tree structure by a chromosome, we choose a coding in which each gene represents the path from root to a menu item as depicted by Fig.2.

The number of genes is not necessarily equal to the number of action (i.e. terminal) items. In some cases, an action could be accessed by different paths. Therefore the chromosome is as long as the sum of allowed occurrences of each action. For example, if an action is allowed twice, there is a need for two genes to that action, each representing a different path. The mapping between genes and actions is kept by an association table. When the path is empty, the action item is associated to the root (e.g. gene N in figure); if the path is null, that action item occurrence is not considered in the menu layout (e.g. gene 1). Such a

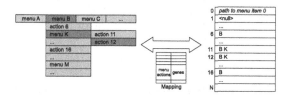

Fig. 2. Chromosome with mapping to the menu layout

chromosome structure is more robust to genetic operations than others, allowing a better control of action items, whose best placement in the menu layout is the ultimate goal of the optimization algorithm. The algorithm is based on three genetic operations:

- selection: a tournament selection has been preferred in order to be less sensitive to the fitness scaling
- crossover: single point crossover
- mutation: gene mutation with a random choice of insertion, deletion and modification of single items.

In particular, the algorithm adopts elitism by random substitution with the best individuals. Tournament is implemented by selecting the best individual after t pairwise comparisons, as described in [14]. Crossover and mutation are described in Fig.3.

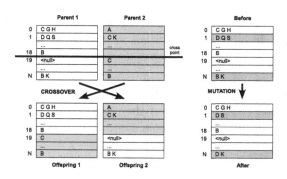

Fig. 3. Crossover and mutation

The menu layout is built by adding paths in the order they occur in the chromosome genes. So, the actual order of items in a certain menu depends on the order they occur in the chromosome, given the same path to them. For instance, if A-B-L precedes A-B-M, L will come first in the menu A-B, otherwise the opposite. A permutation of the mapping entries would allow to obtain different placements. However, the mapping is fixed and not processed by genetic operators. The reason is that the initial path building is purely random. Thus, different placement are considered by the initial production of alternatives. Considering the mapping permutation would not be beneficial, adding only an additional degree of freedom to control by the genetic algorithm.

3.2 Fitness Function

The fitness function of an individual x is aimed to model the trade-off between accessibility and preference compliance. Thus it is defined as convex combination

$$fitness(x) = \sigma \cdot H(x) + (1 - \sigma) \cdot C(x) \qquad (2)$$

where $\sigma \in [0,1]$, $H(x)$ is the degree of accessibility, and $C(x)$ is the degree of constraints' compliance. In particular, $H(x)$ is defined as

$$H(x) = e^{k(1-H_{HAI}(x))} \qquad (3)$$

where $H_{HAI}(x)$ is defined by Eq.(1). The constant k controls the exponential decay. Instead the degree of preference compliance is defined as the weighted mean

$$C(x) = \frac{\sum\limits_{i=1}^{m} \bar{p}_i c_i(x)}{\sum\limits_{i=1}^{m} \bar{p}_i} \qquad (4)$$

where m is the number of preferences, $\bar{p}_i = 1 - p_i$ is the constraint importance, and $c_i(x)$ is the compliance of x to the preference c_i. Therefore, we assumed a compensation between optimization criteria.

The problem of finding an optimal menu layout consists in placing all action items by maximizing accessibility and preference compliance. Preferences can be of different kinds. In our experimentation we considered the following types:

- **Path ordering** (*ancestor*, *successor*): defines a ordering relation between *ancestor* and *successor* along a path
- **Menu ordering** (*predecessor, follower*): defines a ordering relation between *predecessor* and *follower* whenever they coexist within the same menu
- **Number of menu items** (*menu, min, max*): defines the *min* and *max* number of items present in *menu*
- **Occurrence** (*item, min, max*): defines the *min* and *max* number of occurrences of *item*
- **Level** (*item, min, max*): defines the *min* and *max* level for *item*
- **Menu belonging** (*item, menu*): *item* should belong to *menu*

Each preference has a priority $p_i \in [1,5]$, where 1 is the highest priority (i.e. very important), 5 the lowest (i.e. not very important). The degree of compliance of x to each preference is computed as

$$c_i(x) = 1 - \frac{v_i(x)}{mv_i(x)} \qquad (5)$$

with $v_i(x)$ giving the number of criterion violations of x, and $mv_i(x)$ the maximum number of possible violations.

Level	Items Min - Max	Priority
Root	4-5	2
Level 1	2-4	3
Level 2	2-5	3
Level 3	1-2	4

Ordering	Priority
A-B-C-D	2
F-G	3

Menu	Level Min - Max	Priority	Repetition Min - Max	Priority
A	1-1	2	1-1	2
B	1-1	2	1-1	2
C	1-1	2	1-1	2
D	1-2	3		
E	1-3	4		
F			0-1	3
G			0-1	3
H	2-2	3	0-1	3
I			0-1	3
J	2-4	3	1-3	2
K			1-2	4
L	3-4	4	1-2	4

Fig. 4. Preferences. The top-left table provides the number of items at each level, the middle-left table provides two path ordering preferences, the top-right table provides for each menu the desired level and the number of occurrences, the bottom table provides menu belonging preferences specifying the priority of each preference.

4   An Example of Application

As an example of application let us compose a menu layout made of 25 action items $Z1..Z25$ and 12 submenus $A..L$. Layout generation is driven by 60 preferences, outlined in Figure 4. The algorithm was setup with standard parameters: tournaments = 1, crossover probability = 0.8, mutation probability = 0.02, elitism = 2.[1] Figure 5 and Figure 6 show the layout of the best individual respectively after 10 and 1000 generations. Figure 5 depicts a layout with fitness = 0.7949. Indeed, this structure does not satisfy some preferences. In particular,

- the number of children in the root is more than 5
- the tree does not meet any level preference
- menus A and B should not have any repetition

On the other side, A, B and C are in the right order on the menu bar (at level 1) as expected. Instead layout presented in Figure 6 meets better the preference set, thus its fitness value is 0.9717.

When some preferences become mandatory (constraints), the problem becomes harder to solve. In Figure 7 we show a solution layout in this case. In particular, all preferences with priority 2 have been considered mandatory.

The low fitness value (0.566) is due to conflicts between constraints and some other preferences, as it can be easily noticed by observing tables in Figure 4. Obviously, when constraints are in contradictory, there is no solution to the optimization problem. Therefore, it is important to choose the preference and constraint set appropriately, in order to avoid conflicts and contradictions. This is the situation depicted in Figure 8.

---

[1] Parameters have been chosen by a simple qualitative analysis, according to common values adopted for them, without any in-depth quantitative analysis for their optimization.

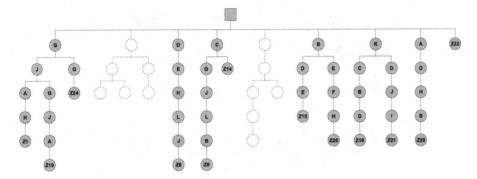

Fig. 5. Layout of the best individual after 10 generations (fitness 0.7949)

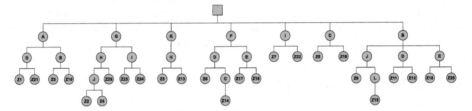

Fig. 6. Layout of the best individual after 1000 generations (fitness 0.9717)

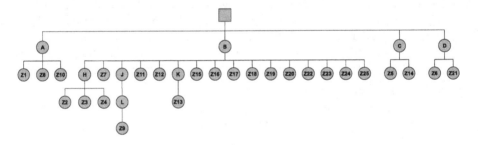

Fig. 7. Layout with constraints (fitness = 0.566)

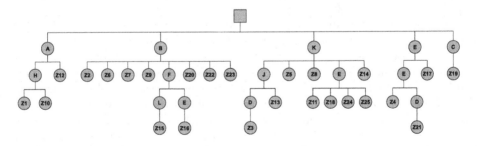

Fig. 8. Layout with compatible constraint and preference sets (fitness = 0.9952)

In this case we defined a legal individual when the menu bar has 4 or 5 items, and A, B, C are on the menu bar with no repetition. Furthermore we imposed that item $Z1$ has to be an action of menu A. These conditions are expressed by 10 constraints: 3 level constraints, 3 occurrence constraints, 1 number of children constraint, 2 path ordering constraints, 1 belonging constraint. The algorithm run 500 generations on population with 1000 individuals. At the end, legal individuals were 102 (i.e. 898 illegal). Fitness of the best legal individual was 0.9952, with $H = 0.9498$.

We can note that action $Z10$ is in A, action $Z21$ in menu D as expected by the menu belonging preferences. Furthermore, some menu (namely F,G,H,I) are allowed to occur more than once, whilst L no more than twice. We can verify that in layout of Figure 8 these preferences are fully satisfied. In particular, F and H occur once, whilst G and L never. Moreover, the number of children of level 2 are between 2 and 5, and between 1 and 2 at level 3.

These performances are not episodic, as it could be argued. We run the algorithm several times with a different number of preferences in order to study quantitatively the convergence. In Figure 9, we report the median of best fitness with a different cardinality of the preference set and population size (100, 200, 500, 1000 individuals). Preference sets of different cardinality (15, 30, 45, 60 preferences) have been chosen with the same distribution of priorities, so that analysis is independent on this factor.

We can notice that the algorithm reached high values of fitness in all cases, although the behavior differs qualitatively according to the number of preferences considered at a time, thus according to the problem difficulty. So, if in the case of 15 preferences, convergence is reached pretty soon, an increasing time is required

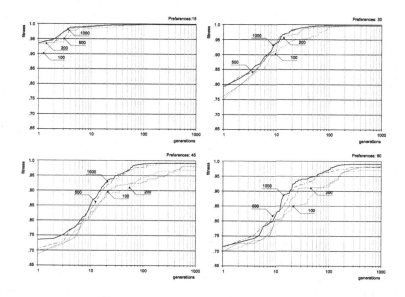

Fig. 9. Algorithm convergence

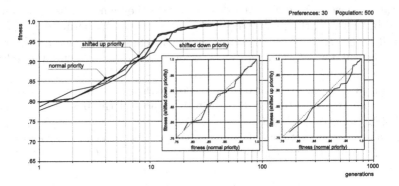

Fig. 10. Algorithm convergence with priority shift

in the case of 30, 45 and 60 preferences. Also population size has an impact on convergence when the number of preferences increases. Indeed, we can observe how the algorithm is not able to converge properly in the case of 60 preferences with 100 and 200 individuals. Another point of interest is the convergence of the algorithm when priority changes. In Figure 10 is outlined the median fitness of the best individual when preference priority is increased (1-3) and decreased (3-5) against the nominal case (2-4).

Obviously the fitness value cannot be the same, as the priority ratios change. However, we can notice how convergence is not heavily affected by a shift of priority, thus resulting robust tho this situation. This means that priority magnitude does not represent a sensible aspect to take into consideration.

5   Conclusions and Future Work

In this paper we presented a genetic algorithm for optimizing the layout of a GUI menu system, keeping into the account accessibility, user preferences and standard guidelines. The resulting solution can be used as a robust starting point aimed to be refined by software engineers. Experimentation provided very encouraging results, proving the ability of a simple genetic algorithm in converging towards solutions with high fitness, also in presence of mandatory constraints. Moreover, the algorithm has been proven to scale the number of constraints, and to be robust to constraint preference variations. However, we aim to investigate two main directions in the future. The first is how to integrate mandatory constraints into the solution generation, so there will be no illegal individuals to deal with. Genetic programming seems to be a promising solution to this problem. The second is about how to find out implicit preferences by analyzing the usage of generated menus, instead of forcing the user to make explicit all her/his preferences. In this case, making the genetic algorithm interactive poses additional interesting questions to be answered regarding to how to sample the search space and to gather user feedback.

Acknowledgments. This work was partially supported by MIUR Project Automatic System For The Visually Impaired (SAPI), n.1642-2006.

References

1. Walker, N., Smelcer, J.B.: A comparison of selection times from walking and pull-down menus. In: Proceedings of ACM CHI 1990 Conference on Human Factors in Computing Systems, pp. 221–225 (1990)
2. Hick, W.E.: On the rate of gain of information. Quarterly Journal of Experimental Psychology 4, 11–26 (1952)
3. Fitts, P.M.: The information capacity of the human motor system in controlling the amplitude of movement. J. Exp. Psychol. 47, 381–391 (1954)
4. Cockburn, A., Gutwin, C., Greenberg, S.: A predictive model of menu performance. In: CHI 2007: Proceedings of the SIGCHI conference on Human factors in computing systems, pp. 627–636. ACM, New York (2007)
5. Accot, J., Zhai, S.: Beyond fitts' law: models for trajectory-based hci tasks. In: CHI 1997: Proceedings of the SIGCHI conference on Human factors in computing systems, pp. 295–302. ACM, New York (1997)
6. Hollink, V., van Someren, M.: Validating navigation time prediction models for menu optimization. In: Althoff, K.D., Schaaf, M. (eds.) LWA. Hildesheimer Informatik-Berichte, University of Hildesheim, Institute of Computer Science, vol. 1, pp. 47–52 (2006)
7. Bernard, M.L.: Examining a metric for predicting the accessibility of information within hypertext structures. PhD thesis, Wichita, KS, USA (2002) Adviser-Charles G. Halcomb
8. Botafogo, R.A., Rivlin, E., Shneiderman, B.: Structural analysis of hypertexts: identifying hierarchies and useful metrics. ACM Trans. Inf. Syst. 10(2), 142–180 (1992)
9. Ahlström, D.: Modeling and improving selection in cascading pull-down menus using fitts' law, the steering law and force fields. In: CHI 2005: Proceedings of the SIGCHI conference on Human factors in computing systems, pp. 61–70. ACM, New York (2005)
10. Ahlström, D., Alexandrowicz, R., Hitz, M.: Improving menu interaction: a comparison of standard, force enhanced and jumping menus. In: CHI 2006: Proceedings of the SIGCHI conference on Human Factors in computing systems, pp. 1067–1076. ACM, New York (2006)
11. Apple Computer Inc.: Apple human interface guidelines. Technical report, Apple Computer Inc. (2006)
12. Inc, S.M.: Java look and feel design guidelines: advanced topics. Addison-Wesley Longman Publishing Co., Inc., Boston (2001)
13. Oliver, A., Regragui, O., Monmarché, N., Venturini, G.: Genetic and interactive optimization of web sites. In: The 11th International World Wide Web Conference, Honolulu, Hawaii, USA, pp. 7–11 (2002)
14. Goldberg, D.E.: Genetic Algorithms in Search, Optimization and Machine Learning. Addison Wesley, Reading (1989)

A Path Relinking Approach with an Adaptive Mechanism to Control Parameters for the Vehicle Routing Problem with Time Windows

Hideki Hashimoto[1] and Mutsunori Yagiura[2]

[1] Graduate School of Informatics, Kyoto University, Kyoto, Japan
hasimoto@amp.i.kyoto-u.ac.jp
[2] Graduate School of Information Science, Nagoya University, Nagoya, Japan
yagiura@nagoya-u.jp

Abstract. We propose a path relinking approach for the vehicle routing problem with time windows. The path relinking is an evolutionary mechanism that generates new solutions by combining two or more reference solutions. In our algorithm, those solutions generated by path relinking operations are improved by a local search whose neighborhood consists of slight modifications of the representative neighborhoods called 2-opt*, cross exchange and Or-opt. To make the search more efficient, we propose a neighbor list that prunes the neighborhood search heuristically. Infeasible solutions are allowed to be visited during the search, while the amount of violation is penalized. As the performance of the algorithm crucially depends on penalty weights that specify how such penalty is emphasized, we propose an adaptive mechanism to control the penalty weights. The computational results on well-studied benchmark instances with up to 1000 customers revealed that our algorithm is highly efficient especially for large instances. Moreover, it updated 41 best known solutions among 356 instances.

1 Introduction

The vehicle routing problem with time windows (VRPTW) is the problem of minimizing the total traveling distance of a number of vehicles, under capacity and time window constraints, where every customer must be visited exactly once by a vehicle. The capacity constraint signifies that the total load on a route cannot exceed the capacity of the assigned vehicle. The time window constraint signifies that each vehicle must start the service at each customer in the period specified by the customer. The VRPTW has a wide range of applications such as bank deliveries, postal deliveries, school bus routing and so on, and it has been a subject of intensive research focused mainly on heuristic and metaheuristic approaches. See the survey by Bräysy, Dullaert and Gendreau [1] for evolutionary algorithms.

We propose a path relinking approach for the VRPTW. The path relinking [2] is an evolutionary mechanism that generates new solutions by combining two or more reference solutions. Our algorithm invokes a path relinking operation for

J. van Hemert and C. Cotta (Eds.): EvoCOP 2008, LNCS 4972, pp. 254–265, 2008.

generating new candidate solutions, which are then improved by a local search whose neighborhood consists of slight modifications of the representative neighborhoods called 2-opt*, cross exchange and Or-opt. To reduce the computation time for searching these neighborhoods, we propose a neighbor list that prunes the neighborhood search heuristically. In our algorithm, infeasible solutions are allowed to be visited during the search, while the amount of violation is penalized. The amount of violation for the capacity constraint is estimated by the amount of capacity excess. To estimate the amount of violation of time window constraints of each route, we consider the total amount of traveling time to be shortened to satisfy the constraints. We also incorporate in our algorithm a frequency-based penalty, in which a customer who often appears in an infeasible route of locally optimal solutions is penalized to direct the search to make those routes with many heavily penalized customers feasible. As the evaluation of these penalties takes time if naively implemented, we propose an efficient algorithm, which enables us to evaluate each neighborhood solution in $O(1)$ time. We also propose an adaptive mechanism to control the weights of these penalties. Finally we report computational results on well-studied benchmark instances with up to 1000 customers. The results show the high competence of our algorithm against existing methods; it updates 41 best known results among 356 instances within a reasonable amount of computation time.

2   Problem Definition

Here we formulate the vehicle routing problem with time windows. Let $G = (V, E)$ be a complete directed graph with vertex set $V = \{0, 1, \ldots, n\}$ and edge set $E = \{(i, j) \mid i, j \in V, i \neq j\}$, and $M = \{1, 2, \ldots, m\}$ be a vehicle set. In this graph, vertex 0 is the depot and other vertices are customers. Each customer i and each edge $(i, j) \in E$ are associated with:

 i. a fixed quantity a_i (≥ 0) of goods to be delivered to i,
 ii. a time window $[e_i, l_i]$,
 iii. a traveling time t_{ij}(≥ 0) and a traveling distance c_{ij}(≥ 0) from i to j.

We assume $a_0 = 0$ and $e_0 = 0$ without loss of generality. Each vehicle has an identical capacity u.

Let σ_k denote the route traveled by vehicle k, where $\sigma_k(h)$ denotes the hth customer in σ_k, and let

$$\sigma = (\sigma_1, \sigma_2, \ldots, \sigma_m).$$

Note that each customer i is included in exactly one route σ_k, and is visited by vehicle k exactly once. We denote by n_k the number of customers in σ_k. For convenience, we define $\sigma_k(0) = 0$ and $\sigma_k(n_k + 1) = 0$ for all k (i.e., each vehicle $k \in M$ departs from the depot and comes back to the depot). Moreover, let s_i be the start time of service at customer i (by exactly one of the vehicles) and s_k^a be the arrival time of vehicle k at the depot. Note that each vehicle is allowed to wait at customers before starting services.

Let us introduce 0-1 variables $y_{ik}(\boldsymbol{\sigma}) \in \{0,1\}$ for $i \in V \setminus \{0\}$ and $k \in M$ by

$$y_{ik}(\boldsymbol{\sigma}) = 1 \iff i = \sigma_k(h) \text{ holds for exactly one } h \in \{1, 2, \ldots, n_k\}.$$

That is, $y_{ik}(\boldsymbol{\sigma}) = 1$ holds if and only if vehicle k visits customer i. The traveling distance of a vehicle k is expressed as $d(\sigma_k) = \sum_{h=0}^{n_k} c_{\sigma_k(h),\sigma_k(h+1)}$. Then the problem we consider in this paper is formulated as follows:

$$\text{minimize} \quad \sum_{k \in M} d(\sigma_k) \tag{1}$$

$$\text{subject to} \quad \sum_{k \in M} y_{ik}(\boldsymbol{\sigma}) = 1, \qquad\qquad\qquad i \in V \setminus \{0\} \tag{2}$$

$$\sum_{i \in V \setminus \{0\}} a_i y_{ik}(\boldsymbol{\sigma}) \leq u, \qquad\qquad k \in M \tag{3}$$

$$t_{0,\sigma_k(1)} \leq s_{\sigma_k(1)}, \qquad\qquad\qquad k \in M \tag{4}$$

$$s_{\sigma_k(i)} + t_{\sigma_k(i),\sigma_k(i+1)} \leq s_{\sigma_k(i+1)}, \quad 1 \leq i \leq n_k - 1, \ k \in M \tag{5}$$

$$s_{\sigma_k(n_k)} + t_{\sigma_k(n_k),0} \leq s_k^{\mathrm{a}} \leq l_0, \qquad k \in M \tag{6}$$

$$e_i \leq s_i \leq l_i \qquad\qquad\qquad\qquad i \in V \setminus \{0\} \tag{7}$$

$$y_{ik}(\boldsymbol{\sigma}) \in \{0,1\}, \qquad\qquad\qquad i \in V \setminus \{0\}, \ k \in M. \tag{8}$$

Constraint (2) means that every customer $i \in V \setminus \{0\}$ must be served exactly once by a vehicle. Constraint (3) means a capacity constraint for vehicle k. Constraints (4)–(6) require that each vehicle cannot serve a customer before arriving at the customer. Constraint (7) is a time window constraint for each customer. Note that essential decision variables in this formulation are routes σ_k, since the values of $y_{ik}(\boldsymbol{\sigma})$ are automatically determined from $\boldsymbol{\sigma}$, and finding appropriate values for s_i and s_k^{a}, if any, is easy when $\boldsymbol{\sigma}$ is fixed.

3   Local Search

In this section, we describe our local search (LS). Our LS searches a visiting order $\boldsymbol{\sigma} = (\sigma_1, \sigma_2, \ldots, \sigma_m)$, which can be infeasible with respect to the capacity and time window constraints. The algorithm evaluates each route σ_k by a function $p(\sigma_k)$, which is the sum of its traveling distance $d(\sigma_k)$ and the penalty for violation of constraints if σ_k is infeasible, and it evaluates a solution $\boldsymbol{\sigma}$ by $\sum_{k \in M} p(\sigma_k)$. The details of function $p(\sigma_k)$ will be discussed in Section 4. Our LS starts from an initial solution $\boldsymbol{\sigma}$ and repeats replacing $\boldsymbol{\sigma}$ with a better solution (with respect to $\sum_{k \in M} p(\sigma_k)$) in its neighborhood $N(\boldsymbol{\sigma})$ until no better solution is found in $N(\boldsymbol{\sigma})$. To define the neighborhood $N(\boldsymbol{\sigma})$, we use the 2-opt*, cross exchange and Or-opt neighborhoods with slight modifications. For the 2-opt* and cross exchange neighborhoods, we propose a neighbor list to prune the neighborhood search heuristically. A similar technique was successfully applied to the traveling salesman and vehicle routing problems [3, 4], in which the list is determined only on the basis of distance; therefore it is not appropriate to apply the existing method directly to the VRPTW. In Section 3.1, we describe the neighbor lists that take into account the time windows, and in Section 3.2, the details of the neighborhoods are described.

3.1 Neighbor List

We consider a neighbor list for each customer i, which is a set of customers preferable to visit immediately after i. Each customer j that can be visited after i (i.e., $e_i + t_{ij} \leq l_j$) is evaluated by $\max\{t_{ij}, e_j - l_i\}$. When a vehicle visits j immediately after i, it takes at least $\max\{t_{ij}, e_j - l_i\}$ time between the start times of i and j. Hence, if this value is small, it is preferable to visit j immediately after i. The algorithm computes these values once at the beginning and stores the best N_{nlist} (a parameter) customers as a neighbor list of i. We set $N_{\text{nlist}} = 20$ in the experiments.

3.2 Neighborhoods

We use the 2-opt*, cross exchange and Or-opt neighborhoods with slight modifications, wherein we restrict the 2-opt* and cross exchange neighborhoods by using the neighbor lists.

The 2-opt* neighborhood was proposed in [5], which is a variant of the 2-opt neighborhood [6] for the traveling salesman problem. A 2-opt* operation removes two edges from two different routes (one from each) to divide each route into two parts and exchanges the second parts of the two routes. Our algorithm searches only those solutions obtainable by a 2-opt* operation in which at least one of the newly added edges is in the neighbor list. The size of this neighborhood is $O(N_{\text{nlist}}n)$.

The cross exchange neighborhood was proposed in [7]. A cross exchange operation removes two paths from two routes (one from each) of different vehicles, whose length (i.e., the number of customers in the path) is at most L^{cross} (a parameter), and exchanges them. Our algorithm searches only those solutions obtainable by a cross exchange operation in which a newly added edge linking the former part of a route and the path from another route is in the neighbor list. The size of this neighborhood is $O((L^{\text{cross}})^2 N_{\text{nlist}}n)$. We set $L^{\text{cross}} = 3$ in the experiments.

The cross exchange and 2-opt* operations always change the assignment of customers to vehicles. We also use an intra-route neighborhood to improve individual routes, which is a variant of the Or-opt neighborhood used for the traveling salesman problem [8]. An intra-route operation removes a path of length at most $L^{\text{intra}}_{\text{path}}$ (a parameter) and inserts it into another position of the same route, where the position is limited within length $L^{\text{intra}}_{\text{ins}}$ (a parameter) from the original position. The size of the intra-route neighborhood is $O(L^{\text{intra}}_{\text{path}} L^{\text{intra}}_{\text{ins}} n)$. We set $L^{\text{intra}}_{\text{path}} = 3$ and $L^{\text{intra}}_{\text{ins}} = 10$ in the experiments.

Figure 1 is an illustration of the neighborhoods. In Figure 1, squares represent the depot (which is duplicated at each end) and small circles represent customers

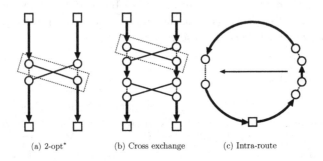

(a) 2-opt*   (b) Cross exchange   (c) Intra-route

Fig. 1. Neighborhood operations in our local search

in the routes. A thin line represents a route edge and a thick line represents a path (i.e., more than two customers may be included). The dotted boxes mean that edges in them are in the neighbor lists.

Our LS searches the above intra-route, 2-opt* and cross exchange neighborhoods, in this order. Whenever a better solution is found, the LS immediately accepts it (i.e., we adopt the first admissible move strategy) and resumes the search from the intra-route neighborhood.

4   Evaluation Function $p(\sigma_k)$

We first define the function $p(\cdot)$ to evaluate a route σ_k. For convenience, throughout this section, we assume that vehicle k visits customers $1, 2, \ldots, n_k$ in this order and let customer n_k+1 represent the arrival at the depot (i.e., $s_{n_k+1} = s_k^a$). The function we adopt is

$$p(\sigma_k) = \begin{cases} d(\sigma_k), & \text{if } \sigma_k \text{ is feasible} \\ d(\sigma_k) + \alpha p_c(\sigma_k) + \beta p_t(\sigma_k) + \sum_{h=1}^{n_k} \gamma_h, & \text{otherwise,} \end{cases} \quad (9)$$

where $p_c(\sigma_k)$ is the amount of capacity excess (i.e., $p_c(\sigma_k) = \max\{0, \sum_{i=1}^{n_k} a_i - u\}$) and $p_t(\sigma_k)$ is the minimum total amount of traveling times to be shortened to satisfy the constraints; i.e.,

$$p_t(\sigma_k) = \min\left\{\sum_{h=1}^{n_k+1} \tau_h \;\middle|\; \begin{array}{l} s_0 \geq 0, \; s_{h-1} + t_{h-1,h} - \tau_h \leq s_h, \\ \tau_h \geq 0, \; e_h \leq s_h \leq l_h, \; h = 1, \ldots, n_k+1 \end{array}\right\}.$$

In function p, α, β and γ_i for each $i \in V$ are parameters, which are controlled adaptively (see Section 5). Parameters α and β are controlled reflecting the difficulties in satisfying the capacity constraint and the time window constraint, respectively. Parameter γ_i reflects the difficulty in visiting customer i by a feasible route. In the evaluation of (9), each traveling time can be shortened by an arbitrary amount (i.e., the resulting traveling time $t_{h-1,h} - \tau_h$ can be negative) to satisfy time window constraints while the shortened amount is penalized as $p_t(\sigma_k)$. This idea of defining p_t was proposed by Nagata [9]. The algorithm computes $p(\sigma_k)$ by each term separately. In the rest of this section, we focus on the computation of $p_t(\sigma_k)$, since the other terms can be efficiently computed by using standard data structures (e.g., [10,11,12,13]).

A key observation to the efficient computation is that each route σ_k of a neighborhood solution is a recombination of a few paths of the current solution. Hence we consider a speeding up approach that stores some useful information of paths from the depot to customers and those from customers to the depot, among those paths of the current routes. For each customer h in a new route σ_k, let $\mathcal{F}_h$ (resp., $\mathcal{B}_h$) be some data structure that contains the information of the path (of σ_k) from the depot to h (resp., from h to the depot). Note that $\mathcal{F}_h$ and $\mathcal{B}_h$ signify the information of the paths of the new route σ_k. For example, if σ_k is generated by a 2-opt* operation, and the path from the depot to h and the path from $h + 1$ to the depot are from the current solution, then $\mathcal{F}_h$ and $\mathcal{B}_{h+1}$ are available from the stored information when they are used to compute $p(\sigma_k)$. On the other hand, for the cross exchange and intra-route neighborhoods, $\mathcal{F}_h$ and $\mathcal{B}_h$ for customers h in inserted paths need to be recomputed, because in the new route σ_k the path from the depot to such an h and that from h to the depot are different from those in the current route. What is important in this approach is to execute the followings efficiently for a given σ_k:

1. construction of $\mathcal{F}_{h+1}$ from $\mathcal{F}_h$ (the forward computation),
2. construction of $\mathcal{B}_h$ from $\mathcal{B}_{h+1}$ (the backward computation), and
3. computation of $p_t(\sigma_k)$ from $\mathcal{F}_h$ and $\mathcal{B}_{h+1}$.

It is not hard to show that each neighborhood solution can be evaluated in $O(T)$ time, if the above operations can be done in $O(T)$ time for any h ($0 \leq h \leq n_k$). However, to accomplish this, the neighborhood need to be searched in an appropriate search order. The detailed description of such a search order is explained in Ibaraki et al. [10]. This strategy has also been used to devise efficient algorithms for a variety of VRPTW variants [11,12,13]. Below we show that the forward and backward computation can be done in $O(1)$ time and the computation of $p_t(\sigma_k)$ from $\mathcal{F}_h$ and $\mathcal{B}_{h+1}$ can also be done in $O(1)$ time. Hence the algorithm can evaluate each neighborhood solution in $O(1)$ time.

Let f_h be the minimum total amount of traveling times to be shortened to satisfy the time window constraints for customers $1, 2, \ldots, h$ when vehicle k visits them along the route. Let s_h^{f} be the start time of service at h that attains f_h together with $s_1^{\mathrm{f}}, \ldots, s_{h-1}^{\mathrm{f}}$, and let $\mathcal{F}_h = (f_h, s_h^{\mathrm{f}})$. Then the forward computation can be done by:

$$s_{h+1}^{\mathrm{f}} = \min\left\{l_{h+1}, \max\{s_h^{\mathrm{f}} + t_{h,h+1}, e_{h+1}\}\right\} \tag{10}$$

$$f_{h+1} = f_h + \max\{s_h^{\mathrm{f}} + t_{h,h+1}, e_{h+1}\} - s_{h+1}^{\mathrm{f}}. \tag{11}$$

In (10), if $l_{h+1} < \max\{s_h^{\mathrm{f}} + t_{h,h+1}, e_{h+1}\}$ holds, the traveling time is shortened to satisfy the time window constraint and this amount is added to f_{h+1} in (11).

The backward computation can be done similarly. Let b_h be the minimum total amount of traveling times to be shortened to satisfy the time window constraints for customers $h, h+1, \ldots, n_k + 1$ when vehicle k starts from h and returns to the depot along the route. Let s_h^b be the start time of service at h that attains b_h together with $s_{h+1}^b, \ldots, s_{n_k+1}^b$, and let $\mathcal{B}_h = (b_h, s_h^b)$. Then the backward computation can be done by:

$$s_h^b = \max \left\{ \min\{l_h, s_{h+1}^b - t_{h,h+1}\}, e_h \right\} \tag{12}$$

$$b_h = b_{h+1} + s_h^b - \min\{l_h, s_{h+1}^b - t_{h,h+1}\}. \tag{13}$$

We can compute $p_t(\sigma_k)$ from $\mathcal{F}_h = (f_h, s_h^f)$ and $\mathcal{B}_{h+1} = (b_{h+1}, s_{h+1}^b)$ by

$$s_{h+1}^f = \min \left\{ l_{h+1}, \max\{s_h^f + t_{h,h+1}, e_{h+1}\} \right\} \tag{14}$$

$$p_t(\sigma_k) = f_h + b_{h+1} + \max\{0, s_{h+1}^f - s_{h+1}^b\}. \tag{15}$$

5   Adaptive Mechanism to Control Parameters

In this section, we describe an adaptive mechanism to control the parameters α, β and γ_i for each customer i. The algorithm (in which the local search (LS) is executed many times) updates these parameters whenever the LS outputs a locally optimal solution. We set their initial values to $\alpha = 1000$, $\beta = 1000$ and $\gamma_i = 100$ in the experiments.

Let $p_c^{\text{sum}}(\sigma) = \sum_{k \in M} p_c(\sigma_k)$ and $p_t^{\text{sum}}(\sigma) = \sum_{k \in M} p_t(\sigma_k)$, and let $p_c^{\min}$ (resp., $p_t^{\min}$) be the minimum $p_c^{\text{sum}}(\sigma)$ (resp., $p_t^{\text{sum}}(\sigma)$) of the solutions in the current reference set R of good solutions, where rules for maintaining R are described in Section 6. Let P_c (resp., P_t) be the number of moves, during the last call to the LS, to a solution σ whose $p_c^{\text{sum}}(\sigma)$ (resp., $p_t^{\text{sum}}(\sigma)$) is less than $p_c^{\min}$ (resp., $p_t^{\min}$) or equals to 0. Let N_{total} be the total number of moves during the last call to LS, and let $N_c = N_{\text{total}} - P_c$ and $N_t = N_{\text{total}} - P_t$. We use parameters δ_{inc}, δ_{dec}, $\delta_{\text{inc}}^{\text{cust}}$ and $\delta_{\text{dec}}^{\text{cust}}$, and in the experiments, we set $\delta_{\text{inc}} = 0.05$, $\delta_{\text{dec}} = 0.1$, $\delta_{\text{inc}}^{\text{cust}} = 0.1$ and $\delta_{\text{dec}}^{\text{cust}} = 0.01$. If the LS found, during last call, a solution σ that satisfied $p_c^{\text{sum}}(\sigma) < p_c^{\min}$ and $p_t^{\text{sum}}(\sigma) < p_t^{\min}$, the parameters α and β are decreased by

$$\alpha := \left(1 - \frac{P_c}{\max\{P_c, P_t\}} \delta_{\text{dec}}\right)\alpha, \qquad \beta := \left(1 - \frac{P_t}{\max\{P_c, P_t\}} \delta_{\text{dec}}\right)\beta.$$

Even if the LS did not find such a solution, if $N_c = 0$ (resp, $N_t = 0$) holds, α (resp., β) is decreased by the same equation. Otherwise they are increased by

$$\alpha := \left(1 + \frac{N_c}{\max\{N_c, N_t\}} \delta_{\text{inc}}\right)\alpha, \qquad \beta := \left(1 + \frac{N_t}{\max\{N_c, N_t\}} \delta_{\text{inc}}\right)\beta.$$

In the locally optimal solution, if a route violates the capacity or time window constraint, γ_i of each customer i in the route is increased by $\gamma_i := (1 + \delta_{\text{inc}}^{\text{cust}})\gamma_i$. For each customer i who is in a feasible route, γ_i is decreased by $\gamma_i := (1 - \delta_{\text{dec}}^{\text{cust}})\gamma_i$.

6   Path Relinking Approach

Let R be a reference set of solutions. Initially R is prepared by applying the LS to randomly generated solutions. Then it is updated by reflecting outcome of the LS. During the search, the algorithm always keeps the size of R to ρ (a parameter). We set $\rho = 10$ in the experiments. Good solutions with respect to p are kept in R, excluding at most two solutions: One which achieves $p_c^{\min}$ and the other which achieves $p_t^{\min}$. After a feasible solution is found (i.e., $p_c^{\min} = 0$ and $p_t^{\min} = 0$), the best feasible solution is always stored as a member of R. Other solutions in R are maintained as follows. Whenever the LS stops, the locally optimal solution σ_{lopt} is exchanged with the worst (with respect to p) solution σ_{worst} in R (excluding the above solutions), provided that σ_{lopt} is not worse than σ_{worst} and is different from all solutions in R.

A path relinking operation is applied to two solutions σ_A (initiating solution) and σ_B (guiding solution) randomly chosen from R, where a random perturbation is applied to σ_B with probability $1/2$ before applying the path relinking (for the purpose of keeping the diversity of the search), and the resulting solution is redefined to be σ_B. We use a cyclic operation, which exchanges partial paths between different routes cyclically, as a random perturbation. Note that a cyclic operation with more than two routes is different from any neighborhood operation we use for the LS, and hence the local search does not get the solution back by one move. In the path relinking operation, we focus on route edges which are used in vehicle routes of a solution. Let $dist(\sigma, \sigma')$ be the number of different route edges between two solutions σ and σ'. It is not difficult to see that the distance $dist(\sigma, \sigma')$ between two different solutions σ and σ' can be shortened by at least one by applying an appropriate 2-opt* operation or intra-route operation to σ. The path relinking operation generates a sequence of solutions $(\sigma_A = \sigma_1, \sigma_2, \ldots, \sigma_q, \ldots, \sigma_B)$ by repeating the following procedure starting from $q = 1$ until $\sigma_q = \sigma_B$ holds: Let σ_{q+1} be the best solution with respect to p among those that satisfy $dist(\sigma_{q+1}, \sigma_B) < dist(\sigma_q, \sigma_B)$ and obtainable from σ_q by a 2-opt* or intra-route operation, and then let $q := q + 1$.

We call a solution σ_q locally minimal in the sequence if $p(\sigma_q) < \min\{p(\sigma_{q-1}), p(\sigma_{q+1})\}$ holds. Let S be the best π (a parameter) solutions among the locally minimal solutions in the sequence. Every solution in S is used as an initial solution of the LS. We set $\pi = 20$ in the experiments. The next path relinking is initiated whenever all solutions in S are exhausted as the starting solutions for the local search.

The proposed algorithm is summarized in Algorithm 1. The algorithm stops when it reaches a given time limit. In Algorithm 1, a call to the local search starting from a solution σ is denoted by $LS(\sigma)$, whose output is the obtained locally optimal solution.

7   Computational Experiments

We conducted computational experiments to evaluate the proposed algorithm. The algorithm was coded in C and run on a PC (Xeon, 2.8 GHz, 1 GB memory).

Algorithm 1. Path Relinking Approach

1: Construct the neighbor lists.
2: Let R be ρ randomly generated solutions. For each $\sigma \in R$, let $\sigma_{\mathrm{lopt}} := LS(\sigma)$ and then let $R := (R \setminus \{\sigma\}) \cup \sigma_{\mathrm{lopt}}$.
3: Let $S := \emptyset$.
4: **while** the stopping criterion is not satisfied **do**
5:     **while** $S = \emptyset$ **do**
6:         Randomly choose two solutions σ_A and σ_B from R ($\sigma_A \neq \sigma_B$).
7:         With probability $1/2$, apply a cyclic operation to σ_B.
8:         Apply the path relinking operation to σ_A and σ_B, and then let S be the set of best π locally minimal solutions in the generated sequence.
9:     **end while**
10:    Randomly choose $\sigma \in S$, and let $S := S \setminus \{\sigma\}$ and $\sigma_{\mathrm{lopt}} := LS(\sigma)$.
11:    Update the penalty weights.
12:    Choose the worst $\sigma_{\mathrm{worst}} \in R$ among those that satisfy (1) σ_{worst} is not the unique feasible solution in R, (2) $\exists \sigma_c \in R \setminus \{\sigma_{\mathrm{worst}}\}$, $p_c^{\mathrm{sum}}(\sigma_c) \leq p_c^{\mathrm{sum}}(\sigma_{\mathrm{worst}})$ and (3) $\exists \sigma_t \in R \setminus \{\sigma_{\mathrm{worst}}\}$, $p_t^{\mathrm{sum}}(\sigma_t) \leq p_t^{\mathrm{sum}}(\sigma_{\mathrm{worst}})$.
13:    **if** $p(\sigma_{\mathrm{lopt}}) \leq p(\sigma_{\mathrm{worst}})$ and σ_{lopt} is different from all solutions in R **then**
14:        $R := (R \setminus \{\sigma_{\mathrm{worst}}\}) \cup \sigma_{\mathrm{lopt}}$
15:    **end if**
16: **end while**
17: Output the incumbent solution and stop.

The parameter setting of the algorithm was determined by preliminary experiments on several instances, in which we observed that the performance of the algorithm was not sensitive to parameter values.

We used Solomon's benchmark instances [14] and Gehring and Homberger's benchmark instances [15]. There are 356 instances in total, and all of them have been widely used in the literature. In Solomon's instances, the number of customers is 100, and in Gehring and Homberger's instances, which are the extended instances from Solomon's instances, the number of customers is from 200 to 1000. The customers are distributed in the plane and the distances between customers are measured by Euclidean distances. For these instances, the number of vehicles m is also a decision variable, and the objective is to find a solution with the minimum vehicle number and the total traveling distance in the lexicographical order (i.e., a solution is better than another (1) if its vehicle number is smaller or (2) if the vehicle numbers are the same but the distance is smaller).

As our algorithm deals with the problem with a fixed number of vehicles, we first set the number of vehicles in each instance to the known smallest number to the best of our knowledge, and repeat the followings. If the algorithm found a feasible solution and the number of vehicles is larger than a lower bound $\lceil \sum_{i \in V} a_i / u \rceil$, we ran the algorithm again after decrementing the number of vehicles by one. On the other hand, if the algorithm was not able to find a feasible solution, we ran the algorithm again after incrementing the number of vehicles by one. Among the 356 instances, the algorithm found a feasible solution in the first run for every instance except for six instances. Among the remaining

Table 1. Comparison of our results with the existing methods for benchmark instances

References		100	200	400	600	800	1000
Hashimoto et al.	CNV	405	**692**	**1381**	2069	2746	3430
(in press) [12]	CTD	57,282	171,223	406,646	847,470	1,444,513	2,204,728
	P4 2.8GHz	17	33	67	100	133	167
Ibaraki et al.	CNV	407	694	1387	2070	2750	3431
(in press) [11]	CTD	57,545	170,484	398,938	825,172	1,421,225	2,155,374
	P4 2.8GHz	17	33	67	100	133	167
Bräysy et al.	CNV	–	695	1391	2084	2776	3465
(2004) [16]	CTD	–	172,406	399,132	820,372	1,384,306	2,133,376
	AMD 700MHz	–	3×2	3×8	3×16	3×26	3×40
Prescott-Gagnon et al.	CNV	405	694	1385	2071	2745	3432
(2007) [17]	CTD	**57,240**	168,556	389,011	800,797	1,391,344	2,096,823
	Opt 2.3GHz	5×30	5×53	5×89	5×105	5×129	5×162
Pisinger and Ropke	CNV	405	694	1385	2071	2758	3438
(2007) [18]	CTD	57,322	169,042	393,210	807,470	1,358,291	2,110,925
	P4 3GHz	10×2	10×8	5×16	5×18	5×23	5×27
Mester and Bräysy	CNV	–	694	1389	2082	2765	3446
(2005) [19]	CTD	–	168,573	390,386	796,172	1,361,586	2,078,110
	P 2GHz	–	8	17	40	145	600
Le Bouthillier et al.	CNV	405	694	1389	2086	2761	3442
(2005) [20]	CTD	57,360	169,959	396,612	809,494	1,443,400	2,133,645
	5×P 850MHz	12	10	20	30	40	50
Le Bouthillier and Crainic	CNV	407	694	1390	2088	2766	3451
(2005) [21]	CTD	57,412	173,061	408,281	836,261	1,475,281	2,225,366
	5×P 850MHz	12	10	20	30	40	50
Gehring and Homberger	CNV	406	696	1392	2079	2760	3446
(2001) [22]	CTD	57,641	179,328	428,489	890,121	1,535,849	2,290,367
	4×P 400MHz	5×14	3×2	3×7	3×13	3×23	3×30
Homberger and Gehring	CNV	408	699	1397	2088	2773	3459
(2005) [15]	CTD	57,422	180,602	431,089	890,293	1,516,648	2,288,819
	P 400MHz	5×17	3×2	3×5	3×10	3×18	3×31
Ours	CNV	405	694	1383	**2068**	**2737**	**3420**
	CTD	57,484	169,070	392,507	800,982	1,367,971	2,085,125
	Xeon 2.8GHz	17	33	67	100	133	167

six instances, it was able to find feasible solutions with one more vehicle for five instances and with two more vehicles for the one. The time limit for each run of the algorithm for 100, 200, 400, 600, 800 and 1000-customer instances are 1000, 2000, 4000, 6000, 8000 and 10000 seconds, respectively. This setting of the time limit is the same with [11, 12].

Table 1 shows the comparison of our results with those obtained by existing methods. A number in the first row shows the number of customers. Our results are denoted by "Ours." For each method, we provide the cumulative number of vehicles (CNV), the cumulative total distance (CTD), the CPU, and the average computation time in minutes for solving an instance. In the notation of the CPU, "P," "P4," and "Opt" mean Pentium, Pentium 4 and Opteron, respectively. Marks "×" in the second column mean the number of CPUs (e.g., "4×P 400MHz" means four CPUs of Pentium 400MHz), and those in other columns mean the number of runs (e.g., "5×30" means five runs each with 30 minutes of computation time). A number in bold in rows CNV indicates that the value is the best among all the algorithms in the table and there is no tie. When there are ties for the best CNV, the corresponding distance value that is the smallest among those ties is indicated by boldface.

From Table 1, the CNV obtained by our algorithm is much smaller than those of the other methods for large instances with 600 customers or more, and the computation time spent by our algorithm seems to be reasonable; e.g., for instances with $n = 1000$, the computation times spent by recent algorithms by Hashimoto et al. [12], Ibaraki et al. [11], Prescott-Gagnon et al. [17], Pisinger and Ropke [18], and Mester and Bräysy [19] are similar to or sometimes larger than ours even if the difference of CPUs are taken into consideration. Moreover, our algorithm updated 41 best known solutions among the 356 instances.[1] This indicates that our algorithm is highly efficient.

8   Conclusion

We proposed a path relinking approach for the vehicle routing problem with time windows with an adaptive mechanism to control parameters. The generated solutions in the path relinking are improved by a local search. In the local search, each neighborhood solution is evaluated in $O(1)$ time and the neighborhood search is pruned heuristically by the neighbor list. During the search, infeasible solutions are allowed to be visited while the amount of violation is penalized. We also proposed an adaptive mechanism to control the penalty weights. The computational results on representative benchmark instances indicate that the proposed algorithm is highly efficient, and furthermore, the algorithm updated 41 best known solutions among 356 instances.

References

1. Bräysy, O., Dullaert, W., Gendreau, M.: Evolutionary algorithms for the vehicle routing problem with time windows. Journal of Heuristics 10, 587–611 (2004)
2. Glover, F., Laguna, M., Marti, R.: Scatter search and path relinking: Advances and applications. In: Glover, F., Kochenberger, G.A. (eds.) Handbook of Metaheuristics, pp. 1–35. Kluwer Academic Publishers, Dordrecht (2003)
3. Johnson, D.S., McGeoch, L.A.: The traveling salesman problem: A case study in local optimization. In: Aarts, E.H.L., Lenstra, J.K. (eds.) Local Search in Combinatorial Optimization, pp. 215–310. John Wiley and Sons, Chichester (1997)
4. Park, N., Okano, H., Imai, H.: A path-exchange-type local search algorithm for vehicle routing and its efficient search strategy. Journal of the Operations Research Society of Japan 43(1), 197–208 (2000)
5. Potvin, J.Y., Kervahut, T., Garcia, B.L., Rousseau, J.M.: The vehicle routing problem with time windows part I: tabu search. INFORMS Journal on Computing 8(2), 158–164 (1996)
6. Lin, S.: Computer solutions of the traveling salesman problem. Bell System Technical Journal 44, 2245–2269 (1965)

---

[1]   We compared our solutions with those reported in the papers [9,11,12,17,21,23] and on the SINTEF website (www.top.sintef.no/vrp/benchmarks.html), as of November 11, 2007. (Note that the SINTEF website includes the results of [18, 19, 22].) Our results of individual instances are available from the authors' web site (http://www.al.cm.is.nagoya-u.ac.jp/~yagiura/papers/vrptwEvoCOP08_abst.html).

7. Taillard, É., Badeau, P., Gendreau, M., Guertin, F., Potvin, J.Y.: A tabu search heuristic for the vehicle routing problem with soft time windows. Transportation Science 31(2), 170–186 (1997)
8. Reiter, S., Sherman, G.: Discrete optimizing. Journal of the Society for Industrial and Applied Mathematics 13(3), 864–889 (1965)
9. Nagata, Y.: Effective memetic algorithm for the vehicle routing problem with time windows: Edge assembly crossover for the VRPTW. In: Proceedings of the Seventh Metaheuristics International Conference (MIC 2007) (2007)
10. Ibaraki, T., Imahori, S., Kubo, M., Masuda, T., Uno, T., Yagiura, M.: Effective local search algorithms for routing and scheduling problems with general time-window constraints. Transportation Science 39(2), 206–232 (2005)
11. Ibaraki, T., Imahori, S., Nonobe, K., Sobue, K., Uno, T., Yagiura, M.: An iterated local search algorithm for the vehicle routing problem with convex time penalty functions. Discrete Applied Mathematics (in press)
12. Hashimoto, H., Yagiura, M., Ibaraki, T.: An iterated local search algorithm for the time-dependent vehicle routing problem with time windows. In: Discrete Optimization (in press)
13. Hashimoto, H., Ibaraki, T., Imahori, S., Yagiura, M.: The vehicle routing problem with flexible time windows and traveling times. Discrete Applied Mathematics 154, 2271–2290 (2006)
14. Solomon, M.M.: Algorithms for the vehicle routing and scheduling problems with time window constraints. Operations Research 35(2), 254–265 (1987)
15. Homberger, J., Gehring, H.: A two-phase hybrid metaheuristic for the vehicle routing problem with time windows. European Journal of Operational Research 162, 220–238 (2005)
16. Bräysy, O., Hasle, G., Dullaert, W.: A multi-start local search algorithm for the vehicle routing problem with time windows. European Journal of Operational Research 159, 586–605 (2004)
17. Prescott-Gagnon, E., Desaulniers, G., Rousseau, L.M.: A branch-and-price-based large neighborhood search algorithm for the vehicle routing problem with time windows. Technical report, GERAD (Group for Research in Decision Analysis) (2007)
18. Pisinger, D., Ropke, S.: A general heuristic for vehicle routing problems. Computers and Operations Research 34, 2403–2435 (2007)
19. Mester, D., Bräysy, O.: Active guided evolution strategies for large-scale vehicle routing problems with time windows. Computers and Operations Research 32, 1593–1614 (2005)
20. Le Bouthillier, A., Crainic, T.G., Kropf, P.: A guided cooperative search for the vehicle routing problem with time windows. IEEE Intelligent Systems 20(4), 36–42 (2005)
21. Le Bouthillier, A., Crainic, T.G.: A cooperative parallel meta-heuristic for the vehicle routing problem with time windows. Computers and Operations Research 32, 1685–1708 (2005)
22. Gehring, H., Homberger, J.: A parallel two-phase metaheuristic for routing problems with time-windows. Asia-Pacific Journal of Operational Research 18, 35–47 (2001)
23. Bent, R., Van Hentenryck, P.: Randomized adaptive spatial decoupling for large-scale vehicle routing with time windows. In: AAAI, pp. 173–178. AAAI Press, Menlo Park (2007)

Reactive Stochastic Local Search Algorithms for the Genomic Median Problem

Renaud Lenne[1,2], Christine Solnon[1], Thomas Stützle[2], Eric Tannier[3], and Mauro Birattari[2]

[1] LIRIS, UMR CNRS 5205, Université de Lyon 1, Lyon, France
{renaud.lenne,christine.solnon}@liris.cnrs.fr
[2] IRIDIA-CoDE, Université Libre de Bruxelles, Brussels, Belgium
{stuetzle,mbiro}@ulb.ac.be
[3] INRIA Rhône-Alpes, LBBE, UMR CNRS 5558, Université de Lyon 1, France
eric.tannier@inria.fr

Abstract. The genomic median problem is an optimization problem inspired by a biological issue: it aims to find the chromosome organization of the common ancestor to multiple living species. It is formulated as the search for a genome that minimizes a rearrangement distance measure among given genomes. Several attempts have been reported for solving this NP-hard problem. These range from simple heuristic methods to a stochastic local search algorithm inspired by `WalkSAT`, a well-known local search algorithm for the satisfiability problem in propositional logic.

The main objective of this research is to develop improved algorithmic techniques for tackling the genomic median problem and to provide new state-of-the-art solutions. In particular, we have developed an algorithm that is based on tabu search and iterated local search and that shows high performance. To alleviate the dependence of the algorithm performance on a single fixed parameter setting, we have included a reactive scheme that automatically adapts the tabu list length of the tabu search part and the perturbation strength of the iterated local search part. In fact, computational results show that we have developed a new very high-performing stochastic local search algorithm for the genomic median problem and we also have found a new best solution for a real-world case.

1 Introduction

Genome rearrangements are large-scale evolutionary events that modify the organization of the genomes. Chromosomes may be fissioned, fusioned, large segments can be translocated or inverted. Given the genome of living species, the reconstruction of rearrangement scenarios has been the subject of a huge amount of literature these last years. It aims to understand what rearrangement events took place and when they occurred in evolution, and it is a promising way for phylogenetic inference [1,2].

The Genomic Median Problem (GMP) is a crucial step in genome rearrangement problems. While for only two genomes, a scenario with a minimum number

J. van Hemert and C. Cotta (Eds.): EvoCOP 2008, LNCS 4972, pp. 266–276, 2008.

of rearrangements can be reconstructed by the way of polynomial methods for many variants of rearrangements (see for example [3,4]), the problem is NP-hard for already three genomes [5]: it consists in searching for a fourth genome that minimizes the distance to three given genomes, in terms of the number of rearrangements.

The classical phylogenetics methods to construct an ancestral genome are based on pieces of sequences, thus making the reconstruction of the organisation of the genome impossible. One of the objectives of the GMP is to find this organisation, making better construction of ancestral genomes. It could also be used as a hint for phylogeny, for example by using the founded median as an entry-point for a phylogenetic algorithm (like in [2]).

There have been various attempts at algorithmically solving the problem. Exact solutions exist for the special case where there is only one chromosome and a rather small instance [5,6]. Incomplete approaches, ranging from rather simple heuristics [6,7] to more complex local search algorithms [8,9] have been proposed. These approaches produce solutions that are often of good quality but that are not necessarily optimal and for larger instances there may be significant gaps to optimal solutions. In addition, compared to the currently available local search techniques, the approaches are rather simple and therefore one can conjecture that there is certainly room for improving their performance.

Motivated by these observations, we propose a new stochastic local search algorithm for the GMP, based on tabu search [10] and iterated local search [11]. A first goal is to improve upon the performances of current state-of-the-art algorithms in terms of run-times required to reach specific bounds on solution quality and to find better quality solutions, thus providing new state-of-the-art solutions that may be of biological relevance. A second goal is to study the influence of the parameters on the solution process with respect to different instances; based on preliminary experiments, we have added a reactive scheme to automatically adapt crucial parameters, which for our algorithm are the tabu list length and the perturbation size, during search.

The paper is organized as follows. Section 2 formally describes the GMP and existing approaches to solve this problem. Section 3 introduces our new stochastic local search approach, based on a combination of tabu search and iterated local search. Section 4 studies the influence of the parameters on the solution process, showing that the best parameter setting varies from one instance to another. Section 5 introduces a reactive scheme for automatically adapting parameters during search. Section 6 experimentally compares static and reactive versions of our algorithm, and also compares our algorithm with state-of-the-art approaches.

2   Problem Definition and Existing Work

A genome will be defined as a graph, in which some edges are directed (the orthologous markers), and some not (the links between the genes).

Representation of genomes by graphs. A genome is seen as a group of chromosomes. A chromosome is seen as a list of oriented genes or markers. A genome composed of k chromosomes defined on a set of n markers is represented by a graph $G = (V, E)$ such that

- V associates two vertices i^- and i^+ with every marker $i \in [1; n]$ and two anonymous vertices with every chromosome $j \in [1; k]$;
- E is composed of two parts E_m and E_c such that
  - for every marker i, E_m contains a *directed* edge (i^-, i^+)
  - for every chromosome c_j containing $|c_j|$ markers, E_c contains $|c_j| + 1$ *non directed* edges

  such that each chromosome corresponds to a path in G which endpoints are anonymous vertices and which alternates directed edges —corresponding to oriented markers— and non directed edges —linking markers.

Let us consider for example a genome composed of 6 markers and the 3 following chromosomes: $c_1 = <\overrightarrow{1}, \overleftarrow{5}>$, $c_2 = <\overrightarrow{2}, \overrightarrow{4}>$, and $c_3 = <\overrightarrow{3}, \overleftarrow{6}>$. This genome is represented by the following graph.

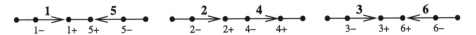

Note that chromosomes have no directions so that paths are not directed. For example, the first chromosome c_1 is equivalent to $<\overrightarrow{5}, \overleftarrow{1}>$.

Genomic distance. A *rearrangement* in a genome is an operation that deletes two non directed edges (a, b) and (c, d), and replace them by (a, c) and (b, d). It is the "double-cut-and-join" operation described in [4,12], or the "2-break rearrangement" of [13]. It simulates chromosome fissions, fusions, translocations, inversions and transpositions.

Let us consider the genome of our previous example. An example of rearrangement consists in replacing edges $(2^+, 4^-)$ and $(3^+, 6^+)$ by edges $(2^+, 3^+)$ and $(4^-, 6^+)$, thus changing chromosomes c_2 to $<\overrightarrow{2}, \overleftarrow{3}>$, and c_3 to $<\overleftarrow{4}, \overleftarrow{6}>$, as displayed in the following graph.

A rearrangement transforms one genome into another. Given two genomes G_1 and G_2 defined on the same set of markers, there is always a way to transform G_1 into G_2 by a sequence of rearrangements. The minimum number of rearrangements can be computed in linear time [4,12]. This number is called the *genomic distance* between G_1 and G_2, and is noted $d(G_1, G_2)$.

The genomic distance $d(G_1, G_2)$ is computed with respect to a non directed graph $G_{1,2} = (V_{1,2}, E_{1,2})$ which is obtained by merging the two graphs $G_1 = (V_1, E_1)$ and $G_2 = (V_2, E_2)$ as follows:

- $V_{1,2} = V_1 \cup V_2$. Note that vertices associated with markers are shared by both G_1 and G_2 as they are defined on a same set of markers. However, anonymous vertices corresponding to chromosome endpoints are different in G_1 and G_2.
- $E_{1,2}$ is the union of all *non directed* edges of G_1 and G_2.

The genomic distance is defined, as in [4,12], by $d(G_1, G_2) = n + p - c(G_{1,2})) + \frac{\#(G_{1,2})}{2}$ where n is the number of markers, p is the number of chromosomes in both genomes, $c(G_{1,2})$ is the number of different paths and cycles in $G_{1,2}$, and $\#(G_{1,2})$ the number of paths which endpoints are anonymous vertices that come from the same genome.

The Genomic Median Problem (GMP). Given three genomes G_1, G_2, G_3 on the same set of markers, the goal of the GMP is to determine a genome G_M that minimizes the sum $d(G_1, G_M) + d(G_2, G_M) + d(G_3, G_M)$.

The reversal median problem, which uses the number of reversal as a distance, was proven to be NP-hard [5] for the special case when the genomes are composed of one unique chromosome. The problem which we handle here uses a different distance, as defined in [4,12] but most of the time the solution coincide. Also, the proof of NP-Completeness is valid for our definition of the distance (which results in a problem called Cycle Median Problem in [5]), since it proves that, for unichromosomal genomes, the problem of minimizing our distance formula is NP-Complete.

Various methods for solving the *GMP* have been proposed. These comprise exact solvers or simple heuristics that work on the particular case of the unichromosomal genomes like GRAPPA [5,6] with some heuristic improvements [7]. For the general problem case with multiple chromosomes, there is a rather simplistic search method in MGR-MEDIAN [8], which uses a greedy constructive algorithm. The best performance results so far have been reported for an algorithm called MedRByLS [9]; this is a local search algorithm inspired by WalkSAT [14], a well-known local search algorithm for the satisfiability problem in propositional logic. We based our algorithm on the same neighborhood and the same data structures as used in MedRByLS.

Neighborhood considered in MedRByLS. To find a genome G_M that minimizes the sum of the distances to three given genomes G_1, G_2 and G_3, MedRByLS iterately modifies a genome G_M by performing local moves. Each move corresponds to a rearrangement in G_M, i.e., the exchange of two non directed edges of G_M, and is evaluated with respect to the three graphs $G_{M,1}$, $G_{M,2}$, and $G_{M,3}$.

At each step, the size of the neighborhood is in $\mathcal{O}(n^2)$, where n is the number of markers. As genomes may have several hundreds of markers, the neighborhood is reduced in [9] with respect to the following principle: a rearrangement is considered only if, in one of the three graphs $G_{M,1}$, $G_{M,2}$, and $G_{M,3}$, it breaks a cycle or a path into two cycles or paths such that one of these two cycles or paths is elementary, i.e., a cycle of length 2 or a path of length 3. Therefore, moves can only increase the number of elementary cycles or paths.

Basic principle of the local search algorithm of MedRByLS. `MedRByLS` follows the random walk framework initially introduced in `WalkSAT` for the SAT problem [14]. It starts local search from an initial genome G which may either be provided by the user or randomly chosen within the set $\{G_1, G_2, G_3\}$. Then, it iteratively chooses a move uniformly at random from the neighborhood defined above; the move is applied if it decreases the sum of the distances; otherwise it is accepted with a small probability p.

3   Tabu Search and Iterated Local Search for the GMP

The random walk framework considered by `MedRByLS` is a rather basic one which may have difficulties in escaping from local minima. Hence, we propose to consider more advanced local search approaches.

We have first re-implemented `MedRByLS`, using the same data structures based on graphs and the same neighborhood definition. This allows a direct comparison of our new algorithm to the original `MedRByLS` using a same implementation of the data structures. For this comparison, we verified that our re-implementation matches the performances of the original version.

As a next step, we enhanced the local search by a simple tabu search scheme. For the search diversification of the resulting tabu search algorithm, we integrated it into the iterated local search framework by adding appropriate perturbations and acceptance criteria. This resulted in an algorithm that we called `MedITaS` (for Median solver by Iterated Tabu Search). More precisely, it consists of the two following main algorithmic components.

3.1   Tabu Search (TS) Algorithm

At first, a simple tabu search (TS) algorithm was implemented. This algorithm forbids the reversal of the last t local search moves (where t is the tabu tenure), that is, the last changed nodes. In order to do this, we use an array of n integers representing the n nodes and, for each node, we put in this array the iteration when it was last changed. This simplifies the task of guessing if a move is tabu or not (that is, if it has been changed in the last cycles). Different from many other simple tabu search algorithms, ours is based on a *first-improvement* pivoting rule because the neighbourhood is very large and, thus, a full scan of it would be too time-consuming. In our experiments we have used a default initial tabu list length of 50.

3.2   Iterated Local Search (ILS) Algorithm

After some preliminary tests, TS seemed to stagnate frequently in poor quality solutions. To overcome this problem, we integrated TS into an iterated local search (ILS) algorithm. ILS uses solution perturbations when the search is deemed to be stuck in plateau-moves or in a basin of attraction to generate new starting solutions for the local search. To know if the search is stuck, we consider the resampling ratio [15], which is computed in constant time thanks to a double hash-table, and which corresponds to the percentage of solutions that are revisited with respect to the number of computed solutions. If this ratio is too high, it means that the search keeps visiting the same few solutions and has to be perturbated to escape the basin of attraction. We also look at the solution value. If this value keeps being constant for a long time, we consider being stuck in plateau-moves. In this case a perturbation could lead to avoid looking for non-interesting solutions.

The perturbation uses a rearrangement of k edges instead of 2. This means that k edges are deleted and replaced by k other edges sharing the same extremities. Finally, an acceptance criterion decides whether either the solution before the perturbation or the one after is kept for the next iteration of the ILS algorithm; in the latter case, the tabu list is emptied. The implemented acceptance criterion accepts a new solution if it is better than the previous one; otherwise, the previous solution has a user-defined probability of beeing kept (in our test, we used the default probability of 0.2).

4   Tuning of MedITaS Parameters

In order to test the influence of parameters on the solution process, we conducted a simple experiment. At first, we randomly generated 20 instances equally split on two different levels of hardness (with respect to the definition of the phase transition by [9]) but with the same size (500 markers). The first set of 10 instances, labeled as *easy*, has a ratio of *number of markers* to *number of rearrangements* of 0.5. The second set of 10 instances, labeled as *hard*, has a value of 1.0 for the same ratio. Each of these sets has been used to off-line tune the algorithm through F-Race [16,17], a tool for the automatic tuning of algorithm parameters.

For the tabu tenure of TS, the best setting resulting from these experiments was of 76 for *easy* instances whereas it was of 86 for *hard* instances. For the strength k of the perturbation of ILS, the best setting was of 17 for *easy* instances whereas it was of 2 for *hard* instances. Hence, the best settings strongly differ between the two benchmark sets, especially for the perturbation.

5   Reactive Search

Experiments reported in the previous section show that both ILS and TS are sensitive to parameter settings and that the best parameter settings are very different from one instance class to another. In addition, in further experiments we have noted that within one instance class, the best parameter settings depend further on the individual instance. Hence, we decided to extend the basic version of MedITaS using a reactive version of TS (which adaptatively changes the tabu tenure) and ILS (working on the perturbation strength). This reactive scheme is inspired from reactive search [18]. It uses the resampling ratio, which is explained in Section 3.2, to determine if one of the parameters needs to be changed.

The reaction mechanism works as follows. If there is too much resampling (by default, after 3 recalculated solutions), the tabu list length is increased (by default, the size is increased by 10). At the opposite, if there is no resampling for a long time (the default value is 500 moves) the tabu list length is shorten by a parameterized number (by default, the tabu list is shortened by 1). A similar mechanism is used for tuning the ILS part. If the solution that is returned after the perturbation and the subsequent local search is an already visited solution, the reactive algorithm increases the strength k of the perturbation, because it seems that the perturbation did not succeed in escaping from the basin of attraction. Here again, the size of the increase and the decrease can be parameterized; as default, we use the value 1. Finally, we should remark that the settings of the parameters that direct the reaction mechanism, at least in principle, should also be tuned. However, here we essentially stick to the default values used, since, as also argued in [18], the parameters steering the reaction mechanism should be reasonably robust.

6   Results

In order to test the efficiency of our algorithms, we ran multiple comparisons. All runs were made on the same machine having a Dual-Core AMD Opteron2216 HE (2 processors at 2.4GHz) and 4GB of RAM; only one core is used for each execution since our algorithm is implemented as a fully sequential one.

6.1   Comparison between Off-Line Tuned and Reactive Algorithms

At first, we generated randomly 20 instances with a ratio of *number of markers* to *number of rearrangements* of 1.0 (which seems, according to the results from [9], to be in the phase transition) of 500 markers. This set has been split in two: 10 instances have been used to off-line tune the algorithm through F-Race [16,17], as before; 10 other instances have been used as a test set. The algorithm to tune is the non-reactive version of MedITaS. We have then compared the results of the off-line tuned version to the reactive version of our MedITaS algorithm starting either with the default initial parameter values or with the parameter values that have been determined by the automatic tuning. For the comparison, we have run those three algorithms for 20 independent trials on each of the 10 test instances and 60 seconds per trial.

Figure 1 plots the cumulative distribution of the frequency of finding a bound on the median to be reached on the 10 test instances. The plot shows that, when comparing the fine-tuned version to the pure reactive version with initial default parameter values, the former gets its first high-quality solutions quicker than the latter. However, after 40 seconds or so, the reactive version reaches higher empirical frequencies. Also, better trade-offs are obtained by initializing the reactive algorithm with the fine-tuned parameter settings: doing so gives good results as quickly as the non-reactive version, and, for higher computation times, it performs similar to the default reactive algorithm.

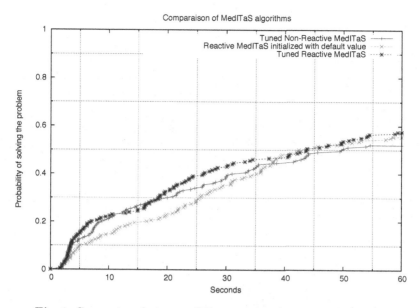

Fig. 1. Comparison between off-line tuned and reactive algorithms

	after 15 seconds		after 30 seconds		after 60 seconds	
	NR tuned	R default	NR tuned	R default	NR tuned	R default
R default	0 / 10	-	6 / 10	-	9 / 10	-
R tuned	8 / 10	0 / 10	8 / 10	3 / 10	9 / 10	8 / 10

Fig. 2. Number of significant differences among the ten instances (the number indicates the number x of times over 10 instances, that is x / 10) as resulting by applying pairwise comparisons using the `Wilcoxon rank sum test`. NR and R respectively stand for Non Reactive and Reactive variants of MedITas; R either starts from fine-tuned or default parameter settings whereas NR always uses fine-tuned parameter settings.

To check for the statistical significance of the results, we did pairwise comparisons using the `Wilcoxon rank sum test` where the p-values were adjusted using the `holm` method on each of the 10 instances separately for correcting for the effect of multiple comparisons. We then counted how often the differences between two algorithms are significant (where the p-value is larger than 0.1). We run those comparison at 15, 30 and 60 seconds run-time.

From Figure 2, we can clearly see that the fine-tuned reactive search performs quite always significantly better than the non-reactive version. We can also see that for a short run-time, neither of the reactive version initialized with default value or the fine-tuned non-reactive one is significantly better than the other.

6.2 Comparison to `MedRByLS`

In another experiment, we compared the performance of our reactive `MedITaS` algorithm (starting with default parameters) to `MedRByLS`, which from a solution

Instance	MedITas avg min max (sdv)	MedRByLS avg min max (sdv)	Instance	MedITas avg min max (sdv)	MedRByLS avg min max (sdv)
1	**541.0** 539 **543** (1.0)	543.0 **538** 568 (6.9)	2	**540.4** 539 **542** (1.0)	542.7 539 569 (6.9)
3	**563.7** 561 **566** (1.5)	565.8 561 590 (6.9)	4	**566.9** 565 **568** (0.8)	570.2 566 600 (8.6)
5	**588.9** 587 **591** (1.3)	593.3 587 624 (8.7)	6	**593.9** 591 **597** (1.7)	597.6 592 631 (9.4)
7	**609.8** 609 **613** (1.1)	613.9 **608** 642 (8.2)	8	**610.2** 608 **612** (1.1)	614.4 608 648 (9.3)
9	**635.1** 634 **637** (1.0)	639.1 634 674 (9.7)	10	**637.1** 635 **639** (1.2)	641.2 635 676 (10.1)
11	**658.5** 656 **661** (1.4)	662.8 656 697 (10.1)	12	**653.4** 651 **655** (1.1)	658.8 652 692 (9.6)
13	**669.1** 668 **672** (1.0)	674.7 668 710 (10.6)	14	**669.3** 667 **671** (1.2)	674.1 667 709 (10.7)
15	**675.2** 674 **677** (1.0)	679.8 **673** 716 (10.7)	16	**684.5** 682 **688** (1.6)	690.8 683 724 (10.7)
17	**693.1** 691 **695** (1.1)	698.9 692 737 (11.5)	18	**692.5** 690 **695** (1.4)	698.8 691 735 (11.2)
19	**707.2** 706 **709** (0.9)	713.0 **705** 753 (11.6)	20	**705.7** 703 **709** (1.4)	712.1 703 751 (11.6)
21	**722.3** 720 **725** (1.1)	728.6 720 770 (12.6)	22	**716.2** 714 **718** (1.0)	722.9 715 764 (12.4)

Fig. 3. Comparison between `MedITaS` and `MedRByLS` on randomly generated instances

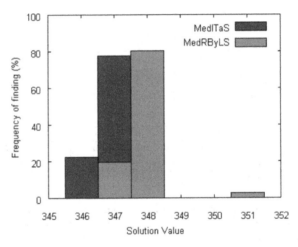

Fig. 4. Comparison between `MedITaS` and `MedRByLS` on a real-world instance (the human-mouse-rat comparison)

quality point of view is currently the state-of-the-art algorithm [9]. For this comparison, we used 22 randomly generated instances of different difficulties (with respect to the definition of the phase transition by [9]) but with the same size (500 markers). The set has 11 levels of hardness and 2 instances per level. On this set we run our `MedITaS` algorithm and our implementation of the basic local search algorithm `MedRByLS` from [9] for 20 independent trials on each instance and 40 seconds per trial. The comparison of the best solution qualities reached by both algorithms on each instance is given in Figure 3. From this figure, we can see that `MedITaS` always gives solution qualities that are at least as good as `MedRByLS` (often the differences are also statistically significant) and that the gap between the two algorithms tends to increase as instances become harder. Also, the standard deviation of `MedITaS` is very low and remain constant as the instances become harder, as opposite to `MedRByLS` which give an higher deviation on harder instances.

6.3   Real World Instance

In another experiment we used a real-world instance: the human-mouse-rat comparison, which was also used in [9]. This instance is made of 424 markers and the best median so far had a value of 346. We ran each algorithm 35 times for a computation time limit of 60 seconds. From these runs, we generated the graph in Figure 4, which represents the histogram of the frequency of finding certain solution qualities with the two main algorithms (MedITaS, in its reactive and fine-tuned version, and MedRByLS).

Figure 4 shows that MedITaS finds solutions that are at least as good as those found by MedRByLS and always of a very good quality (of 347 or better), while MedRByLS sometimes fails to find good ones: on some runs it returned a solution of value 351. We should also notice that MedRByLS has a quite low probability (less than 20%) of finding a solution of 347 or better. Finally, it should be also mentioned that in our experiments MedITaS found a new best solution for this instance with an evaluation function value of 345.

7   Discussion

Our implementation of Iterated Tabu Search gave very promising results. First, we have seen that the reactive version of our algorithm can handle relatively well a wide range of different instances without having the need to be off-line tuned. But we have also shown that a reactive search starting with fine-tuned parameters performs slightly better and a lot quicker than starting from a default value.

Then, we have seen that MedITaS always gives at least as good or better results, in the same computation time, than the former best algorithm (MedRByLS). We also found a new best solution for the human-mouse-rat common ancestor.

The developed algorithmic techniques perform significantly better than previously available ones from a solution quality point of view. But from a biological point of view, the distance used here (as the one used in all previous attempts at solving the problem) does not seem to reflect the biological reality of the evolution process, as it is also explained in [19]). Thus, a research on a biologically more relevant distance has to be envisaged. Also, we noted in our experiments that there were a lot of medians with the exactly same value. It could be a good idea to do some comparison between them trying to extract some valuable information on the most probable characteristics of the real ancestor.

Acknowledgements

This article is based on [20], a shorter paper by the same authors. The authors would like to thank to Yannet Interian for her kind help in any questions regarding the implementation of her algorithm. Renaud Lenne is supported by a FRIA Grant. Thomas Stützle acknowledges support from the Belgian FNRS of which he is a Research Associate. Eric Tannier is funded by the Agence Nationale pour la Recherche (ANR), projects REGLIS and GENOMICRO.

References

1. Moret, B., Tang, J., Warnow, T.: Reconstructing phylogenies from gene content and gene-order data. In: Gascuel, O. (ed.) Mathematics of Evolution and Phylogeny, pp. 321–352. Oxford Univ. Press, Oxford
2. Bernt, M., Merkle, D., Middendorf, M.: Using median sets for inferring phylogenetic trees. Bioinformatics 23, e129–e135 (2007)
3. Hannenhalli, S., Pevzner, P.A.: Transforming cabbage into turnip: Polynomial algorithm for sorting signed permutations by reversals. Journal of the ACM 46(1), 1–27 (1999)
4. Yancopoulos, S., Attie, O., Friedberg, R.: Efficient sorting of genomic permutations by translocation, inversion and block interchange. bioinformatics 21(16), 3340–3346 (2005)
5. Caprara, A.: The reversal median problem. INFORMS Journal on Computing 15, 93–113 (2003)
6. Moret, B., Siepel, A., Tang, J., Liu, T.: Inversion medians outperform breakpoint medians in phylogeny reconstruction from gene-order data (2002)
7. Arndt, W., Tang, J.: Improving inversion median computation using commuting reversals and cycle information. In: Tesler, G., Durand, D. (eds.) RECMOB-CG 2007. LNCS (LNBI), vol. 4751, pp. 30–44. Springer, Heidelberg (2007)
8. Bourque, G., Pevzner, P.: Genome-scale evolution: Reconstructing gene orders in the ancestral species. Genome Res. 12(1), 26–36 (2002)
9. Interian, Y., Durrett, R.: Genomic midpoints: Computation and evolutionary implications (Submitted, 2007)
10. Glover, F., Laguna, F.: Tabu Search. Kluwer Academic Publishers, Norwell, MA, USA (1997)
11. Stutzle, T.: Iterated local search for the quadratic assignment problem (1999)
12. Bergeron, A., Mixtacki, J., Stoye, J.: A unifying view of genome rearrangements. In: Bücher, P., Moret, B.M.E. (eds.) WABI 2006. LNCS (LNBI), vol. 4175, pp. 163–173. Springer, Heidelberg (2006)
13. Alekseyev, M.A., Pevzner, P.A.: Multi-break rearrangements and chromosomal evolution. Theoretical Computer Science (to appear, 2007)
14. Selman, B., Kautz, H., Cohen, B.: Noise strategies for improving local search. In: Proceedings of the 12th National Conference on Artificial Intelligence, pp. 337–343. AAAI Press / The MIT Press, Menlo Park, CA, USA (1994)
15. van Hemert, J., Bäck, T.: Measuring the searched space to guide efficiency: The principle and evidence on constraint satisfaction. In: Guervós, J.J.M., Adamidis, P.A., Beyer, H.-G., Fernández-Villacañas, J.-L., Schwefel, H.-P. (eds.) PPSN 2002. LNCS, vol. 2439, pp. 23–32. Springer, Heidelberg (2002)
16. Birattari, M.: race: Racing methods for the selection of the best. R package version 0.1.56 (2005)
17. Birattari, M.: The Problem of Tuning Metaheuristics as Seen from a Machine Learning Perspective. PhD thesis, Université Libre de Bruxelles, Brussels, Belgium (2004)
18. Battiti, R., Protasi, M.: Reactive local search for the maximum clique problem. Algorithmica 29(4), 610–637 (2001)
19. Eriksen, N.: Reversal and transposition medians. Theoretical Computer Science 374(1-3) (2007)
20. Lenne, R., Solnon, C., Sttzle, T., Tannier, E., Birattari, M.: Effective Stochastic Local Search Algorithms for the Genomic Median Problem. In: Doctoral Symposium on Engineering Stochastic Local Search Algorithms (SLS-DS). IRIDIA Technical Report Series, pp. 1–5 (2007)

Solving Graph Coloring Problems Using Learning Automata

Noureddine Bouhmala[1] and Ole-Christoffer Granmo[2]

[1] Vestfold University College, Norway
[2] University of Agder, Grimstad, Norway

Abstract. The graph coloring problem (GCP) is a widely studied combinatorial optimization problem with numerous applications, including time tabling, frequency assignment, and register allocation. The growing need for more efficient algorithms has led to the development of several GCP solvers. In this paper, we introduce the first GCP solver that is based on Learning Automata (LA). We enhance traditional Random Walk with LA-based learning capability, encoding the GCP as a Boolean satisfiability problem (SAT). Extensive experiments demonstrate that the LA significantly improve the performance of RW, thus laying the foundation for novel LA-based solutions to the GCP.

Keywords: SAT, Learning automata, Combinatorial optimization, Graph coloring problem.

1 Introduction

In the family of graph coloring problems (GCP), an undirected graph $G(V, E)$ is given, where V is a set of vertices, and E is a set of vertice-pairs called edges. With k colors $\{1...k\}$ to assign, a k-coloring of G refers to a mapping $C : V \rightarrow \{1...k\}$ such that if $C(p) = C(q)$ then $(p, q) \notin E$. That is, vertices that are directly connected by an edge cannot be assigned the same color. There exist two variants of this problem. In the *optimization variant*, the goal is to find the chromatic number $\chi(G)$ which is the minimum k for which there exists a k-coloring of G. In the *decision variant*, the task is to decide whether a coloring of G exists for a particular number of colors k. All these problems are known to be NP-complete, so it is unlikely that a polynomial-time algorithm exists that solves any of these problems. In this paper, we present the first (as far as we know) heuristic technique that combines Learning Automate (LA) [1] with Random Walk (RW) [2], with the purpose of solving the decision variant of the GCP.

The GCP is a well studied problem and has several applications including frequency assignment [3], register allocation [4], pattern matching [5], timetabling [6], and the solution of sparse linear systems [7]. Encoding the GCP as Boolean satisfiability (SAT) and solving it using efficient SAT algorithms has caused considerable interest. The SAT problem, which also is known to be NP-complete [8], can be defined as follows. A propositional formula $\Phi = \bigwedge_{j=1}^{m} C_j$ with m clauses

J. van Hemert and C. Cotta (Eds.): EvoCOP 2008, LNCS 4972, pp. 277–288, 2008.
© Springer-Verlag Berlin Heidelberg 2008

and n Boolean variables is given. Each Boolean variable $x_i, i \in \{1, \ldots, n\}$, takes one of the two values, **True** or **False**. Each clause C_j is a disjunction of Boolean variables and has the form:

$$C_j = \left(\bigvee_{k \in I_j} x_k \right) \vee \left(\bigvee_{l \in \bar{I}_j} \bar{x}_l \right),$$

where $I_j, \bar{I}_j \subseteq \{1, \ldots.n\}$, $I \cap \bar{I}_j = \emptyset$, and $\bar{x}_i$ denotes the negation of x_i.

The task is to determine whether there exists an assignment of truth values to the variables under which Φ evaluates to **True**. Such an assignment, if it exists, is called a satisfying assignment for Φ, and Φ is called satisfiable. Otherwise, Φ is said to be unsatisfiable. Note that since we have two choices for each of the n Boolean variables, the size of the search space S becomes $|S| = 2^n$. That is, the size of the search space grows exponentially with the number of variables.

Most SAT solvers use a Conjunctive Normal Form (CNF) representation of the formula Φ. In CNF, the formula is represented as a conjunction of clauses, each clause is a disjunction of literals, and a literal is a Boolean variable or its negation. For example, $P \vee Q$ is a clause containing the two literals P and Q. The clause $P \vee Q$ is satisfied if either P is **True** or Q is **True**. When each clause in Φ contains exactly k literals, the resulting SAT problem is called k-SAT.

This paper proposes a new heuristic solution to the GCP, encoded as SAT. In essence, the traditional Random Walk (RW) strategy is enhanced with learning capability, in the form of Learning Automata. Learning Automata have been used to model biological systems [9], and have attracted considerable interest in the last decade because they can learn the optimal actions when operating in (or interacting with) unknown stochastic environments. Furthermore, they combine rapid and accurate convergence with low computational complexity.

Our paper is organized as follows. In Sect. 2 we review various algorithms for solving SAT-encoded graph coloring problems. Sect. 3 explains the basic concepts of Learning Automata and introduces our new approach — the Learning Automata Random Walk (LARW). In Sect. 4, we analyse the results from testing LARW on an extensive set of GCP instances. Finally, in Sect. 5 we present a summary of the paper and provide pointers to further work.

2   Previous Work and Recent Developments

The GCP has been extensively studied due to its simplicity and applicability. The simplicity of the problem coupled with its intractability makes it an ideal platform for exploring new algorithmic techniques. This has led to the development of several algorithms for solving graph coloring problems which usually fall into two main categories: *exact* systematic search (SS) algorithms and stochastic local search algorithms (SLS). Exact systematic search algorithms are guaranteed to return a solution to a problem if one exists and prove it insoluble otherwise. Exact algorithms include specialized branch-and-bound algorithms [10]. Also, techniques based on integer programming formulations of the GCP have been studied [11]. The most popular and efficient systematic search algorithms for SAT are based on the Davis-Putnam (DP) [12] procedure which enumerates all possible variable assignments by means of a binary search tree.

These algorithms can be very effective on specific classes of graphs, however, when problems scales up, their solution effectiveness typically degrades in an exponential manner. Indeed, due to their combinatorial explosive nature, large and complex SAT problems are hard to solve using systematic search algorithms. One way to overcome the combinatorial explosion is to give up completeness.

Local search algorithms are based on what is perhaps the oldest optimization method – *trial and error*. Typically, they start with an initial assignment of truth values to variables, randomly or heuristically generated. Satisfiability can then be formulated as an iterative optimization problem in which the goal is to minimize the number of unsatisfied clauses. Thus, the optimum is obtained when the value of the objective function equals zero, which means that all clauses are satisfied. During each iteration, a new value assignment is selected from the "neighborhood" of the present one, by performing a "move". Most local search algorithms use a 1-flip neighborhood relation, which means that two truth value assignments are considered to be neighbors if they differ in the truth value of *only* one variable. Performing a move, then, consists of switching the present value assignment with one of the neighboring value assignments, e.g., if the neighboring one is better (as measured by the objective function). The search terminates if no better neighboring assignment can be found. Note that choosing a fruitful neighborhood, and a method for searching it, is usually guided by intuition – theoretical results that can be used as guidance are sparse. Popular local search algorithms include GSAT [13], GSAT with random walk [2], Walk-SAT [14]. Recently, new algorithms, such as [15], have emerged using history-based variable selection strategies in order to avoid flipping the same variable.

3   Learning Automata for SAT-Encoded GCPs

We base our work on the principles of Learning Automata (LA) [1]. LA have been used to model biological systems [9], and have attracted considerable interest in the last decade because they can learn the optimal actions when operating in (or interacting with) unknown stochastic environments. Furthermore, they combine rapid and accurate convergence with low computational complexity. Although the GCP has not been addressed from a LA point of view before, LA solutions have recently been proposed for several other combinatorial optimization problems. In [16,17] a so-called Object Migration Automaton is used for solving the classical equipartitioning problem. An order of magnitude faster convergence is reported compared to the best known algorithms at that time. A similar approach has also been discovered for the Graph Partitioning Problem [18]. LA has furthermore been used to tackle Stochastic Knapsack Problems [19]. Finally, the list organization problem has successfully been addressed by LA schemes and have been found to converge to the optimal arrangement with probability arbitrary close to unity [20]. Inspired by the success of the above solution schemes, we will in the following propose a LA based solution scheme for SAT-encoded GCPs.

3.1 A Learning SAT Automaton

Generally stated, a LA performs a sequence of actions on an *environment*. The environment can be seen as a generic *unknown* medium that responds to each action with some sort of reward or penalty, perhaps *stochastically*. Based on the responses from the environment, the aim of the LA is to find the action that minimizes the expected number of penalties received. Because we treat the environment as unknown, we will here only consider the definition of our LA.

Our LA can be defined in terms of a quintuple [1]:

$$\{\underline{\Phi}, \underline{\alpha}, \underline{\beta}, \mathcal{F}(\cdot, \cdot), \mathcal{G}(\cdot, \cdot)\}.$$

Above, $\underline{\Phi} = \{\phi_1, \phi_2, \dots, \phi_s\}$ is the set of internal automaton states. $\underline{\alpha} = \{\alpha_1, \alpha_2, \dots, \alpha_r\}$ is the set of automaton actions. And, $\underline{\beta} = \{\beta_1, \beta_2, \dots, \beta_m\}$ is the set of inputs that can be given to the automaton. An output function $\alpha_t = \mathcal{G}[\phi_t]$ determines the next action performed by the automaton given the current automaton state. Finally, a transition function $\phi_{t+1} = \mathcal{F}[\phi_t, \beta_t]$ determines the new automaton state from: (1) the current automaton state and (2) the response of the environment to the action performed by the automaton.

Based on the above generic framework, the crucial issue is to design automata that can learn the optimal action when interacting with the environment. Several designs have been proposed in the literature, and the reader is referred to [1] for an extensive treatment. In this paper we target the SAT-encoded GCP, and our goal is to design a team of LA that seeks the solution of GCP instances. We build upon the work of Tsetlin and the linear two-action automaton [9,1]. Briefly stated, for each literal in the SAT-encoded GCP instance that is to be solved, we construct an automaton with

- States: $\underline{\Phi} = \{-N - 1, -N, \dots, -1, 0, \dots, N - 2, N - 1\}$.
- Actions: $\underline{\alpha} = \{\textbf{\textit{True}}, \textbf{\textit{False}}\}$.
- Inputs: $\underline{\beta} = \{reward, penalty\}$.

Fig. 1 specifies the $\mathcal{G}$ and $\mathcal{F}$ matrices of our automaton (cf. general definition of LA above). The $\mathcal{G}$ matrix can be summarized as follows. If the automaton state is positive, then action **True** will be chosen by the automaton. If, on the other hand, the state is negative, then action **False** will be chosen. Note that since

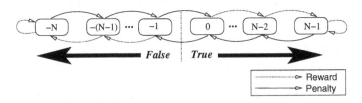

Fig. 1. The state transitions and actions of our Learning Automaton for SAT-encoded GCP

we initially do not know which action is optimal, we set the initial state of our automaton randomly to either '-1' or '0'.

The state transition matrix $\mathcal{F}$ determines how learning proceeds. As seen in the figure, providing a *reward* input to the automaton *strengthens* the currently chosen action, essentially by making it less likely that the other action will be chosen in the future. Correspondingly, a *penalty* input *weakens* the currently selected action by making it more likely that the other action will be chosen later on. In other words, the automaton attempts to incorporate past responses when deciding on a sequence of actions.

3.2 Learning Automata Random Walk (LARW)

Overview: In addition to the definition of the LA, we must define the environment that the LA interacts with. Simply put, the environment is a SAT-encoded GCP instance as defined in Sect. 1. Each variable of the SAT problem instance is assigned a dedicated LA, resulting in a team of LA. The task of each LA is to determine the truth value of its corresponding variable, with the aim of satisfying all of the clauses where that variable appears. In other words, if each automaton reaches its own goal, then the overall SAT problem at hand has also been solved.

Pseudo-code: With the above perspective in mind, we will now present the details of the LARW that we propose. Fig. 2 contains the complete pseudo-code for solving SAT-encoded GCP, using a team of LA. As seen from the figure, the LARW corresponds to an ordinary Random Walk (RW), however, both satisfied and unsatisfied clauses are used in the search. Furthermore, the assignment of truth values to variables is indirect, governed by the states of the LA. At the core of the LARW is a punishment/rewarding scheme that guides the team of LA towards the optimal assignment. In the spirit of automata based learning, this scheme is incremental, and learning is performed gradually, in small steps. To elaborate, in each iteration of the algorithm, we randomly select a single clause. A variable is randomly selected from that clause, and the corresponding automaton is identified. If the clause is unsatisfied, the automaton is punished. Correspondingly, if the clause is satisfied, the automaton is rewarded, however, only if the automaton makes the clause satisfied.

Remark 1: Like a two-action Tsetlin Automaton, our proposed LA seeks to minimize the expected number of penalties it receives. In other words, it seeks finding the truth assignment that minimizes the number of unsatisfied clauses among the clauses where its variable appears.

Remark 2: Note that because multiple variables, and thereby multiple LA, may be involved in each clause, we are dealing with a game of LA [1]. That is, multiple LA interact with the same environment, and the response of the environment depends on the actions of several LA. In fact, because there may be conflicting goals among the LA involved in the LARW, the resulting game is competitive. The convergence properties of general competitive games of LA

Procedure learning_automata_random_walk()

Begin
    /* Initialization */
    For i := 1 **To** n **Do**
        /* The initial state of each automaton is set to either '-1' or '1' */
        state[i] = random_element($\{-1, 0\}$);
        /* And the respective literals are assigned corresponding truth values */
        If state[i] == -1 **Then** x_i = *False* **Else** x_i = *True*;

    /* Main loop */
    While Not stop($\mathcal{C}$) **Do**
        /* Draw unsatisfied clause randomly */
        C_j = random_unsatisfied_clause($\mathcal{C}$);
        /* Draw clause literal randomly */
        i = random_element($I_j \cup \bar{I}_j$);
        /* The corresponding automaton is penalized for choosing the "wrong" action */
        If $i \in I_j$ **And** state[i] $< N - 1$ **Then**
            state[i]++;
            /* Flip literal when automaton changes its action */
            If state[i] == 0 **Then**
                flip(x_i);
        Else If $i \in \bar{I}_j$ **And** state[i] $> -N$ **Then**
            state[i]--;
            /* Flip literal when automaton changes its action */
            If state[i] == -1 **Then**
                flip(x_i);

        /* Draw satisfied clause randomly */
        C_j = random_satisfied_clause($\mathcal{C}$);
        /* Draw clause literal randomly */
        i = random_element($I_j \cup \bar{I}_j$);
        /* Reward corresponding automaton if it */
        /* contributes to the satisfaction of the clause */
        If $i \in I_j$ **And** state[i] ≥ 0 **And** state[i] $< N - 1$ **Then**
            state[i]++;
        Else If $i \in \bar{I}_j$ **And** state[i] < 0 **And** state[i] $> -N$ **Then**
            state[i]--;
    EndWhile
End

Fig. 2. Learning Automata Random Walk Algorithm

have not yet been successfully analyzed, however, results exists for certain classes of games, such as the Prisoner's Dilemma game [1]. In our case, the LA involved in the LARW are non-absorbing, i.e., every state can be reached from every other state with positive probability. This means that the probability of reaching the solution of the SAT problem instance at hand is equal to 1 when running the game infinitively. Also note that the solution of the SAT problem corresponds to a Nash equilibrium of the game.

Remark 3: In order to maximize speed of learning, we initialize each LA randomly to either the state '-1' or '0'. In this initial configuration, the variables will be flipped relatively quickly because only a single state transition is necessary for a flip. Accordingly, the joint state space of the LA is quickly explored in this configuration. However, as learning proceeds and the LA move towards their boundary states, i.e., states '-N' and 'N-1', the flipping of variables calms down. Accordingly, the search for a solution to the SAT problem instance at hand becomes increasingly focused.

4   Experimental Results

4.1   Benchmark Instances

As a basis for the empirical evaluation of LA, we selected a benchmark test suite of 3-colorables graphs that shows phase transition. All the instances are known to be hard and difficult to solve and are available from the SATLIB website (http://www.informatik.tu-darmstadt.de/AI/SATLIB). All the benchmark instances used in this experiment are satisfiable instances and have been used widely in the literature.

Note that due to the randomization of LARW, the number of flips required for solving a problem instance varies widely between different runs. Therefore, for each problem instance, we run LARW and RW 100 times with a cutoff parameter (maxflips) which is high enough (10^7) to guarantee a success rate close to 100%.

4.2   Search Trajectory

The manner in which each LA converges to an assignment is crucial for better understanding the LA SAT Game's behavior. In Fig. 3, we show how the best and current assignment progresses during the search using a random 3-SAT problem with 150 variables and 645 clauses taken from the SAT benchmark library. The plot located on the left of the figure suggests that problem solving with LARW happens in two phases. In the first phase, which corresponds to the early part of the search (the first 5% of the search), LARW behaves as a hill-climbing method. In this phase, which can be described as a short one, up to 95% of the clauses are satisfied. The best assignment climbs rapidly at first, and then flattens off as we mount the plateau, marking the start of the second phase. The plateau spans a region in the search space where flips typically leave the best assignment unchanged. The long plateaus becomes even more pronounced as the number of flips increases. More specifically, the plateau appears when trying to satisfy the last few remaining clauses.

To further investigate the behavior of LARW once on the plateau, we looked at the corresponding average state of the LA as the search progresses. The plot located on the right in Fig. 3 shows the resulting observations. At the start of plateau, search coincides in general with an increase in the average state. The longer the plateau runs, the higher the average state becomes. An automaton with high average state needs to perform a series of actions before its current state changes to either -1 or 0, thereby making the flipping of the corresponding variable possible. The transition between each plateau corresponds to a change

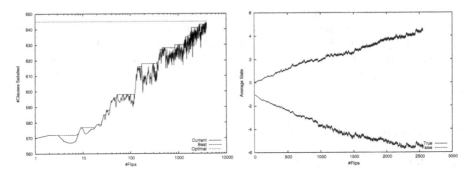

Fig. 3. (Left)LARW's search space on a 150 variable problem with 645 clauses (uf150-645). Along the horizontal axis we give the number of flips, and along the vertical axis the number of satisfied clauses. (Right) Average state of automaton. Horizontal axis gives the number of flips, and the vertical axis shows the average state of automaton.

to the region where a small number of flips gradually improves the score of the current solution ending with an improvement of the best assignment. The search pattern brings out an interesting difference between LARW and the standard use of SLS. In the latter, one generally stops the search as soon as no more improvements are found. This can be appropriate when looking for a near-solution. On the other hand, when searching for a global maximum (i.e., a satisfying assignment) stopping when no flips do no yield an immediate improvement is a poor strategy.

4.3 Run-Length-Distributions (RLDs)

As an indicator of the behavior of the algorithm on a single instance, we choose the median cost when trying to solve a given instance in 100 trials, and using an extremely high cutoff parameter setting of $Maxsteps = 10^7$ in order to obtain a maximal number of successful tries. The reason behind choosing the median cost rather than the mean cost is due to the large variation in the number of flips required to find a solution. To get an idea of the variability of the search cost, we analyzed the cumulative distribution of the number of search flips needed by both LARW and RW for solving single instances. Due to non-deterministic decisions involved in the algorithm (i.e.,initial assignment, random moves), the number of flips needed by both algorithms to find a solution is a random variable that varies from run to run. More formally, let k denotes the total number of runs, and let $f'(j)$ denotes the number of flips for the j-th successful run (i.e, run during which a solution is found) in a list of all successful runs, sorted according to increasing number of flips, then the cumulative empirical RLD is defined by $\hat{P}(f'(j) \leq f) = \frac{|\{j|f'(j)\leq f\}|}{k}$. For practical reasons we restrict our presentation here to the instances corresponding to small, medium, and large sizes from the underlying test-set.

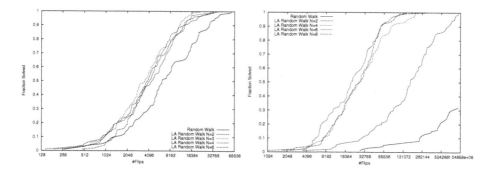

Fig. 4. (Right) LARW Vs RW: cumulative distributions for a 90-variables *graph coloring* problems with 300 clauses(flat90-300). (Left) LARW cumulative distribution for a 150-variable *graph coloring* problem with 545 clauses (flat375-1403). Along the horizontal axis we give the number of flips, and along the vertical axis the fraction of problems solved for different values of N.

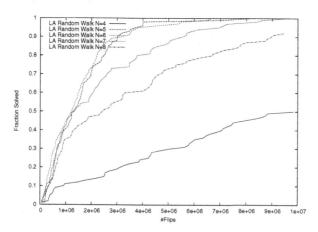

Fig. 5. (Right) LARW Vs RW : cumulative distributions for a 375-variable *graph coloring* problems with 1403 clauses (flat375-1403). Along the horizontal axis we give the number of flips, and along the vertical axis the fraction of problems solved for different values of N.

Fig. 4 and Fig. 5 shows RLDs obtained by applying RW and LARW to individual SAT-encoded graph coloring problem instances. As can be seen from Fig. 4, we observe that on the small size instance, the two algorithms show no cross-over in their corresponding RLDs. This provides evidence for the superiority of LARW compared to RW (i.e, N = 1) as it gives consistently higher success probabilities, regardless of the number of search steps. On the medium size instance, we observe a stagnation behavior with a low asymptotic solution probability corresponding to a value around 0.3. As can be easily seen, both methods show the existence of an initial phase below which the probability for finding a solution is 0. Both methods start the search from a randomly chosen assignment which typically violates many clauses. Consequently, both methods need some time to reach the first local optimum which possibly could be a feasible solution. The plot in Fig. 5 shows that the performance of RW for the large instance (flat375-1403) is even far more dramatic as the probability of finding a feasible solution within the required number of steps is 0. The value of the distance between the minimum and the maximum number of search steps needed for finding a feasible solution using RW is higher compared to that of LARW and increases with the hardness of the instance. The learning automaton mechanism pays off as the instance gets harder, and the probability of success gets higher as N increases and reaches an optimal value.

4.4   Mean Search Cost

In this section, we focus on the behavior of the two algorithms using 100 instances from a test-set of small, medium instances. We chose not to include the plot for the large instance (flat375-1403) because RW was incapable of solving it during the 100 trials. For each instance the median search cost (number of local search steps) is measured and we analyze the distribution of the mean search cost over the instances from each test-set. The different plots show the cumulative hardness distributions produced by 100 trials on 100 instances from a test-set.

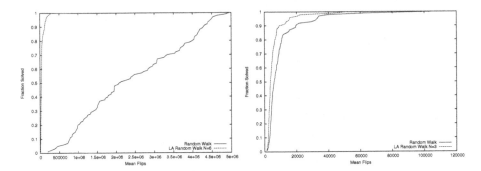

Fig. 6. Hardness distribution across test-set flat150-545. (Left) Hardness distribution across test-set for flat90-300. Along the horizontal axis we give the median number of flips per solution , and along the vertical axis the fraction of problems solved.

Several observations can be made from the plots in Fig. 6 which shows the hardness distributions of the two algorithms for SAT-encoding graph coloring problem instances. There exists no cross-overs in the plots of both figures which makes LARW the clear winner. The RW shows a higher variability in search cost compared to LARW between the instances of each test-set. The distributions of the two algorithms confirms the existence of instances which are harder to solve than others. In particular, as can be seen from the long tails of these distributions, a substantial part of problem instances are dramatically harder to solve with RW than with LARW. The harder the instance, the higher the difference between the average search costs of two algorithms (a factor of approx.

up to 50). The finite automaton learning mechanism employed in LARW offers an efficient way to escape from highly attractive areas in the search space of hard instances leading to a higher probability success as well as reducing the average number of local search steps to find a solution. The empirical hardness distribution of SAT-encoded graph coloring problems in Fig. 6 shows that it was rather easy for both algorithms to find a feasible solution in each trial across the test set flat90-300 with LARW showing on average a lower search cost within a given probability compared to RW. The plot reveals the existence of some instances on which RW suffers from a strong search stagnation behavior. The plot located on the left of Fig. 6 shows a striking poor average performance of RW compared to LARW on the test set flat150-545. Conversely, LARW shows a consistent ability to find solutions across the instances on this test set. For LARW, we observe a small variability in search cost indicated by the distance between the minimum and the maximum number of local search steps needed to find a solution. The differences in performance between these two algorithms can be characterized by a factor of ca. 10 in the median. The performance differences observed between the two algorithms for small size instances are still observed and very significant for medium size instances. This suggests that the finite learning automaton is considerably more effective for larger instances.

5   Conclusions

In this work, we have introduced a new approach for solving GCP based on combining Learning Automata with Random Walk. Thus, in order to get a comprehensive picture of the new algorithm's performance, we used a set of problems consisting of randomly generated SAT-encoded graph coloring instances. All the selected problem instances are located in the so-called phase transition and have been widely used by different authors in the context of evaluating the performance of metaheuristics. RW suffers from stagnation behaviour which directly affects its performance. This same phenomenon is, however, observed with LARW only for large instances. Based on the analysis of RLD's, we observe that the probability of finding a solution within any arbitrary number of search steps is higher for LARW compared to that of RW. To get an idea of the variability of the solution cost between the instances of the test sets, we analysed the cumulative distribution of the mean search cost. Results indicated that the harder the instance, the higher the difference between the mean search costs of the two algorithms. The difference can be several order of magnitude in favour of LARW.

References

1. Narendra, K.S., Thathachar, M.A.L.: Learning Automata: An Introduction. Prentice Hall, Englewood Cliffs (1989)
2. Selman, B., Kautz, H.A., Cohen, B.: Noise Strategies for Improving Local Search. In: Proceedings of AAAI 1994, pp. 337–343. MIT Press, Cambridge (1994)
3. Gamst, A.: Some lower bounds for a class of frequency assignment problems. IEEE Transactions of Vehicular Technology 35, 8–14 (1986)

4. Chow, F., Hennessy, J.: The priority-based coloring approach to register allocation. ACM Transactions on Programming Languages and Systems 12, 501–536 (1990)
5. Ogawa, H.: Labeled point pattern matching by delaunay triangulation and maximal cliques. Pattern Recognition 19, 35–40 (1996)
6. Werra, D.D.: An introduction to timetabling. European Journal of Operations Research 19, 151–162 (1985)
7. Gebremedhin, A., Manne, F., Pothen, A.: What color is your jacobian? graph coloring for computing derivatives. SIAM Review 47, 629–705 (2005)
8. Cook, S.: The complexity of theorem-proving procedures. In: Proceedings of the Third ACM Symposuim on Theory of Computing, pp. 151–158 (1971)
9. Tsetlin, M.L.: Automaton Theory and Modeling of Biological Systems. Academic Press, London (1973)
10. Caramia, M., Dell'Olmo, P.: Bounding vertex coloring by truncated multistage branch and bound. Networks 44, 231–242 (2004)
11. Mehrotra, A., Trick, M.A.: A column generation approach for graph coloring. IN-FORMS Journal on Computing 8, 344–354 (1996)
12. Davis, M., Putnam, H.: A computing procedure for quantification theory. Journal of the ACM 7, 201–215 (1960)
13. Selman, B., Levesque, H., Mitchell, D.: A new method for solving hard satisfiability problems. In: Proceedings of AAA 1992, pp. 440–446. MIT Press, Cambridge (1992)
14. McAllester, D., Selman, B., Kautz, H.: Evidence for Invariants in Local Search. In: Proceedings of AAAI 1997, pp. 321–326. MIT Press, Cambridge (1997)
15. Glover, F.: Tabu search-part 1. ORSA Journal on Computing 1, 190–206 (1989)
16. Oommen, B.J., Ma, D.C.Y.: Deterministic learning automata solutions to the equipartitioning problem. IEEE Transactions on Computers 37(1), 2–13 (1988)
17. Gale, W., Das, S., Yu, C.: Improvements to an Algorithm for Equipartitioning. IEEE Transactions on Computers 39, 706–710 (1990)
18. Oommen, B.J., Croix, E.V.S.: Graph partitioning using learning automata. IEEE Transactions on Computers 45(2), 195–208 (1996)
19. Granmo, O.C., Oommen, B.J., Myrer, S.A., Olsen, M.G.: Learning Automata-based Solutions to the Nonlinear Fractional Knapsack Problem with Applications to Optimal Resource Allocation. IEEE Transactions on Systems, Man, and Cybernetics Part B (2006)
20. Oommen, B.J., Hansen, E.R.: List organizing strategies using stochastic move-to-front and stochastic move-to-rear operations. SIAM Journal on Computing 16, 705–716 (1987)

Author Index

Lecture Notes in Computer Science

Sublibrary 1: Theoretical Computer Science and General Issues

Vol. 4746: A. Bondavalli, F. Brasileiro, S. Rajsbaum (Eds.), Dependable Computing. XV, 239 pages. 2007.

Vol. 4743: P. Thulasiraman, X. He, T.L. Xu, M.K. Denko, R.K. Thulasiram, L.T. Yang (Eds.), Frontiers of High Performance Computing and Networking ISPA 2007 Workshops. XXIX, 536 pages. 2007.

Vol. 4742: I. Stojmenovic, R.K. Thulasiram, L.T. Yang, W. Jia, M. Guo, R.F. de Mello (Eds.), Parallel and Distributed Processing and Applications. XX, 995 pages. 2007.

Vol. 4739: R. Moreno Díaz, F. Pichler, A. Quesada Arencibia (Eds.), Computer Aided Systems Theory – EUROCAST 2007. XIX, 1233 pages. 2007.

Vol. 4736: S. Winter, M. Duckham, L. Kulik, B. Kuipers (Eds.), Spatial Information Theory. XV, 455 pages. 2007.

Vol. 4732: K. Schneider, J. Brandt (Eds.), Theorem Proving in Higher Order Logics. IX, 401 pages. 2007.

Vol. 4731: A. Pelc (Ed.), Distributed Computing. XVI, 510 pages. 2007.

Vol. 4728: S. Bozapalidis, G. Rahonis (Eds.), Algebraic Informatics. VIII, 291 pages. 2007.

Vol. 4726: N. Ziviani, R. Baeza-Yates (Eds.), String Processing and Information Retrieval. XII, 311 pages. 2007.

Vol. 4719: R. Backhouse, J. Gibbons, R. Hinze, J. Jeuring (Eds.), Datatype-Generic Programming. XI, 369 pages. 2007.

Vol. 4711: C.B. Jones, Z. Liu, J. Woodcock (Eds.), Theoretical Aspects of Computing – ICTAC 2007. XI, 483 pages. 2007.

Vol. 4710: C.W. George, Z. Liu, J. Woodcock (Eds.), Domain Modeling and the Duration Calculus. XI, 237 pages. 2007.

Vol. 4708: L. Kučera, A. Kučera (Eds.), Mathematical Foundations of Computer Science 2007. XVIII, 764 pages. 2007.

Vol. 4707: O. Gervasi, M.L. Gavrilova (Eds.), Computational Science and Its Applications – ICCSA 2007, Part III. XXIV, 1205 pages. 2007.

Vol. 4706: O. Gervasi, M.L. Gavrilova (Eds.), Computational Science and Its Applications – ICCSA 2007, Part II. XXIII, 1129 pages. 2007.

Vol. 4705: O. Gervasi, M.L. Gavrilova (Eds.), Computational Science and Its Applications – ICCSA 2007, Part I. XLIV, 1169 pages. 2007.

Vol. 4703: L. Caires, V.T. Vasconcelos (Eds.), CONCUR 2007 – Concurrency Theory. XIII, 507 pages. 2007.

Vol. 4700: C.B. Jones, Z. Liu, J. Woodcock (Eds.), Formal Methods and Hybrid Real-Time Systems. XVI, 539 pages. 2007.

Vol. 4699: B. Kågström, E. Elmroth, J. Dongarra, J. Waśniewski (Eds.), Applied Parallel Computing. XXIX, 1192 pages. 2007.

Vol. 4698: L. Arge, M. Hoffmann, E. Welzl (Eds.), Algorithms – ESA 2007. XV, 769 pages. 2007.

Vol. 4697: L. Choi, Y. Paek, S. Cho (Eds.), Advances in Computer Systems Architecture. XIII, 400 pages. 2007.

Vol. 4688: K. Li, M. Fei, G.W. Irwin, S. Ma (Eds.), Bio-Inspired Computational Intelligence and Applications. XIX, 805 pages. 2007.

Vol. 4684: L. Kang, Y. Liu, S. Zeng (Eds.), Evolvable Systems: From Biology to Hardware. XIV, 446 pages. 2007.

Vol. 4683: L. Kang, Y. Liu, S. Zeng (Eds.), Advances in Computation and Intelligence. XVII, 663 pages. 2007.

Vol. 4681: D.-S. Huang, L. Heutte, M. Loog (Eds.), Advanced Intelligent Computing Theories and Applications. XXVI, 1379 pages. 2007.

Vol. 4672: K. Li, C. Jesshope, H. Jin, J.-L. Gaudiot (Eds.), Network and Parallel Computing. XVIII, 558 pages. 2007.

Vol. 4671: V.E. Malyshkin (Ed.), Parallel Computing Technologies. XIV, 635 pages. 2007.

Vol. 4669: J.M. de Sá, L.A. Alexandre, W. Duch, D.P. Mandic (Eds.), Artificial Neural Networks – ICANN 2007, Part II. XXXI, 990 pages. 2007.

Vol. 4668: J.M. de Sá, L.A. Alexandre, W. Duch, D.P. Mandic (Eds.), Artificial Neural Networks – ICANN 2007, Part I. XXXI, 978 pages. 2007.

Vol. 4666: M.E. Davies, C.J. James, S.A. Abdallah, M.D. Plumbley (Eds.), Independent Component Analysis and Signal Separation. XIX, 847 pages. 2007.

Vol. 4665: J. Hromkovič, R. Královič, M. Nunkesser, P. Widmayer (Eds.), Stochastic Algorithms: Foundations and Applications. X, 167 pages. 2007.

Vol. 4664: J. Durand-Lose, M. Margenstern (Eds.), Machines, Computations, and Universality. X, 325 pages. 2007.

Vol. 4661: U. Montanari, D. Sannella, R. Bruni (Eds.), Trustworthy Global Computing. X, 339 pages. 2007.

Vol. 4649: V. Diekert, M.V. Volkov, A. Voronkov (Eds.), Computer Science – Theory and Applications. XIII, 420 pages. 2007.

Vol. 4647: R. Martin, M.A. Sabin, J.R. Winkler (Eds.), Mathematics of Surfaces XII. IX, 509 pages. 2007.

Vol. 4646: J. Duparc, T.A. Henzinger (Eds.), Computer Science Logic. XIV, 600 pages. 2007.

Vol. 4644: N. Azémard, L. Svensson (Eds.), Integrated Circuit and System Design. XIV, 583 pages. 2007.

Vol. 4641: A.-M. Kermarrec, L. Bougé, T. Priol (Eds.), Euro-Par 2007 Parallel Processing. XXVII, 974 pages. 2007.

Vol. 4639: E. Csuhaj-Varjú, Z. Ésik (Eds.), Fundamentals of Computation Theory. XIV, 508 pages. 2007.

Vol. 4638: T. Stützle, M. Birattari, H. H. Hoos (Eds.), Engineering Stochastic Local Search Algorithms. X, 223 pages. 2007.

Vol. 4630: H.J. van den Herik, P. Ciancarini, H.H.L.M.(J.) Donkers (Eds.), Computers and Games. XII, 283 pages. 2007.

Vol. 4628: L.N. de Castro, F.J. Von Zuben, H. Knidel (Eds.), Artificial Immune Systems. XII, 438 pages. 2007.

Vol. 4627: M. Charikar, K. Jansen, O. Reingold, J.D.P. Rolim (Eds.), Approximation, Randomization, and Combinatorial Optimization. XII, 626 pages. 2007.